LASER-PLASMA INTERACTIONS 3

Proceedings of the Twenty Ninth Scottish
Universities Summer School in Physics
ST. ANDREWS August 1985

A NATO Advanced Study Institute

edited by

M B HOOPER

Published by the
Scottish Universities Summer School in Physics

SUSSP PUBLICATIONS

Edinburgh University Physics Department
King's Buildings, Mayfield Road
Edinburgh

Further copies of this book may be obtained
directly from the above address

ISBN 0 905945 12 3

Printed in Great Britain
by The Camelot Press, Southampton

SUSSP PROCEEDINGS

1	1960	Dispersion Relations
2	1961	Fluctuation, Relaxation and Resonance in Magnetic Systems
3	1962	Polarons and Excitons
4	1963	Strong Interactions and High Energy Physics
5	1964	Nuclear Structure and Electromagnetic Interactions
6	1965	Phonons in Perfect and Imperfect Lattices
7	1966	Particle Interactions at High Energies
8	1967	Mathematical Methods in Solid State and Superfluid Theory
9	1968	Physics of Hot Plasma
10	1969	Quantum Optics
11	1970	Hadronic Interactions of Electrons and Photons
12	1971	Atoms and Molecules in Astrophysics
13	1972	Electronic/Structural Properties of Amorphous Semiconductors
14	1973	Phenomenology of Particles at High Energies
15	1974	The Helium Liquids
16	1975	Non-Linear Optics
17	1976	Fundamentals of Quark Models
18	1977	Nuclear Structure Physics
19	1978	The Metal Non-Metal Transition in Disordered Systems
20	1979	Laser-Plasma Interactions
21	1980	Gauge Theories and Experiments at High Energies
22	1981	Magnetism in Solids
23	1982	Lasers: Physics, Systems and Techniques
24	1982	Laser-Plasma Interactions 2
25	1983	Quantitative Electron Microscopy
26	1983	Statistical and Particle Physics
27	1984	Fundamental Forces
28	1985	Superstrings and Supergravity
29	1985	Laser-Plasma Interactions 3
30	1985	Synchrotron Radiation
31	1986	Disordered Systems
32	1987	Computational Physics
33	1987	Laboratory and Astrophysical Plasma Spectroscopy

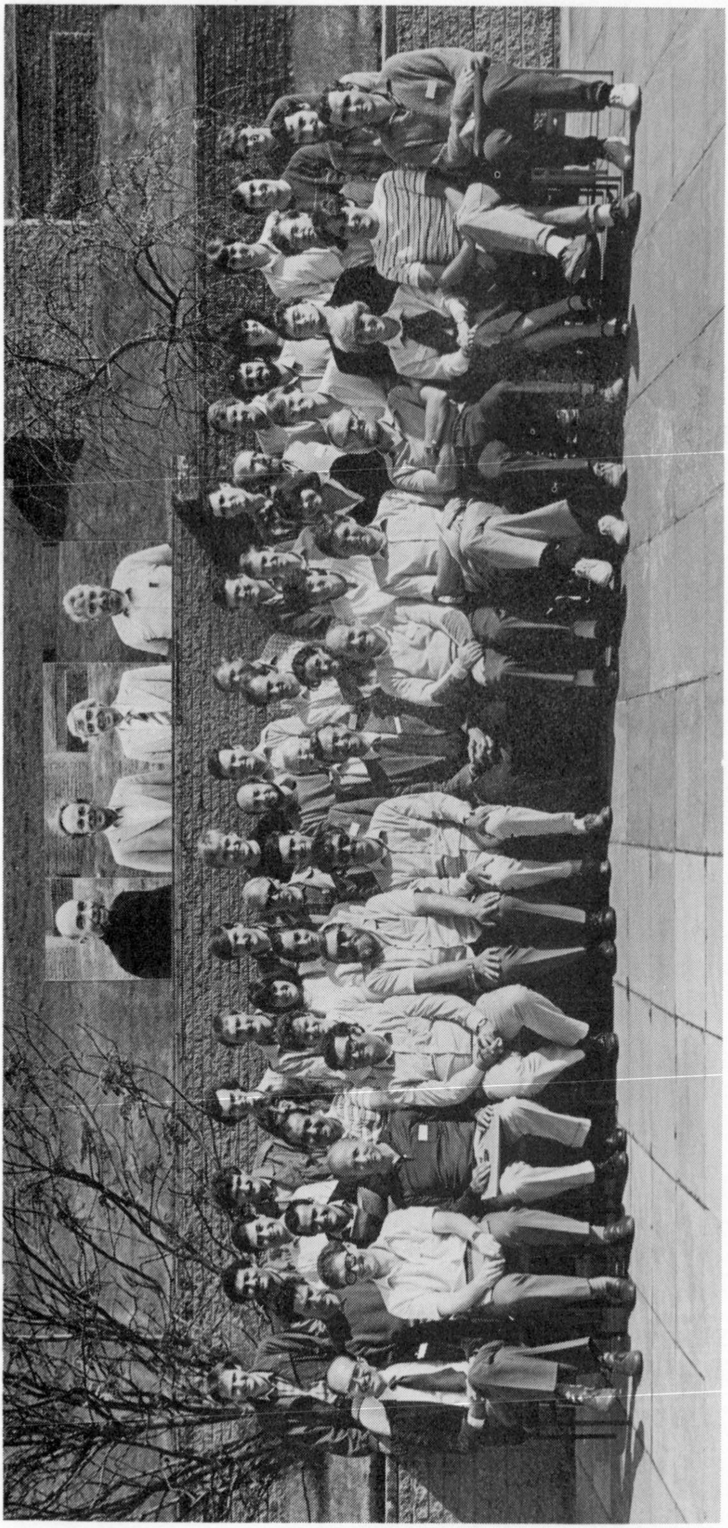

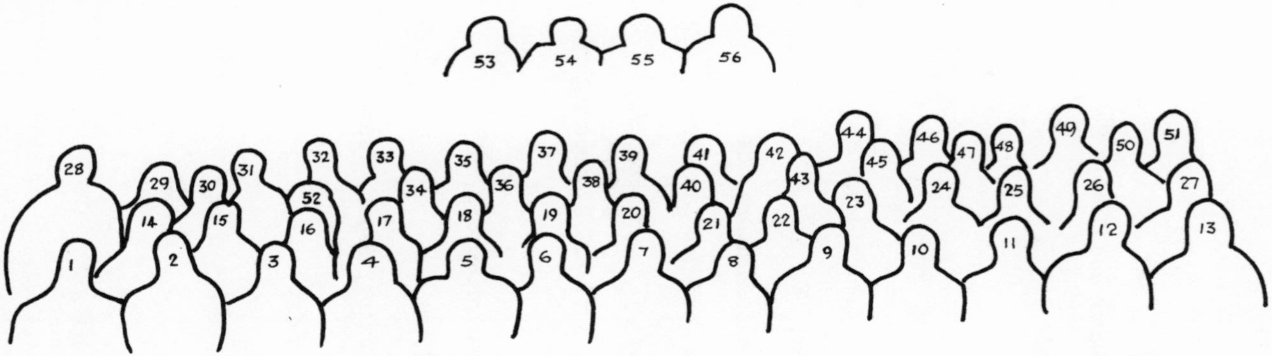

1. Dr M B Hooper
2. Dr R G Evans
3. Dr R More
4. Dr D Shvarts
5. Dr H E R Preston
6. Dr R A Cairns
7. Dr J J Sanderson
8. Prof P Mulser
9. Prof G J Pert
10. Dr M H Key
11. Mrs S Smart
12. Dr L Wood
13. Mr J Farny
14. Dr D Giulietti
15. Mr I Deha
16. Mr L Boufendi
17. Miss S L Thomson
18. Dr L M Small
19. Mr M J D Henshaw
20. Miss A M Woods
21. Mrs G Nave
22. Mrs A Klisnick
23. Mrs P Monier
24. Mr P A Norreys
25. Mr B Faral
26. Mr O Larroche
27. Mr N Fried
28. Mr J Mika
29. Mr B A Shiwai
30. Dr L Nocera
31. Mr R Weber
32. Mr D A Diver
33. Mr A Evans
34. Miss P M Evans
35. Mr S Hüller
36. Dr L Drska
37. Mr A Rogoyski
38. Mr A G Cartlidge
39. Mr W P Lewis
40. Mr A W Taylor
41. Dr T Bar-Noy
42. Dr Y Kaufman
43. Mr E M Robertson
44. Mr D Ofer
45. Dr J B Weill
46. Mr S H Batha
47. Mr J P Marangos
48. Ms K Stein
49. Mr D A Perkins
50. Dr J E Andrew
51. Mr S J Davidson
52. Mr C D Jackson
53. Dr J Meyer-ter-Vehn
54. Dr W L Kruer
55. Dr T A Hall
56. Dr F O'Neill

EXECUTIVE COMMITTEE

Sanderson, Dr J J	*St Andrews (Director)*
Cairns, Dr R A	*St Andrews (Secretary)*
Hooper, Dr M B	*Strathclyde (Editor)*
Preston, Dr H E R	*Strathclyde (Treasurer and Social Secretary)*
Wood, Dr L	*St Andrews (Steward)*

LECTURERS

Evans, Dr R G	*Rutherford Appleton Laboratory, UK*
Hall, Dr T A	*University of Essex, UK*
Key, Dr M H	*Rutherford Appleton Laboratory, UK*
Kruer, Dr W L	*Lawrence Livermore National Laboratory, USA*
Meyer-ter-Vehn, Dr J	*Max Planck Institut für Quantenoptik, W Germany*
More, Dr R	*Lawrence Livermore National Laboratory, USA*
Mulser, Prof P	*Technische Hochschule Darmstadt, W Germany*
O'Neill, Dr F	*Rutherford Appleton Laboratory, UK*
Pert, Prof G J	*University of Hull, UK*
Shvarts, Dr D	*Nuclear Research Centre, Negev, Israel*

Participants

Dr J E Andrew, *Aldermaston*

Dr T Bar-Noy, *Negev*

Mr S H Batha, *Rochester*

Mr L Boufendi, *C E N, Algeria*

Mr A G Cartlidge, *Cambridge*

Mr S J Davidson, *Aldermaston*

Mr I Deha, *C E N, Algeria*

Mr D A Diver, *Glasgow*

Dr L Drska, *Prague*

Mr A Evans, *Swansea*

Miss P M Evans, *Bristol*

Mr B Faral, *Paris*

Mr J Farny, *Warsaw*

Mr N Fried, *Jerusalem*

Dr D Giulietti, *Pisa*

Mr M J D Henshaw, *Hull*

Mr C D Jackson, *Essex*

Dr Y Kaufman, *Negev*

Mrs A Klisnick, *Paris-Sud*

Mr O Laroche, *Limeil-Valenton*

Mr W P Lewis, *Swansea*

Mr J P Marangos, *Imperial College*

Mr J Mika, *Berne*

Mrs P Monier, *Paris*

Miss G Nave, *Imperial College*

Dr L Nocera, *St Andrews*

Mr P A Norreys, *Royal Holloway*

Mr D Ofer, *Holon*

Mr D A Perkins, *Aldermaston*

Mr E M Robertson, *Belfast*

Mr A M Rogoyski, *King's College*

Mr B A Shiwai, *Essex*

Dr L M Small, *St Andrews*

Ms K Stein, *Darmstadt*

Mr A W Taylor, *St Andrews*

Miss S L Thomson, *Glasgow*

Mr R Weber, *Berne*

Dr J B Weill, *Limeil-Valenton*

Miss A M Woods, *St Andrews*

DIRECTOR'S PREFACE

The 29th Scottish Universities Summer School in Physics, which was held in St. Andrews from 28 July to 10 August 1985, was the third School to be devoted to the topic of Laser Plasma Interactions. The success of the first of these Schools in 1979, coupled with the widespread interest in this important and rapidly developing research area, led to the establishment of the 1982 and 1985 Schools as SERC supported short courses. Despite this switch of principal financial support from NATO to SERC the Schools have remained under the organizational 'umbrella' of SUSSP and, indeed, SUSSP has retained responsibilit for publication of the Proceedings. This appears to be a very happy combination.

Following the policy of the 1982 School half of the lecturers were drawn from the earlier Schools to give continuity in the more basic topics as well as to best illustrate where new developments and directions are appearing. It is, of course, impossible in a two-week period to cover all aspects of the subject and in each of the Schools the basic lectures have been supplemented by just some of the newer or more peripheral topics- these have changed from School to School. In a slight departure from the pattern of the earlier Schools it was decided to give these topics fewer lectures each so that more of them might be included. Although less basic, such topics are usually of the greatest current interest and there is never any shortage of them. Thus, in this School there were lectures on non-Debye plasmas, rare gas halide and X-UV lasers, and we were brought up to date on non-local transport theory and the prospects for inertial confinement fusion. Regrettably, due to a withdrawal through illness, we had no lectures on spectroscopy and atomic physics but this was compensated to some extent by giving a little more time to other topics. To all of the lecturers go the very sincere thanks of the Organizing Committee. The success of a Summer School lies ultimately in the skill of its lecturers; on this primary consideration the Committee is in no doubt that this was a very successful School.

In thanking the members of the Committee for their hard work in organizing the School it is my great pleasure to pay tribute also to my predecessor. This series of Schools was due to the inspiration and enthusiasm of Professor E.W. Laing and the foundations that he laid as Director of the first two, coupled with the skill and experience of the School's Officers, made my task as Director a very light and pleasant one. My thanks, too, go to Mrs. Sandra Smart, who stepped in as School Secretary at fairly short notice and ran the School Office with great competence and good will, and to the University of St. Andrews, particularly those members of the Physics Department and John Burnet Hall who made us so welcome.

Finally, I should like to express my appreciation to all who attended the School for their support and lively participation in both the academic and social activities of the School. Numbers were almost exactly the same as for the 1982 School with half the membership coming from overseas, thus maintaining the international atmosphere, not only of these, but of all the Scottish Universities Summer Schools in Physics. A small grant from the SUSSP Governing Committee enabled us to provide modest bursaries for a few of the overseas participants and it is a pleasure to acknowledge this and, especially, the continued financial support of the Science and Engineering Research Council.

J.J. Sanderson
St. Andrews

EDITOR'S NOTE

These Proceedings build on the foundation provided by previous volumes in this series, Laser-Plasma Interactions (1979) and Laser-Plasma Interactions 2 (1982). Those lecturers who were covering similar ground to that already covered before were invited to provide updating material, whereas those in newer areas gave fuller, more self-contained articles.

I am very grateful to all the authors for not only giving excellent lectures at the School, but also for following it up by their careful work on these proceedings. My thanks are also due to Mrs. Alison Eadie for her meticulous typing of those scripts not provided in camera-ready form.

As a Summer School, rather than a Research Conference, the aim of the lecturers was to assist the new generation of laser-plasma physicist to grasp different parts of this rapidly developing subject.. We all hop that these Proceedings will help fulfill this role to a wider audience.

Michael Hooper

CONTENTS

PLASMA PROCESSES IN NON-IDEAL PLASMAS

INTRODUCTION TO THE PHYSICS AND APPLICATIONS OF LASER PRODUCED PLASMAS

M H Key
Rutherford Appleton Laboratory, Chilton, Oxfordshire England

Although the physics of laser plasma interaction is intrinsically interesting, it is the wide range of significant applications of laser produced plasmas which makes it worthwhile to pursue this branch of science. Notable among these applications are:

- Laser Fusion
- Study of basic physics of dense non-ideal plasmas
- Applications of intense X-ray sources
- X-ray laser development
- Applications of intense sources of energetic particles
- Beat wave accelerator development

This introduction will briefly outline the concepts and the current state of progress of these applications.

1.
LASER FUSION

Inertially confined fusion (ICF) is in principle realisable by compression and heating of Deuterium-Tritrium fuel to produce conditions illustrated in Fig 1,[1] [and J Meyer ter Vehn ibid].

Thermonuclear burn is initiated in a central region at the temperature (kT ~ 5keV) required for the fusion reaction. A minimum density x radius ($\rho r \sim 0.3$ g cm^{-2}) is required to reabsorb the energy of α particles released in the fusion reaction ($D+T \rightarrow \alpha + n$). The reaction 'ignites' by virtue of extra heating from the α particles. In Fig 1 the pressure in the central region (for a density of 30g cm^{-3}) is 120 Gbar and the thermal energy content is 0.08 MJ.

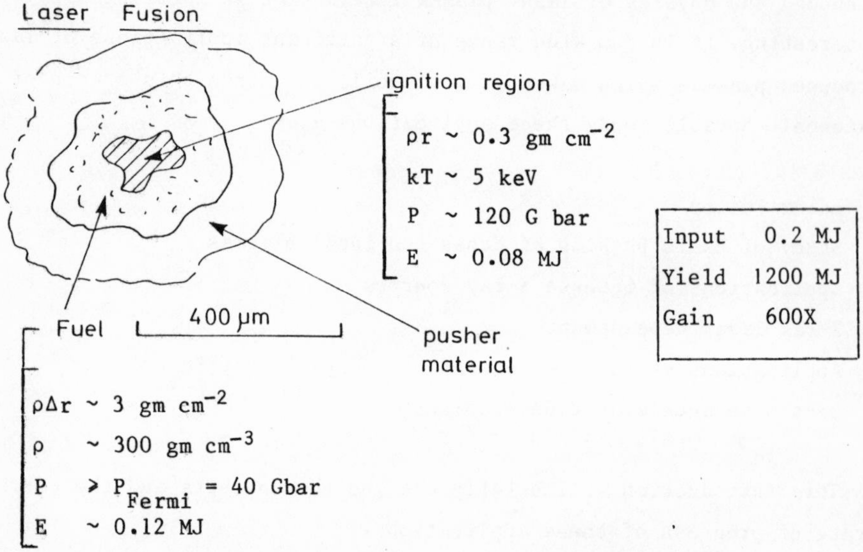

Fig 1 Illustration of typical conditions for high gain inertially confined fusion.

The ignition region is surrounded by relatively cold fuel which is 'burned' as the reaction spreads by α particle heating from the ignition region. The minimum pressure in the fuel, for a given density, is the pressure of Fermi degenerate electrons

$$P_f/1\text{Mbar} = 3.3(\rho/1\text{gm cm}^{-3})^{5/3} \qquad (1)$$

which would be realised in perfectly adiabatic compression of liquid DT ($\rho \sim 0.2$ g cm^{-3}). Any heat input (from energetic plasma electrons or X-rays [W Kruer ibid]) raises the pressure above the Fermi degenerate minimum.

The cold fuel must have a density x radius $\rho r \gtrsim 3$ gcm^{-2} in order to give a sufficiently large inertial confinement time for the thermonuclear burn to occur efficiently. [The product ρr is proportional to the more familiar Lawson number $n\tau$, since the confinement time τ is proportional to r, and the particle number density n is proportional to the mass density ρ. The probability of any nucleus making a fusion collison with another is proportional to $n\tau$ and therefore to ρr]. It follows that the required mass and energy content of the fuel scales as ρ^{-2} and that a higher density leads to a smaller scale fusion burn. The scale which is envisaged for laser fusion is illustrated in Fig 1. The density is $\rho \sim 300$gm cm^{-3} requiring pressure $P \gtrsim P_f = 40$ gigabar and the thermal energy content in the fuel is $E \gtrsim 0.12$ MJ. The diameter of the compressed fuel is 400μm and the thermonuclear energy yield for the system shown in Fig 1 would be 1200MJ. Thus for an initial energy content of 0.2 MJ, the gain G would be 600 x[2].

Useful fusion power could be obtained if the product of the relevant efficiency factors exceeds the above gain factor, ie.

$$G \cdot \varepsilon_{hyd} \cdot \varepsilon_{abs} \cdot \varepsilon_{las} \cdot \varepsilon_{therm} \gtrsim 1 \qquad (2)$$

where ε_{therm} is the thermal cycle efficiency (~ 0.3), ε_{las} is the laser efficiency (~ 0.1), ε_{abs} is the absorption efficiency (~ 0.5) and ε_{hyd} is the hydrodynamic efficiency of conversion of absorbed laser energy into compressed fuel energy (~ 0.1)[4].

The method of compression envisaged is the implosion of a spherical shell containing the DT fuel, as illustrated in Fig 2. The direct drive approach (a) uses laser irradiation of the outer surface of the shell to produce a pressure of the order of 50 Mbar. The indirect drive method (b) uses the trapping of energy in a cavity to generate a similar pressure from trapped plasma and ablation by thermal (soft X-ray) radiation.

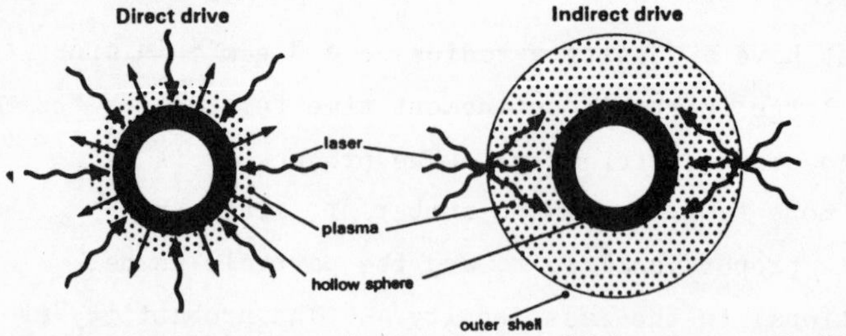

Fig 2 Alternative schemes for laser-driven implosions. In direct drive, a hollow spherical target is illuminated with laser light from all directions and plasma flows out from the irradiated surface. In indirect drive, the target is contained in an outer shell, with laser radiation entering through two tiny ports. The enclosure is filled with plasma and soft X-ray radiation.

With a suitably tailored variation of pressure with time the contents of the shell can be compressed adiabatically(3) while a small region at the centre can be heated to ignition by convergence of a spherical shock wave.

There are many basic physics questions which still need to be answered and these act as a stimulus to the study of laser plasma interactions discussed in the following sections of these proceedings. Referring now to direct drive we can identify:

The maximum ablation pressure P_a - shorter wavelengths are better but what limits are imposed by reduced thermal smoothing? [M H Key ibid]. Higher irradiance is better but how high before hot electrons become a problem, and how does the longer scale length in a full-sized fusion target affect the problem? [W Kruer ibid]. What limits heat flow and how does this affect pressure? [M H Key ibid].

Pressure multiplication in implosions The implosion process multiplies the ablation pressure with an approximate scaling [4]

$$P \sim P_a \, (r/\Delta r) \, G \qquad (3)$$

where $r/\Delta r$ is the radius to wall thickness ratio and G a spherical convergence enhancement. $r/\Delta r$ is limited by the Rayleigh-Taylor instabilty but to what value? [M H Key ibid] and G is limited (at what level?) by symmetry imperfections connected with non uniformity of irradiation, which is mitigated by thermal smoothing and may be exacerbated by thermal instabilities [M H Key ibid].

Preheat and density The final density for a given pressure is determined by the level of preheat by X-rays and hot electrons. Larger targets with thicker walls reduce the problem but the longer scale length of plasma in large targets makes it worse. Hot electrons from Raman, 2 plasmon and filamentation instabilities place upper bounds on I and on λ (at what levels?) [W Kruer ibid]. Preheating by shocks can be suppressed by 'shaping' the ablation pressure as a function of time - can this be done adequately?

Symmetry The irradiation geometry determines uniformity of illumination but thermal smoothing improves symmetry requiring high I and λ, but how high since this constraint is contrary to that imposed by preheating? What role is played by heat flow instabilities and self-focussing?

Ignition Targets can be designed to give a 'hot' core for ignition but can this be hot enough and small enough relative to the rest of the fuel?

Hydrodynamic efficiency This is improved with high values of $r/\Delta r$ but what level is possible bearing in mind the Rayleigh-Taylor limit on $r/\Delta r$? Short wavelength helps by ablating more mass at lower velocity but what is finally achievable?

Absorption Is absorption efficiency adequate in long scale length plasmas or will Brillouin backscatter be a problem? How does shorter wavelength help and is there a satisfactory solution?

Direct/Indirect drive Which is best?

Scalability How many of these questions can be answered in small scale experiments?

Laser fusion experiments to date have been on a scale determined by the available laser power and energy with $E_L \leqslant 10kJ$ and $\rho r \leqslant 10^{-1} g\ cm^{-2}$. Larger scale experiments are now starting at $E_L \approx 50kJ$ with the NOVA laser facility in the USA(5) which should enable the first approach to ignition conditions.

The advantage of short laser wavelength for reducing preheat and giving good absorption efficiency and high drive pressure has been established by numerous experiments [M H Key ibid] and NOVA operates at the 3rd harmonic of the Nd:glass laser wavelength, viz $\lambda = 0.35\mu m$. The efficiency of Nd:glass lasers such as Nova ($\varepsilon_{las} = 0.001$) is much to low for a fusion reactor and much interest attaches to the efficient KrF laser which produces UV radiation, ($\lambda = 0.249\mu m$) with efficiency up to several per cent [F O'Neill ibid].

2.
DENSE NON CLASSICAL PLASMAS

Plasma in which the Coulomb interaction energy between plasma ions is larger than the thermal energy ($\Gamma > 1$) behave differently from 'classical' plasmas in which the opposite is true, and there are interesting parallels with condensed matter studies of liquid salts and with astrophysical analysis of conditions inside stars. In particular multiple ionisation ($Z \gg 1$), simultaneous with high Γ

requires high pressure $P \alpha \Gamma^3 Z^3$, and the only possibilty of studying such plasmas in the laboratory is by laser compression[6].

Areas of current investigation include theoretical analysis of 'non ideal' behaviour; eg pressure ionisation, the plasma electric microfield and consequent changes in K absorption edges, collision rates, energy transport and radiation. A significant question is what is observable? eg absorption spectra[7] and plasma parameters such as P, ρ, T, etc.

The accessible parameter range, see Fig (3), is also being assessed making use of shock compression, shaped pressure impulses and spherical implosions.

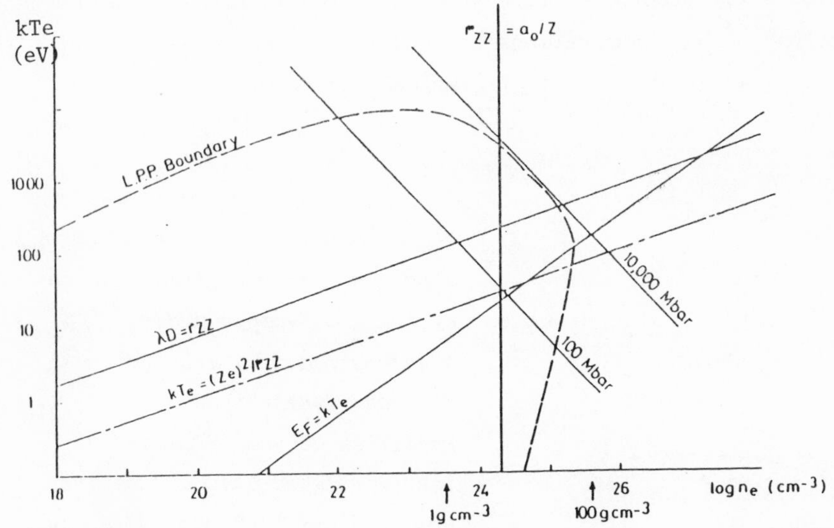

Fig 3 The accessible range of plasma parameters is shown by the boundary for LPP's (laser produced plasmas) against an ordinate of electron temperature kT_e in electron volts and abscissa of electron number density $n_e(cm^{-3})$. Lines of constant pressure are shown at 10,000 and 100 megabars. Non-Debye plasmas are bounded by Debye length λ_D less than the ion separation r_{zz}. Strong Coulomb coupling occurs for interaction energy $(Ze)^2/r_{zz}$ greater than thermal energy kT_e. Electron degeneracy prevails for Fermi energy E_F exceeding thermal energy kT_e. Pressure ionisation occurs for ion separation r_{zz} less than the radius of the first Bohr orbit a_0/Z (shown for Z = 3).

3.
X-RAY SOURCES

Laser produced plasmas are the brightest laboratory source of soft X-ray emission, with single pulse brightness exceeding that of synchrotron radiation sources by up to a million times (Fig 4). The average power of the latter is of course much higher because the pulses occur at > 10MHz but the unique possibility offered by laser produced plasmas is to make measurements using a single pulse with subnanosecond time resolution. The major areas of application are in absorption spectroscopy in general, and in particular to condensed matter research using X-ray absorption fine structure EXAFS[8] and reflection X-ray absorption fine structure REFLEXAFS[9]. Pulsed diffraction[10] has applications in condensed matter studies and microscopic radiography[11] in biological research.

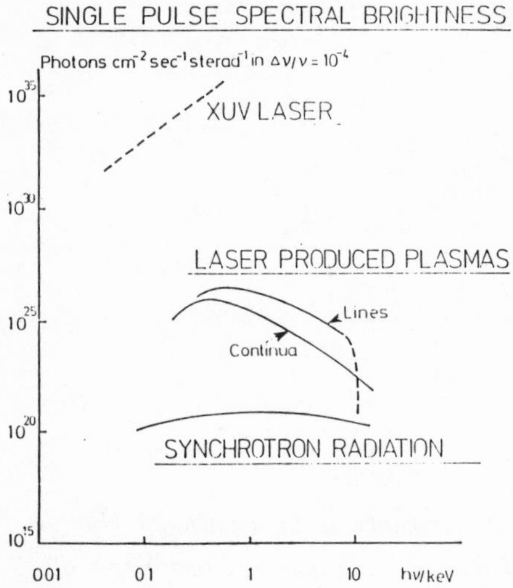

Fig 4 Comparison of the absolute spectral brightness of XUV and X-ray sources. (Synchrotron data are from the SERC Synchrotron Radiation Source (SRS) and laser-produced plasma data are compiled for 10^{12} watt Nd glass lasers.[15]

The spectrum of X-ray emission[12] is readily changed for these various purposes from continuum to narrow line by choice of target material (Fig 5) and up to 50% of the laser intensity can be converted to X-rays using targets of high atomic number.

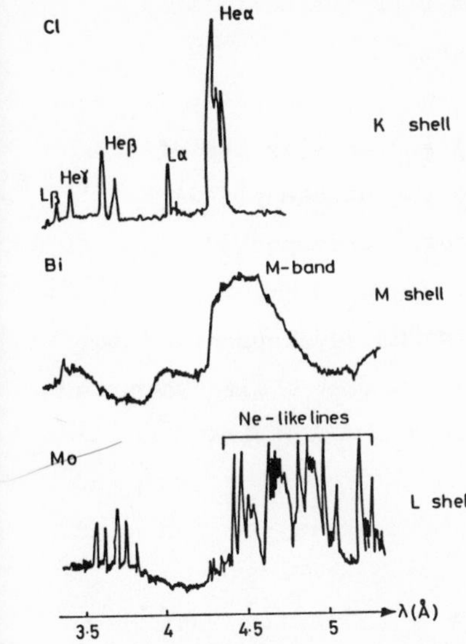

Fig 5 X-ray spectra, from top, a polymer target containing chlorine showing hydrogenic and He-like resonance lines, middle, a bismuth target showing the M band quasi continuum and bottom, Mo target showing the L shell Ne-like line emission. The spectral range is shown in the scale below.

The application to microscopy of living cells is very topical. A subnanosecond flash of X-rays with photon energy below the K absorption edge of oxygen but above the K edge of carbon, is strongly absorbed by carbon based materials in living cells but transmitted by their water environment. The absorption coefficient is high enough for a single cell to produce a radiograph pattern which can be recorded on an X-ray photoresist in contact with the 'in vivo' specimen. Resolution down to 100nm has aleady been demonstrated making it possible for the first time to study the submicron structure of living cells. A closely related possiblity is to replicate microcircuit patterns on the X-ray resist to develop even more miniaturised systems[(13)(14)].

The physics problems in this area are of two kinds, ie the optimisation of the source and the development of the new applications. In the present context the first is more relevant and the key points are to understand and optimise:

Source Brightness requiring understanding of energy transport atomic physics and radiation transfer and particularly of the effect of using shorter laser wavelength.

Spectral range involving ionisation physics and maximising energy density to generate shorter wavelengths.

Duration/size Shorter pulses and small source size are of interest and require understanding of cooling of the plasma by radiation and expansion and of lateral energy transport.

High repetition rate/small scale Successful development of many of these applications depends on providing smaller scale systems of higher pulse repetition rate, eg with medium sized excimer lasers[14].

4.
X-RAY LASERS

The intense thermal fluorescent X-ray emission of laser produced plasmas is the key to their application as amplifying media for X-ray lasers[15]. It is readily shown that the fluorescent brightness of a laser medium having a given amplification coefficient increases with frequency as $\nu^{4.5}$, leading to a need for prodigious fluorescent brightness for lasers of wavelengths significantly shorter than the current 100nm limit. The problems of constructing a laser resonator at soft X-ray wavelengths are severe and research is aimed at creating single or double transit ASE (amplified spontaneous emission) lasers is illustrated in Fig (6). The condition for a laser is that the output intensity should reach saturation, where stimulated emission becomes more probable than spontaneous fluorescence. It is readily shown[15] that the gain needed for this is about exp (14) and that the ouput power is about approximately

$$P/(10^6 \text{ Watt}) = (A/10^{-4} \text{ cm}^{-2})\ (h\nu/100\text{eV})^{4.5} \qquad (4)$$

where A is the output area of the plasma in Fig (6).

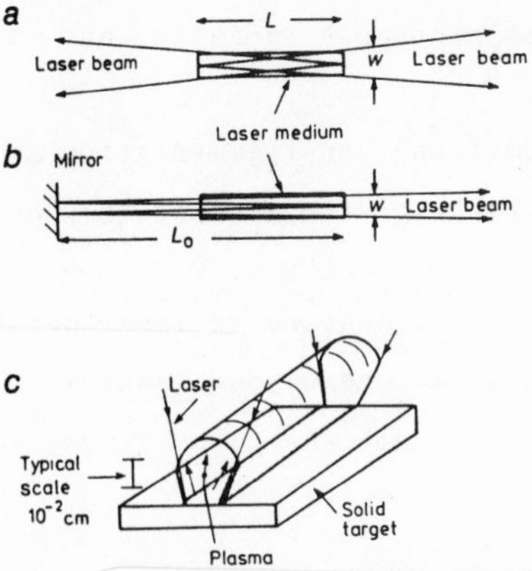

Fig 6 Schematic single-pass (a) and double-pass (b) ASE lasers, (c) Schematic model of plasma production by a laser beam focused on to a line.

The resulting spectral brightness would of course be much greater than the thermal fluorescence of laser produced plasmas as shown in Fig (4).

A major problem is to find ways of creating population inversion in laser produced plasmas and there are several very active lines of investigation being pursued including:

- recombination in transient supercooled plasma, eg in CVI[16]
- collisional excitation of 3p-3s transitions in neon-like ions, eg SeXXV[17]
- resonant photo pumping using intense line emission, eg NaXI pumping NeVIII[18].

Line focus geometry for plasma production is needed as illustrated in Fig (6) and this implies new problems in design of irradiation optics[19] and in target fabrication.

There is much interesting physics in this area, which is extremely topical as a result of recent work on Se XXV showing gain of exp (6.5) in a 12mm length of SeXXV plasma[17].

Physics problems of relevance include:

- Plasma hydrodynamics needed to create specific density and temperature structure and time variation.

- Atomic physics of ionic composition and excited state population.
- Radiative physics of reabsorption of resonance radiation and its effect on excited state population.
- Thermal instability (plasma jet formation) and filamentation as potential destroyers of axial uniformity of the amplifying plasma.

There are also intriguing problems of diagnostics of laser action and in the development of the first applications of XUV lasers.

5. PARTICLE SOURCE APPLICATIONS

Thermonuclear reaction of D and T in implosions creates very high intensity emission of essentially monoenergetic ($\Delta E/E \sim 10^{-2}$) particles with an exceptionally high source brightness, eg with a circa 1kJ, 1nsec laser,

$$\left.\begin{array}{ll} n & (14\ \text{MeV}) \\ \alpha & (3.5\ \text{MeV}) \end{array}\right\} 10^{8} \text{ particles at } 10^{22}\ \text{cm}^{-2}\ \text{sec}^{-1}\ \text{sterad}^{-1}$$

Charged particles, notably α and p, interact strongly with matter and can be used to probe transient events with information being obtained form their energy loss recorded with circa 200:1 resolution by the analysis of etched tracks produced in CR39 polymer[20].

Applications include the study of charged particle energy loss in matter at high pressure and temperature[21] (induced by laser compression). This is of importance in the evaluation of the feasibility of ion beam fusion[22], and the study of density and thickness changes induced by Rayleigh-Taylor instability in laser accelerated targets[23].

The supersonic flow of collionless ions in the ablation plasma is also of exceptional intensity and finds applications on the study of collisionless shocks[24].

Laser plasma physics problems relevant to particle sources include

Thermonuclear yield optimisation Using 'exploding pusher' targets (eg 10^{12} neutrons produced in recent experiments[25]). Which laser wavelength is best?

Particle Energy Spectra Control of energy loss in escaping from the target, arising from the $\rho\Delta r$ of the imploding shell, (or indeed study of the shell thickness and uniformity from the energy loss).

Ion Emission Ionisation physics and knowledge of density and temperature in the flow.

6. BEAT WAVE ACCELERATOR[26]

It has recently become apparent that the electric field strength which can in principle be created in large amplitude plasma wves can exceed perhaps 100 fold that in conventional RF accelerators used in high energy nuclear physics. With HEP research facing a major problem in going to the next level of energy in the TeV range, requiring linear accelerators of unrealistic length (~75Km) a high field accelerator is of real interest.

A plasma in which the plama frequency is equal to the beat frequency ($\omega_{L_1} - \omega_{L_2}$) of 2 laser waves will be driven to large amplitude oscillation, Fig 7. The phase velocity of the wave is close to c, with relativistic γ,

$$\gamma = \omega_L/\omega_p = \omega_L/(\omega_{L_1} - \omega_{L_2}) \quad (5)$$

and maximum field strength,

$$eE_{MAX} = m_e c\,\omega_p . \quad (6)$$

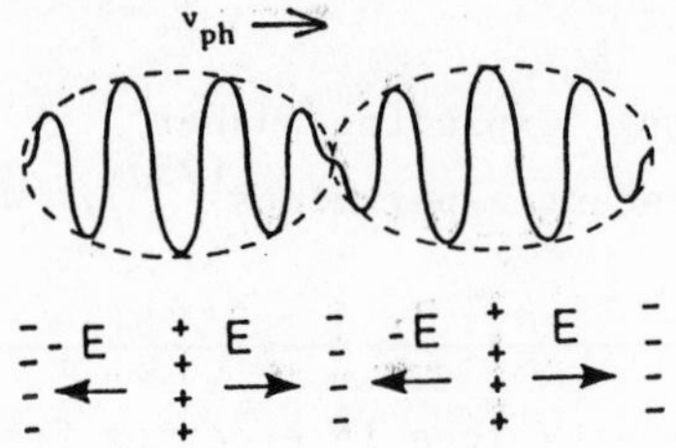

Fig 7 Schematic diagram of a beat pattern and corresponding concentrations of positive and negative charge and associated electric fields in a plasma.

Thus an electron could 'surf ride' on the wave experiencing continuous acceleration until it lost phase with the wave having gained energy of $2\gamma^2 mc^2$. An attractive possibility is to use the 1.06 and 1.05μm wavelengths of Nd YAG and YLF to resonate in a Z pinch plama with $n_e \sim 10^{17}\,cm^{-3}$ with $E_{MAX} \sim 3 \times 10^8$ V/cm and $\gamma \sim 100$.

There is much new and interesting physics here involving basic laser plasma interactions, eg

- the mechanisms governing the saturated amplitude of the plasma wave
- competing non linear effects
- the energy extraction efficiency
- diagnosis of accelerated particles and of the plasma wave itself.

Conclusion

The wide range of interesting and relevant science associated with laser plasma interactions is I hope suggested adequately by these notes and will encourage the reader to delve in more depth into the subject as set out in the subsequent chapters by expert authors which follow.

REFERENCES

1. S. Bodner, J. Fusion Energy, 1, 221 (1981).
2. J. Meyer ter Vehn, Nucl. Fusion, 22, 561 (1982).
3. R. Kidder, Nucl. Fusion, 16, 3 (1976).
4. B. Ahlborn, M. Key, A. Bell, Phys. Fluids, 25, 541 (1982).
5. Lawrence Livermore National Laboratory, Laser Program Annual Report (1983).
6. See R. More ibid.
7. J. Hares et al, Rutherford Appleton Laboratory Report RAL 85-047, A5.1 (1985).
8. R. Eason, D. Bradley, J. Kilkenny, G. Greaves, J. Phys C 17, 5067 (1984).
9. R. Eason, D. Bradley, P. Dobson, J. Hares, Appl. Phys. Letts., 47, 422 (1985).
10. P. Dobson, J. Hares, S. Tabatabaei, J. Lunney, R. Eason, Opt. Comm. (in press).
11. R. Rosser, K. Baldwin, R. Feder, D. Bassett, A. Cole, R. Eason, J. Microscopy, 138, 311 (1985).
12. R. Eason, P. Cheng, R. Feder, A. Michette, R. Rosser, F. O'Neill, Y. Owadano, P. Rumsby, M. Shaw, I. Turcu, Optica Acta (in press).
13. D. Nagel, M. Peckerar, R. Whitlock, J. Greig, R Peckhacer, Elect. Letts., 14, 781 (1978).
14. F. O'Neill, M. Gower, I. Turcu, Y. Owadano, Appl. Optics (in press).
15. M. Key, Nature, 316, 314 (1985).
16. D. Jackoby, G. Pert, L. Shorrock, G Tallents, J. Phys. B, 15, 3557 (1982).
17. D. Mathews et al, Phys. Rev. Letts., 34, 110 (1985).
18. J. Apruzeze, J Davis, K Witney, JAP, 53, 4020 (1982).
19. I. Ross, E. Hodgson, J. Phys. E Sci. Inst., 18, 169 (1985).
20. A. Fews, D. Henshaw, Nucl. Instr. and Methods, 197, 517 (1982).
21. P. Evans, A. Fews, S. Knight, D. Pepler, W. Toner, Rutherford Appleton Laboratory Report RAL 85-047, A1.10 (1985).
22. K. Long, N. Tahir, Phys. Lett., 91A, 451 (1982).

23. P. Evans, A. Fews, A. Cole, C. Edwards, C. Hooker, D. Pepler, W Tòner, J Wark, Rutherford Appleton Laboratory Report RAL 85-047, A2.21 (1985).

24. P. Choi et al, Rutherford Appleton Laboratory Report RAL-85 047, A4.22 (1985).

25. C. Yamanaka, S. Nakai (to be published).

26. T. Tajima, J. Dawson, Phys. Rev. Lett., 43, 268 (1979).
R. Evans, Rutherford Appleton Laboratory Report 84-086-1984 (1984).

LASER LIGHT PROPAGATION AND ABSORPTION

P. Mulser

Institut für Angewandte Physik, Technische Hochschule Darmstadt

1.

COLLISIONAL ABSORPTION

1.1 Momentum Equation

Under the influence of high power radiation dense matter is quickly ionized to such a high degree that the model of a fully ionized plasma may be applied in the following[1]. Let the electromagnetic field

$$\underline{E}(\underline{x},t) = \hat{E}(\underline{x},t)e^{-i\omega t} \tag{1.1}$$

act on the electrons, the velocity distribution of which is indicated by $f(\underline{V})$. Owing to the large ion mass m_i and the relation between electron and ion temperatures $T_i < T_e$ the Lorentz model of immobile ions (in comparison to the thermal speed of the electrons) is an excellent approximation. At the collision parameter b and electron is deflected in the Coulomb field of the ion charge $q = -Ze$ by the amount of

$$\tan\frac{\chi}{2} = \frac{Ze^2}{4\pi\varepsilon_o b\, m_e V^2} \tag{1.2}$$

(Fig.1) thereby transmitting in the direction of $\underline{V}$ the momentum component $m_e V(1-\cos\chi)$ to the ion. From eq. (1.2) the differential

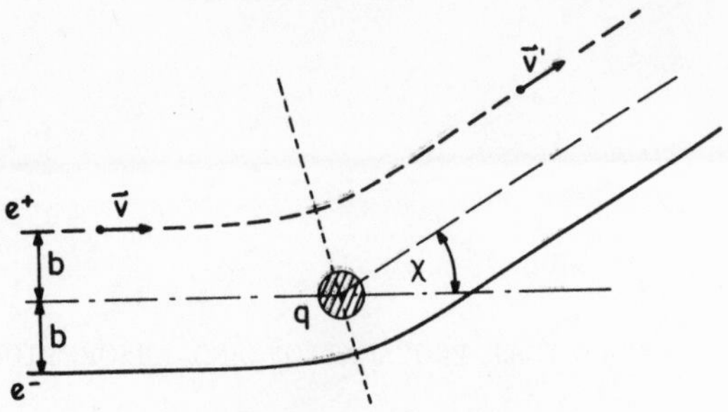

Fig. 1 - Deflection of a positive (-) or negative (——) by a positive ion of charge q = Ze.

cross-section $\sigma(\chi)$ is deduced,

$$\sigma(\chi) = \left(\frac{Ze^2}{4\pi\varepsilon_o . 2m_e V^2}\right)^2 \frac{1}{\sin^4(\chi/_2)} , \quad (1.3)$$

and the average momentum loss to the ions of density n_i per unit time is given by

$$m_e n_i \int_{\underline{V}} \int_{\chi_{min}}^{\pi} \sigma(\chi)\underline{V}(1-\cos\chi)f(\underline{V})(\underline{V}).2\pi \sin\chi \, d\chi \, d\underline{V} =$$

$$= m_e n_i Y \int_{\underline{V}} \frac{1}{V^3} \underline{V} f(\underline{V}) d\underline{V} \; ; \; Y = 4\pi \left(\frac{Ze^2}{4\pi\varepsilon_o m_e}\right)^2 \ln \frac{m_e V^2}{Ze^2/4\pi\varepsilon_o b_{max}} \quad (1.4)$$

The electromagnetic field forces the distribution function to oscillate between $\underline{V} + \hat{V}_{os}$ and $\underline{V} - \hat{V}_{os}$, $\hat{V}_{os} = e\hat{E}/m_e\omega$. Thus f becomes

$$f(\underline{V},t) = f_o(\underline{V} - \underline{V}_{os}(t)), \; \underline{V}_{os} = \hat{V}_{os} e^{-i\omega t} , \quad (1.5)$$

where $f_o(\underline{V})$ is the unshifted distribution of the time instants at which $\underline{V}_{os} = 0$ holds. Under the assumption

$$|\underline{V}_{os}| << \langle V \rangle \quad (1.6)$$

$f(\underline{V}_1 t)$ may be expanded into

$$f(\underline{V}_1 t) = f_o(\underline{V}) - \underline{V}_{os} \nabla_V f_o(\underline{V})$$

to yield the average momentum loss per unit time

$$m_e n_i Y \int \frac{\underline{V}}{V^3} f(\underline{V}) d\underline{V} = -m_e n_i Y \int \frac{\underline{V}}{V^3} (\underline{V}_{os} \nabla f_o) d\underline{V} .$$

Generally (or more precisely, when there are enough electron-electron and electron-ion collisions) f_o is assumed isotropic, $f_o(\underline{V}) = f_o(V)$:

$$-m_e n_i Y \int \frac{\underline{V}}{V^3} (\underline{V}_{os} \nabla_V f_o) d\underline{V} = -m_e n_i Y \underline{V}_{os} \int \frac{\partial f_o}{\partial V} \cos^2\theta . 2\pi \sin\theta d\theta dV =$$

$$-m_e \underline{V}_{os} . 4\pi n_i \frac{Y}{3} \int_0^\infty \frac{\partial f_o}{\partial V} dV = -m_e \underline{V}_{os} \nu_{ei} . \qquad (1.7)$$

For a Maxwellian distribution,

$$f_o(\underline{V}) = \left(\frac{\beta}{\pi}\right)^{\frac{3}{2}} e^{-\beta V^2}, \ \beta = \frac{m_e}{2kT_e}, \qquad (1.8)$$

this directly yields the familiar expression for the collision frequency ν_{ei},

$$\nu_{ei} = 4\pi n_i \frac{Y}{3} f_o(0) = \frac{4(2\pi)^{\frac{1}{2}}}{3} \left(\frac{Ze^2}{4\pi\varepsilon_o m_e}\right)^2 \left(\frac{me}{kT_e}\right)^{\frac{3}{2}} n_i \, ln\Lambda , \qquad (1.9)$$

which generally is deduced from a Fokker-Planck equation(2). $ln\,\Lambda$ is the Coulomb logarithm of eq. (1.4). In cgs.grad units the numerical value of ν_{ei} is

$$\nu_{ei} = 3.6 \times \frac{Z^2 n_i}{T_e^{\frac{3}{2}}} \, ln\,\Lambda . \qquad (1.10)$$

These considerations show that the correct equation of motion for an electron in the oscillatory field is

$$\frac{d\underline{V}_{os}}{dt} + \nu_{ei} \underline{V}_{os} = -\frac{e}{m_e} \hat{E}(\underline{x}_1 t) e^{-i\omega t}, \qquad (1.11)$$

from which the current density

$$\underline{j} = -n_e e \underline{V}_{os} = i\varepsilon_o \frac{\omega_p^2}{\omega} \frac{1}{1+i\frac{\nu_{ei}}{\omega}} \underline{E} , \quad \omega_p^2 = \frac{n_e e^2}{m_e \varepsilon_o} ,$$

is obtained. The absorbed power per unit volume follows from the time averaged Poynting theorem,

$$\overline{\nabla \underline{S}} = -\overline{\underline{j}\underline{E}} = -\frac{\varepsilon_o}{2} \frac{\omega_p^2}{\omega^2} \frac{\nu_{ei}}{1+\left(\frac{\nu_{ei}}{\omega}\right)^2} \hat{E}\hat{E}^* . \qquad (1.12)$$

Generally, in laser plasmas $(\nu_{ei}/\omega)^2$ is very small. From Maxwell's equations the following wave equation is deduced by eliminating $\underline{B}$

$$\nabla \times \nabla \times \underline{E} + \frac{1}{c^2}\frac{\partial^2}{\partial t^2}\underline{E} = -\frac{1}{\varepsilon_o c^2}\frac{\partial}{\partial t}\underline{j} \; . \qquad (1.13)$$

As long as the electron density is smooth, i.e.

$$L = \frac{n_e(\underline{r})}{|\nabla n_e|} << \lambda(\underline{r}) \qquad (1.14)$$

holds with $\lambda(\underline{r})$ the local light wavelength, $\nabla\underline{E} = 0$ can be set and eq. (1.13) transforms, with the expression for $\underline{j}$, into the stationary wave equation

$$\nabla^2\hat{E} + \underline{k}^2 n^2(\underline{x})\hat{E} = 0 \; , \quad k = \omega/c \; , \qquad (1.15)$$

where the refractive index $n(\underline{x})$ is related to the electrical conductivity $\sigma(\underline{x})$ by

$$n = 1 + i\frac{\sigma}{\varepsilon_o\omega} = 1 - \frac{\omega_p{}^2}{\omega^2}\frac{1}{1+i\frac{\nu_{ei}}{\omega}} \; , \quad \sigma = i\varepsilon_o\frac{\omega_p{}^2}{\omega}\frac{1}{1+i\frac{\nu_{ei}}{\omega}} \qquad (1.16)$$

Since under condition (1.14) $\underline{B} = n\underline{k} \times E/\omega$ holds $\underline{S}$ becomes

$$\underline{S} = \tfrac{1}{2}\varepsilon_o \underline{c}\, \mathrm{Re}(n)\, \hat{E}\hat{E}^* \; .$$

By setting $\omega_{p1}^2 = \dfrac{\omega_p{}^2}{1+\left(\frac{\nu_{ei}}{\omega}\right)^2}$, $A = 1-\omega_{p1}{}^2/\omega^2$, $B = \dfrac{\nu_{ei}}{\omega}\dfrac{\omega_{p1}^2}{\omega^2}$

Re(n) and Im(n) are obtained from

$$\mathrm{Re}(n) = \sqrt{\tfrac{1}{2}\{\sqrt{A^2+B^2}+A\}} \; , \quad \mathrm{Im}(n) = \sqrt{\tfrac{1}{2}\{\sqrt{A^2+B^2}-A\}} \qquad (1.17)$$

For $B^2 << A^2$ and $\omega_{p1} < \omega$ holds

$$\mathrm{Re}(n) = \left(1-\frac{\omega_p{}^2}{\omega^2}\right)^{\frac{1}{2}} \; , \quad \mathrm{Im}(n) = \frac{1}{2}\frac{\omega_p{}^2\,\nu_{ei}}{\omega^3\left(1-\frac{\omega_p{}^2}{\omega^2}\right)^{\frac{1}{2}}}$$

The law for collisional absorption can be presented as

$$\overline{\nabla\underline{S}} = -k\frac{\omega_{p1}^2}{\omega^3}\frac{\nu_{ei}}{\mathrm{Re}(n)}\underline{S} \; , \qquad (1.18)$$

with the absorption coefficient α,

$$\alpha = k\,\frac{\omega_{pl}^2}{\omega^3}\,\frac{\nu_{ei}}{\mathrm{Re}(n)} = 2k\,\mathrm{Im}(n)\,. \qquad (1.19)$$

In regions of $L < \lambda(\underline{r})$, for instance near critical surfaces with $\omega_{pl} = \omega$, instead of eq. (1.18) the more general relation eq. (1.12) has to be used and $\underline{S}$ has to be determined from the steady state wave equation (1.15).

1.2 The Collision Frequency ν_{ei}

The formulae for collisional absorption presented in the foregoing section are valid under the following conditions:

a) $\hat{V}_{os} \ll \langle V\rangle$, $(\langle V\rangle = (k\,T_e/m_e)^{\frac{1}{2}} \simeq V_{thermal})$,

b) oscillatory amplitude $\delta_{os} \ll \lambda$,

c) $V_{os}(t)$ does not considerably change during the interaction with the ion.

In addition, condition (1.14) for electron density variations must hold. As far as relation a) is concerned we observe that

$$\hat{E} = (2I/\varepsilon_o c)^{\frac{1}{2}} = 27.5 I^{\frac{1}{2}}\ [\mathrm{V/cm}],\ [I] = \mathrm{W/cm^2},$$

$$\frac{\hat{V}_{os}}{\langle V\rangle} = \frac{2^{\frac{1}{2}}e}{(\varepsilon_o c m_e k_B)^{\frac{1}{2}}}\,\frac{1}{\omega}\left(\frac{I}{T_e}\right)^{\frac{1}{2}} = 7\times10^{-5}\,\frac{\omega_{Nd}}{\omega}\left(\frac{I}{T_e}\right)^{\frac{1}{2}}\ [\mathrm{W/cm^2\ {}^oK}]\,. \qquad (1.20)$$

This means, for instance, that at $I_{Nd} = 10^{14}\ \mathrm{W/cm^2}$ and $T_e = 1\ \mathrm{keV}$ $\hat{V}_{os}/\langle V\rangle = 0.2$, i.e. the inequality is fulfilled, however, generally condition a) is critical. With inequality a) also b) is fulfilled unless the electron is resonant at ω. In the expression for the Coulomb logarithm,

$$\underline{ln}\,\Lambda = ln\,\frac{m_e V^2}{Ze^2/4\pi\varepsilon_o b_{max}}\ ,$$

traditionally the maximum impact parameter b_{max} is identified with the Debye length λ_D

$$\lambda_D = (\varepsilon_o k T_e/n_e e^2)^{\frac{1}{2}} = V_{th}/\omega_p = 6.9\ (T_e/n_e)^{\frac{1}{2}}\ [{}^o\mathrm{K\ cgs}]\ ,$$

and $m_e V^2$ is substituted by $m_e\langle V\rangle^2 = 3kT_e$. In this way Λ becomes

$$\Lambda = \frac{12\pi\varepsilon_o k T_e}{Ze^2/\lambda_D} = \frac{\text{mean kin. energy}}{\text{mean pot. energy}} = 12\pi n_e \lambda_D^{\,3}\ .$$

However, this is only correct for $\omega < \omega_p$. It is easy to see that in an electron-ion collision energy is absorbed effectively from the wave as long as during the interaction time $\tau \simeq 2b/V$ the phase of the $\underline{E}$-field changes by less than 2π. This argument leads to the cut-off parameter and Coulomb logarithm

$$b_{max} = \frac{\pi\langle V\rangle}{\omega}, \quad ln\,\Lambda = ln\,\frac{4\pi^2\varepsilon_o\langle V\rangle^3}{Ze^2\omega} \tag{1.21}$$

The functional dependence of $ln\,\Lambda$ agrees with a more careful investigation in the literature[3]. In the low density region ($\omega_p \ll \omega$) it represents a reduction of collisional absorption.

In eq. (1.4) the integration in χ was extended to π which corresponds to a minimum impact parameter $b_{min} = 2Ze^2/4\pi\varepsilon_0 meV^2 \simeq Ze^2/4\pi\varepsilon_0 kT_e$. With increasing temperature this becomes less than the de Broglie wavelength λ_B,

$$\lambda_B = 2\pi\frac{\hbar}{mV} \simeq \pi\hbar/(m_e kT_e)^{\frac{1}{2}}.$$

Since localization of an electron better than λ_B is not possible the maximum of b_{min} and λ_B has to be chosen for b_{min}.

From the deduction of ν_{ei} in the foregoing section it becomes clear that electron-electron collisions do not contribute to absorption: because of the same displacement δ_{os} all electrons undergo in the wave at neighbouring positions, no momentum transfer to an arbitrary electron occurs on the average. Only at relativistic temperature the situation changes since the displacement $\delta_{os}(\underline{r},t)$ is no longer equal for two electrons at the same position and different speed. Consequently, electron-electron collisions may also contribute to absorption of radiation.

It was calculated that electron-electron collisions contribute to reduce the dc electrical conductivity by a factor between 0.6 for Z=1 and 1.0[4,5]. From this such a correction is sometimes erroneously taken into account, although it has been shown by several authors that this correction reduces to zero for $\omega > \nu_{ei}$[2]. Thus, under conditions a), b), and c) the absorption coefficient for a plasma with Maxwellian electron distribution becomes

$$\alpha = \frac{1}{c}\left(\frac{\omega_{p1}}{\omega}\right)^2 \frac{\nu_{ei}}{\mathrm{Re}(n)} = 1.2 \times 10^{+11}\left(\frac{n_e}{n_{Nd}}\right)^2 \frac{Z}{T_e^{\frac{3}{2}}} \frac{ln\,\Lambda(\omega)}{\kappa\,\mathrm{Re}\,(n)} \;[\mathrm{cgs}\ ^{\circ}\mathrm{K}] \qquad (1.22)$$

κ is the small correction factor $\kappa = 1 + \nu_{ei}^2/\omega^2$ and $n_{Nd} = 10^{21}\ \mathrm{cm}^{-3}$ is the critical electron density for ω_{Nd}. Formula (1.23) is in agreement with that presented in Ref.6.

1.3 Hard sphere model

By collisional damping reversible oscillatory energy is transformed into disordered, thermal energy. In order to get more physical insight in the details of this process and to illustrate the limitations of the collision frequency given by eq. (1.10) it is instructive to study the phenomenon of absorption in a hard sphere model. For this purpose we consider the following situation of Fig.2 : An electron having relative speed $\underline{v}_0$ collides with the corresponding ion at the time instant t_0. After the collision its directed velocity is $\underline{v}'_0$. When no field is present incident and reflected angles are equal, $\alpha' = \alpha$, and the same holds for $|\underline{v}_0|$ and $|\underline{v}'_0|$. However in the presence of an oscillatory field $\underline{E}(\underline{x},t) = \hat{E}\cos(\underline{kx} - \omega t)$ the sum of velocities, $\underline{v}(t) = \underline{v}_0 + \underline{v}_{os}(t)$, is conserved in its magnitude in the collision. That is the origin of α' being different from α ("symmetry breaking").

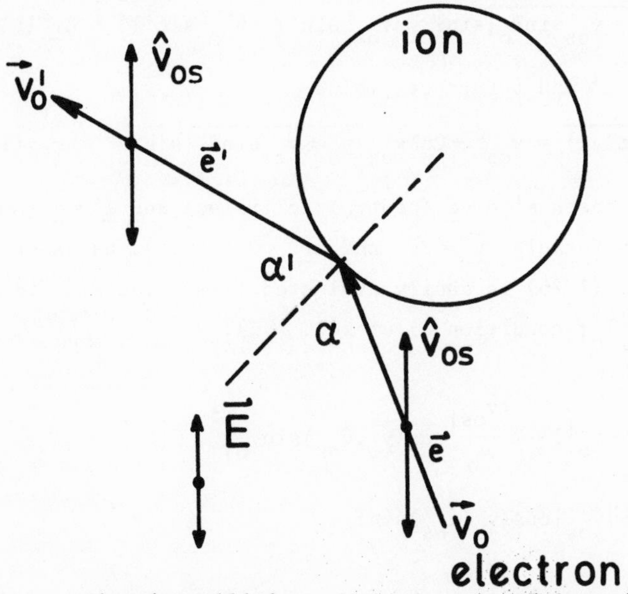

Fig. 2 - Elastic collision of an electron with an ion (hard sphere model). The E-field breaks the symmetry of reflection ($\alpha' \neq \alpha$).

In detail the following situation occurs. Before the collision the speed of the electron is in the direction of the unit vector $\underline{e}$.

$$\underline{v}(t') = \underline{v}_o + \hat{v}_{os}\sin(\underline{kx} - \omega t'),\ \hat{v}_{os} = \frac{e}{m\,\omega}\hat{E}\ ;\ t' < t_o \tag{1.23}$$

In a hard sphere model this expression for $\hat{v}_{os}$ is the correct one. Due to the collision $\underline{v}(t_o)$ is scattered in the direction of $\underline{e}'$, so that the total speed for $t \geq t_o$ is

$$\underline{v}(t) = |\underline{v}_o + \hat{v}_{os}\sin(\underline{kx} - \omega t_o)|\underline{e}' + \hat{v}_{os}\{\sin(\underline{kx} - \omega t) - \sin(\underline{kx} - \omega t_o)\} \tag{1.24}$$

The energy gain of the electron is $\frac{1}{2}m_e\{\underline{v}^2(t) - \underline{v}^2(t')\}$, the difference $\Delta\underline{v}^2$ being given by $(\Psi = \underline{kx} - \omega t)$

$$\begin{aligned}\Delta\underline{v}^2 &= \underline{v}^2(t) - \underline{v}^2(t') = 2\underline{v}_o\hat{v}_{os}(\sin\Psi_o - \sin\Psi') + 2\hat{v}_{os}{}^2\sin^2\Psi_o \\ &+ 2\underline{e}'\hat{v}_{os}|\underline{v}_o + \hat{v}_{os}\sin\Psi_o| \cdot (\sin\Psi - \sin\Psi_o) + \hat{v}_{os}{}^2\sin^2\Psi \\ &- 2\hat{v}_{os}{}^2\sin\Psi_o\sin\Psi - \hat{v}_{os}{}^2\sin^2\Psi'.\end{aligned} \tag{1.25}$$

This has to be averaged over all collision instants t_o occurring with equal probability,

$$\begin{aligned}\frac{1}{\Delta t}\int_{t'}^{t}\Delta\underline{v}^2 dt_o &= -2\underline{v}_o\hat{v}_{os}\sin\Psi' + \hat{v}_{os}{}^2 + 2\underline{e}'\hat{v}_{os}\overline{|\underline{v}_o + \hat{v}_{os}\sin\Psi_o|}\sin\Psi \\ &- 2\underline{e}'\hat{v}_{os}\overline{|\underline{v}_o + \hat{v}_{os}\sin\Psi_o|\sin\Psi_o} + \hat{v}_{os}{}^2\sin^2\Psi - \hat{v}_{os}{}^2\sin^2\Psi' = \Delta\underline{v}^2(t',t).\end{aligned}$$

Time averaging over t' and t further yields

$$\overline{\Delta\underline{v}^2(t',t)} = \hat{v}_{os}{}^2 - 2\underline{e}'\hat{v}_{os}\overline{|\underline{v}_o + \hat{v}_{os}\sin\Psi_o|\sin\Psi_o} \tag{1.26}$$

For the two special cases $\underline{e}' = -\underline{e}$ (central collision) and $\underline{e}' = \underline{e}$ (no collision) from this formula $\overline{\Delta v^2} = 2\hat{v}^2$ and $\Delta\underline{v}^2 = 0$ results, as expected. Only for $|\hat{v}| << v_o$ eq. (1.26) is easily evaluated (weak electric field or high temperature T_e; condition a) of sec. 1.2):

$$|\underline{v}_o + \hat{v}_{os}\sin\Psi_o| - v_o\left(1 + 2\frac{|\hat{v}_{os}|}{v_o}\cos(\underline{v}_o,\hat{v}_{os})\sin\Psi_o\right)^{\frac{1}{2}}$$

$$= v_o + |\hat{v}_{os}|\cos(\underline{v}_o,\hat{v}_{os})\sin\Psi_o.$$

Therefore we get

$$\overline{|\underline{v}_o + v_{os}\sin\Psi_o|\sin\Psi_o} = \tfrac{1}{2}|\hat{v}_{os}|\overline{\cos(\hat{v}_{os},\underline{v}_o)}. \tag{1.27}$$

If $\underline{v}_o$ is isotropic this average is zero and the net mean energy gain in one collision is $\frac{1}{2}m_e\hat{v}_{os}^2$. For hard spheres the differential cross section is constant. Therefore the collision frequency is given by $\nu_{ei} = n_i\sigma\underline{v} = \pi n_i d^2\underline{v}$, $d = 2r$, since the heavy particles of density n_i can be assumed at rest (the differential cross section is $\sigma_\Omega = d^2/4$). When the model is applied to soft Coulomb collisions $\underline{v}_{os}$ has to be determined from eq. (1.11). With this the energy absorbed per unit volume and unit time follows as

$$\frac{\partial\varepsilon_{th}}{\partial t} = \frac{1}{2}n_e m_e \nu_{ei}\left[\mathrm{Re}(\underline{v}_{os})\right]^2 = \frac{\varepsilon_0}{2}\left(\frac{\omega_{p1}}{\omega}\right)^2 \nu_{ei}\hat{E}\hat{E}^* ,$$

which is identical with eq. (1.12). Consequently, also the same absorption coefficient follows as in sec. 1.1. For this purpose one only needs to identify the hard sphere cross section σ with

$$\sigma = \frac{\nu_{ei}}{n_i\underline{v}} = \frac{\nu_{ei}/n_i}{(3kT_e/m_e)^{\frac{1}{2}}} = \frac{4(2\pi)^{\frac{1}{2}}}{3^{\frac{3}{2}}}\left(\frac{Ze^2}{4\pi\varepsilon_0}\right)^2 \frac{ln\,\Lambda(\omega)}{(kT_e)^2} . \qquad (1.28)$$

Numerically this is

$$\sigma = 1.6 \times 10^{-6}\,\frac{Z^2}{T_e^2}\,ln\,\Lambda(\omega)\ \left[\mathrm{cgs}\ {}^oK\right] \qquad (1.29)$$

Finally the question arises how appropriate a hard sphere model is for studying collision frequencies and asymmetry effects induced by a strong oscillatory electric field. In the following two tables some values of the electron thermal velocity, the corresponding distance it covers during one period of oscillation and its de Broglie wavelength λ_B as functions of temperature T_e are reported as well as oscillatory speeds $\hat{v}_{os}$ for different Nd laser intensities.

Table 1

T_e	1eV	10	100	1keV	10
$v_e = (3kT_e/m_e)^{\frac{1}{2}}$ [cms^{-1}]	6×10^7	2×10^8	6×10^8	2×10^9	6×10^9
$\Delta x = \frac{2\pi}{\omega_{Nd}} v_e$ [A]	20	60	200	600	2000
$\lambda_B = h(3m_e kT_e)^{-\frac{1}{2}}$ [A]	10	3	1	0.3	0.1

Table 2

I_{Nd} [W/cm^2]	10^{10}	10^{12}	10^{14}	10^{16}	10^{18}
$v_{os} = \frac{e\hat{E}}{m_e\omega_{Nd}} = 25\, I^{\frac{1}{2}} \left(\frac{\omega_{Nd}}{\omega}\right)$	2.5×10^6	2.5×10^7	2.5×10^8	2.5×10^9	2.5×10^{10}

It results that a hard sphere model represents an acceptable approximation within a wide range of parameters.

1.4 Inverse Bremsstrahlung absorption

To illustrate the physics of absorption of radiation by collisions from a different point of view in this section the absorption coefficient α is deduced from the emission of bremsstrahlung radiation of an optically thin plasma.

The total power P emitted by an accelerated nonrelativistic electron is given by[(7)]

$$P = \frac{2}{3}\frac{e^2}{4\pi\varepsilon_o c^3}\,\dot{\underline{v}}^2(t) \tag{1.30}$$

In a nearly thermal plasma velocity changes $\dot{\underline{v}}(t)$ are mainly produced by electron-ion and electron-electron collisions. An electron-ion pair constitutes a dipole system to a very good approximation as long as the maximum impact parameter b_{max} is small compared to the wavelength emitted. For laser plasmas and frequencies of $\omega \simeq \omega_{Nd}$ this is always the case because of

$$\lambda_{Nd} >> \lambda_D \simeq b_{max} .$$

On the other hand, a colliding electron-electron system represents a quadrupole which at nonrelativistic speeds and under the above conditions gives negligible contribution to radiation emission. The colliding electron-electron pair can also be viewed as two electric dipoles of relative phase difference of π (Fig.3). As a consequence, their dipole emission cancels. Since the principle of detailed balancing applies[(8)] absorption of radiation does not occur in electron-electron encounters either.

The total spectral intensity I_ω emitted by n_e electrons into the unit solid angle is obtained by Fourier-transforming Larmor's formula, eq. (1.30), and averaging over all impact parameters b and the

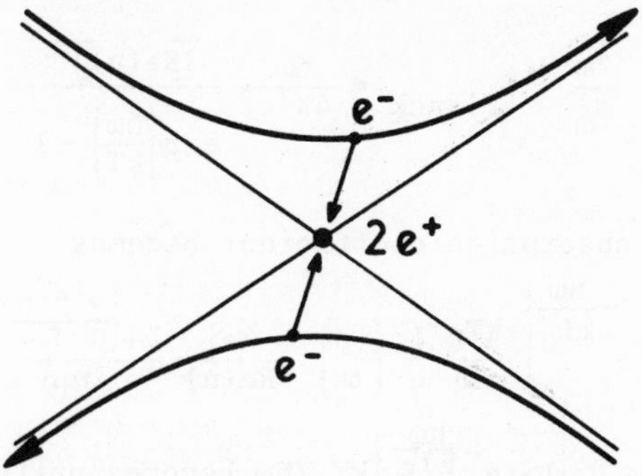

Fig.3 - Two colliding electrons form a pair of dipoles, the radiation of which cancels by interference. $2e^+$: imaginary positive charges in the centre at rest.

corresponding velocity distribution $f(\underline{v})$ (procedure analogous to that of sec. 1.1). In this way for a maxwellian velocity distribution

$$I_\omega = \frac{d^2P}{d\Omega d\omega} = \frac{\mathrm{Re}(n)\omega_p^2}{4\sqrt{3}\pi^2c^3}\nu_{ei}kT_e e^{-\frac{\hbar\omega}{kT_e}} \frac{\sigma(\omega,T_e)}{ln\,\Lambda} \qquad (1.31)$$

is obtained[2,9]. ν_{ei} is the collision frequency from eq. (1.9); $\sigma(\omega,T_e)$ is the Gaunt factor which provides for the necessary quantum corrections in different (ω,T_e) regions. For ω_{Nd} and harmonics $\hbar\omega \ll kT_e$ holds and the ratio between the thermal de Broglie wavelength b_h and the impact parameter b_o for 90° deflection,

$$b_h/b_o = \frac{4\pi\varepsilon_o}{Ze^2}\left(\frac{2kT_e}{m_e}\right)^{\frac{1}{2}}\hbar = 2.4 \times 10^{-3}\; T_e^{\frac{1}{2}}/Z, \quad [T_e] = {}^oK , \qquad (1.32)$$

is such in laser plasmas that the Born-Elwert expression for $\sigma(\omega,T_e)$ applies[2] (since $\hbar\omega \ll kT_e$, $(b_h/b_o)^2 \gg 1$ holds):

$$\sigma(\omega,T_e) = \frac{\sqrt{3}}{\pi}\,ln\left(\frac{4}{G}\frac{kT_e}{\hbar\omega}\right) = 0.55 \times \left[0.81 + ln\,\frac{kT_e}{\hbar\omega}\right] \quad . \qquad (1.33)$$

$G = 1.781$ is the Euler constant. For instance, for $kT_e/\hbar\omega = 10^3$ $\sigma = 4.2$ results. The absorption coefficient α_ω can now be determined from Kirchhoff's law. Bremsstrahlung radiation from an optically thin plasma generally represents such a small energy loss that (i) it can be in thermal equilibrium kinetically although thermal radiative equilibrium is not established and (ii) the spectral radiation intensity $I_\omega(T_e)$ is entirely determined - to a high degree of accuracy - by the kinetic electron distribution. Therefore, the relation holds

$$\frac{I_\omega}{\alpha_\omega} = I_{\text{Planck}} = \frac{\hbar\omega^3}{4\pi^3 c^2}\,\frac{[\text{Re}(n)]^2}{\exp\left(\frac{\hbar\omega}{kT}\right) - 1} \tag{1.34}$$

Consequently, the absorption coefficient becomes

$$\alpha_\omega = \left(1 - e^{-\frac{\hbar\omega}{kT_e}}\right) \frac{kT_E}{\hbar\omega}\,\frac{1}{c}\left(\frac{\omega_p}{\omega}\right)^2 \frac{\nu_{ei}}{\text{Re}(n)}\,\frac{ln\left(\frac{4}{G}\,\frac{kT_e}{\hbar\omega}\right)}{ln\,\Lambda} \quad . \tag{1.35}$$

Because of $\hbar\omega/kT \ll 1$ $\left(1 - e^{-\frac{h\omega}{kT_e}}\right) kT_e/\hbar\omega$ becomes unity. Comparison with eq. (1.2.2) yields

$$\frac{\alpha_\omega}{\alpha} = \frac{1}{ln\,\Lambda}\; ln\left(\frac{4}{G}\,\frac{kT_e}{\hbar\omega}\right) .$$

To give a numerical example, for $\hbar\omega/kT_e = 10^{-3}$ the ratio becomes $\alpha_\omega/\alpha = 7.7/ln\,\Lambda$; for $ln\,\Lambda = 7.7$ α_ω becomes equal to α. In any case the consideration shows that

a) α_ω is, within the uncertainty of $ln\,\Lambda$, practically the same as α, in this way justifying the identification of "inverse bremsstrahlung absorption" with "collisional absorption";

b) the bremsstrahlung intensity $I_\omega(T_e)$ represents the spontaneously emitted amount of radiation, whereas in the absorption coefficient α stimulated re-emission of radiation is included also;

c) at very low temperatures or higher photon energies (e.g. $\hbar\omega/kT_e \simeq 1$) a purely classical calculation of α begins to fail because of $1-e^{-\hbar\omega/kT} \neq \hbar\omega/kT$.

Observation b) needs two further comments. There are several ways of formulating the principle of detailed balancing and Kirchhoff's law. It is convenient to use it in the form of eq. (1.34) where the emitted power density per unit solid angle is indicated by I_ω. This is because experimentally I_ω can be uniquely determined only when the surrounding radiation field is zero; as a consequence I_ω represents the <u>spontaneously</u> emitted radiation. On the other hand, in a measurement of α_ω the "true" or effective absorption coefficient is always determined which is the difference between absorption and stimulated re-emission of radiation between two levels E_1 and E_2 $(E_2 - E_1 = \hbar\omega)$. With their populations n_1 and n_2, the Einstein coefficients $B_{12} = B_{21}$ for stimulated emission, and ρ_ω the spectral radiant energy density

$\rho_\omega = 4\pi I_\omega/c$, α_ω is determined by

$$\alpha_\omega I_\omega = (n_1 B_{21} - n_2 B_{12})\rho_\omega = \frac{4\pi}{c} B_{12}(n_1 - n_2) I_\omega . \tag{1.36}$$

(Refractive index and degeneracy were assumed to be unity.) If κ_ω refers to level excitation by absorption,

$$\kappa_\omega = \frac{4\pi}{c} B_{12} n_1 ,$$

and the levels are in thermodynamic equilibrium (for instance because of collisions; Maxwellian distribution in the case of continuous energy spectrum), $n_2 = n_1 e^{-\frac{\hbar\omega}{kT_e}}$, one obtains

$$\alpha_\omega = \kappa_\omega \left(1 - e^{-\frac{\hbar\omega}{kT_e}}\right) \tag{1.37}$$

κ_ω is often used in theoretical work, however, it is not accessible to direct observation. For correctness it has to be mentioned that the rate equation (1.36) holds as soon as the shortest time of interest is much longer than the transverse relaxation (or phase memory) time T_2 [10]. In conclusion, when Kirchhoff's law is written in the form $I_\omega/\alpha_\omega = I_\omega^{Planck}$, I_ω is the spontaneously emitted radiation and α_ω is the net absorption coefficient. Both I_ω as well as α_ω are directly observable quantities.

The next comment refers to the question whether from Maxwell's equations spontaneous or induced processes are calculated. By considering a driven harmonic oscillator in its classical limit (e.g. an antenna) we conclude that as long as the oscillator radiates in a steady state, Maxwell's equations describe induced emission; when the field is switched off spontaneous emission is also described correctly assuming, of course, a high average quantum state $\langle n \rangle$.

1.5 Laser Field Effects on Collisional Absorption

The formulae for the absorption coefficient α presented in the former sections are valid in the limit of vanishing electric field. However, as may be deduced from the hard sphere model, considerable deviations should appear when v_{os} becomes comparable with v_{th}. The effects we expect in the strong field case are the following:

α) reduction of α because of increasing $\langle v \rangle = \phi(v_{th} + v_{os})$;

β) non-Maxwellian electron velocity distribution, thus affecting the transport coefficients;

γ) anisotropy of velocity distribution induced by $\vec{v}_{os}$.

In contrast to the low field case at high $\vec{E}$-fields the interaction time of a collision will no longer be short compared with $2\pi/\omega$. Strong field corrections of α and the electron distribution function have been considered by several authors in the past. An early semi-quantum mechanical treatment was presented by Rand[11]. A classical, more complete calculation was performed at about the same time by Silin[12]. Essentially identical results with some additional evaluations were reproduced in a simpler and more comprehensive way by Catto and Speziale[13]. Starting from a Fokker-Planck equation of the form

$$\frac{\partial f}{\partial t} - \frac{e}{m}\vec{E}\nabla_v f = A\nabla_v\left(\frac{v^2 I - \vec{v}\vec{v}}{}\nabla_v f\right) + C_{ee}(f),\ A = 2\pi\frac{Z^2e^4n_i}{(4\pi\varepsilon_o)^2m_e}\ ln\,\Lambda\ ,$$

they expanded f in terms of ν/ω into the series $f = f_0 + f_1 + f_2 + \ldots\ldots$ and assumed a Maxwellian distribution for f_o. In this way C_{ee} reduces to zero. The effective collision frequency ν is defined in such a way to satisfy Poynting's theorem,

$$<\nabla\vec{S}> = -<\vec{j}\vec{E}> = \tfrac{1}{2}\,\frac{n_e e^2}{\omega^2}\,\nu\,\vec{E}\vec{E}^* \tag{1.38}$$

Its connection with ν_{ei} from eq. (1.11) is $\nu = \nu_{ei}/(1+\nu_{ei}^2/\omega^2)$, and it is known as soon as $\vec{j}$ is obtained. The latter is calculated from the relation

$$\frac{\partial\vec{j}}{\partial t} = -e\int\vec{v}\,\frac{\partial f}{\partial t}\,d\vec{v}\ .$$

In this way for a linearly polarized wave of the form of eq. (1.1) a collision frequency results which for $\hat{v}_{os}/v_e>1$, $v_e = (kT_e/m_e)^{\frac{1}{2}}$, satisfie the inequality

$$\nu_{ei} > \nu > \nu_{circ} = \left(\frac{\pi}{2}\right)^{\frac{1}{2}}\left(\frac{v_e}{\hat{v}_{os}}\right)^3\nu_{ei}\ . \tag{1.39}$$

ν_{circ} refers to a circularly polarized wave of the same $\hat{v}_{os}$. Once more a different approach was used by Pert[14] to obtain basically similar results, and a nice comparison of different models was made by this author[15].

An even more serious impact of a large $\vec{v}_{os}$ is seen in the distribution function[16]. By comparing the time τ_h of heating the electrons up to the temperature T_e,

$$\tau_h = \tfrac{3}{2}\,n_e kT_e/<\vec{j}\vec{E}> = 3\tau_{ei}\left(\frac{v_e}{\hat{v}_{os}}\right)^2,\ \tau_{ei} = \nu^{-1}\ ,$$

it results that τ_h becomes shorter than the electron-electron collision time τ_{ee} as soon as

$$Z\hat{v}_{os}^{\ 2} \gtrsim 3v_e^{\ 2}$$

holds. After one electron-ion or electron-electron collision $f(\vec{v})$ becomes nearly isotropic for moderate v_{os}/v_e-ratios. However, when driven at constant $\vec{E}$-field strength, f(v) evolves into a self-similar distribution of the form

$$f(v) = \kappa e^{-v^2/5u^5} \quad , \quad u^5 = 5A\hat{v}_{os}^{\ 2}/6 \tag{1.40}$$

which is flatter than a Maxwellian. The concomitant reduction factor in the absorption coefficient α as a function of $x = \hat{v}_{os}^{\ 2}/v_e^{\ 2}$ is reported in Fig. 4a. At x = 1 the reduction amounts to 40%. It has to be mentioned that in all calculations reported so far (11-15) the approximation of zero interaction time τ is made.

The problem of absorption and transport was formulated in nearly full generality by Jones and Lee (17). Starting from three different approaches (Boltzmann equation, Fokker-Planck equation, BBGKY hierarchy) the authors derive differential equations for the isotropic and the anisotropic parts of $f(\vec{v})$. If the anistropic parts are neglected

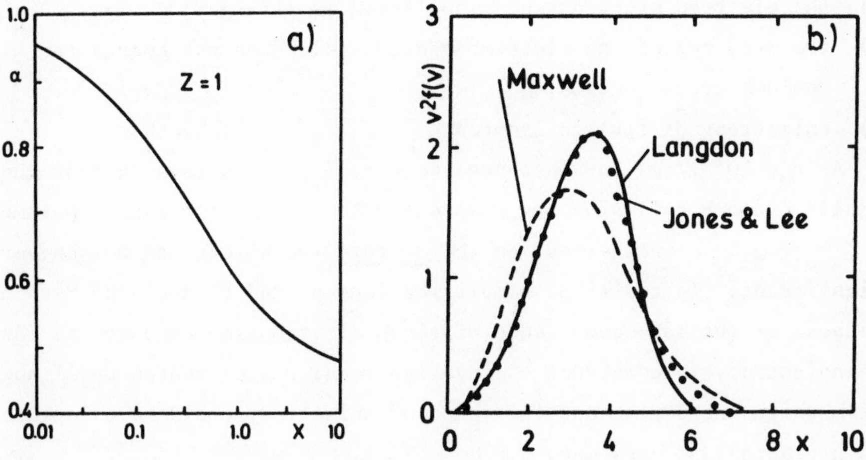

Fig. 4 a) Reduction of the absorption coefficient α as a function of $x = \hat{v}_{os}^{\ 2}/v_e^{\ 2}$ (16).
b) Electric field effect on the electron distribution function $v^2f(v)$ for $x = 2.25$ (17).

self-similar solutions are obtained; in the low field limit Langdon's solution, eq. (1.40), is recovered whereas in the high field limit, $\hat{v}_{os} \gg v_e$, to the authors' surprise a Maxwellian distribution results,

$$f = f_o = \frac{1}{(2\pi)^{\frac{3}{2}} v_e^{\,3}} \, e^{-\frac{v^2}{2v_e^{\,2}}} .$$

However, this is quite natural, since because of the strong heating v_e rises to such final values that $\hat{v}_{os}^{\,2}/v_e^{\,2} \ll 1$ is fulfilled. The authors also find harmonics of ω in $f(\vec{v})$ and memory effects, e.g. owing to the phase change of the $\vec{E}$-field during an electron-ion collision. Numerical evaluations in addition to analytical considerations show a much extended parameter region of validity of Langdon's solution, eq. (1.40). For $x = 2.25$ this is shown in Fig. 4b. The anisotropy effect is present at all field strengths $x > 0$. After a few electron collisions f_{an} follows f_o adiabatically. The altered distribution function also affects the transport coefficients (e.g. heat conduction) of laser plasmas.

In all theoretical evaluations of α under the influence of a strong oscillating field more or less the following simplifications were made up to now:

a) small angle deflections;
b) the electron orbit during a collision is straight;
c) the presence of the electromagnetic field does not change the Coulomb cross-section;
d) anisotropy of $f(\vec{v})$ is ignored.

At $P = 10^{15}$W/cm^2 for instance the electric field is such that the single Coulomb collision may completely be altered for impact parameters $b \lesssim \hat{v}_{os}/\omega_{Nd}$. Nevertheless, on the average the effect may not be very significant. In addition, simplifications a) and b) are also reasonable because of the rare occurrence of close electron-ion encounters. As far as anisotropy is concerned no physical model was presented until now from which the degree and the shape of anisotropy could be calculated in a general and easy way.

2.
LASER LIGHT TRANSPORT

2.1 Light Paths

Before studying laser light absorption by the resonant excitation of electron plasma waves in detail the geometrical propagation of light in the inhomogeneous laser plasma should be briefly discussed. The unmagnetized plasma is capable of sustaining three different types of eigenmodes: a transverse or electromagnetic wave and two longitudinal electron plasma or electrostatic modes, i.e. the high frequency Langmuir and the low frequency ion acoustic wave. Since they all are described by the wave equation (1.13) they are considered here altogether.

From the fluid equations of motion for electrons and ions and their continuity equations the following expressions for the electron and ion current densities are obtained[18],

$$\frac{\partial}{\partial t}\vec{j}_e = e s_e^{\,2}\nabla n_i - \varepsilon_0 s_e^{\,2}\nabla(\nabla\vec{E}) + \varepsilon_0\omega_{pe}^{\;2}\,\vec{E} - \nu_e\vec{j}_e$$
$$- \frac{e}{m_e}\vec{j}_e \times \vec{B} + \frac{e^2}{m_e}(n_e - n_0)\vec{E} - (\vec{j}_e\nabla)\vec{v}_e - \vec{v}_e(\nabla\vec{j}_e) \tag{2.1}$$

$$\frac{\partial}{\partial t}\vec{j}_i = -e s_i^{\,2}\nabla n_i + \varepsilon_0\omega_{pi0}^{\;2}\vec{E} + \frac{e^2}{m_i}(n_i - n_0)\vec{E} - \nu_i\vec{j}_i - (\vec{j}_i\nabla)\vec{v}_i - \vec{v}_i(\nabla\vec{j}_i) \tag{2.2}$$

ν_e, ν_i are appropriate damping coefficients for Landau damping etc. By inserting the linear parts of these expressions in eq. (1.13) one obtains the following electric conductivities and dispersion relations[18],

<u>electromagnetic mode</u>: $\sigma = \sigma_e = \sigma_\perp = i\varepsilon_0 \dfrac{\omega_{pe}^{\;2}}{\omega + i\nu_e}$,

$$\omega^2 = \frac{\omega_{pe}^{\;2}}{1 + i\nu_e/\omega} + c^2k^2 \;;$$

<u>hf electrostatic mode</u>: $\sigma = \sigma_e = \sigma_{\parallel} = i\varepsilon_0 \dfrac{\omega_{pe}^{\;2} + s_e^{\,2}k^2}{\omega + i\nu_e} = i\varepsilon_0\omega$,

$$\omega^2 = \frac{\omega_{pe}^{\;2} + s_e^{\,2}k^2}{1 + i\nu_e/\omega} \;;$$

<u>ion acoustic mode</u>: $\sigma = \sigma_e + \sigma_i = i\varepsilon_0\omega,\; \omega = sk,$ (2.3)

$$\sigma_e = -i\varepsilon_0\frac{s^2}{s_e^{\,2}}\,\frac{\omega_{pe}^{\;2}}{\omega + i\nu_a} = -\frac{\sigma_i}{1 + \gamma_e(k\lambda_D)^2} \;,$$

$$n_i - n_e = n_e \frac{s_e{}^2 k^2}{\omega_{pe}{}^2} = \gamma_e (k\lambda_D)^2 n_0$$

$$(s^2 = s_i{}^2 + \frac{m_e}{m_i} \frac{s_e{}^2}{1+\gamma_e(k_D)^2} ,$$

γ_e adiabatic coefficient).

Neglecting damping the linear wave equation reduces to

$$\nabla \times \nabla \times \vec{E} - \frac{\omega^2}{c^2} \vec{E} = \frac{1}{\varepsilon_0 c^2} \{-\varepsilon_0 \omega_{pe}{}^2 \vec{E} + \varepsilon_0 s_e{}^2 \nabla(\nabla\vec{E}) - e s_e{}^2 \nabla n_i - \frac{\partial \vec{j}_i}{\partial t}\} . \quad (2.4)$$

In a homogeneous medium the wave eq. (1.15) for $\nabla\vec{E} = 0$ follows therefrom, and similarly

$$\nabla \left(i + i\frac{\sigma}{\varepsilon_0 \omega} \right) \vec{E} = \nabla(\varepsilon\vec{E}) = 0$$

for $\nabla \times \vec{E} = i\omega\vec{B} = 0$. In an inhomogeneous medium the optical (or WKB) approximation is valid in regions with gentle density gradients, i.e. $L(\vec{r}) \gg \lambda(\vec{r})$. For monochromatic wave this consists in assuming a smooth phase function $\psi(\vec{x})$ defined as

$$\vec{E}(\vec{x},t) = \hat{E}(\vec{x},t)e^{i\psi(\vec{x})}$$

and keeping only the leading terms in the derivatives of $\vec{E}(\vec{x},t)$:

$$\nabla \times \nabla \times \vec{E} \simeq \nabla\psi \times (\hat{E} \times \nabla\psi) , \; \nabla(\nabla\vec{E}) = \nabla\psi \times (\vec{E} \times \nabla\psi) - \vec{E}(\nabla\psi)^2 .$$

This transforms eq. (2.4) for $\omega > \omega_{pe}$ and $\partial \vec{j}_i/\partial t \simeq 0$, into

$$(\nabla\psi)^2 \vec{E} - \nabla\psi(\vec{E}\nabla\psi) = \frac{\omega^2}{c^2} n^2(\vec{x})\vec{E} - \frac{s_e^2}{c^2} \nabla\psi(\vec{E}\nabla\psi) .$$

For $\vec{E} \perp \nabla\psi$ (electromagnetic wave) and $\vec{E} \parallel \nabla\psi$ (electrostatic wave) it splits into two eikonal equations; a third one results from eqs. (2.3,4) under the condition $\omega \ll \omega_{pe}$. Thus we obtain

$$\vec{E} \perp \nabla\Psi : \quad (\nabla\psi)^2 = (\frac{\omega}{c})^2 n^2(\vec{x}), \Rightarrow \nabla\psi = \pm \vec{k}n(\vec{x}) ,$$

$$\vec{E} \parallel \nabla\psi : \quad (\nabla\psi)^2 = (\frac{\omega}{s_e})^2 n^2(x), \Rightarrow \nabla\psi = \pm \vec{k}_{es} n(\vec{x}) ,$$

$$(\nabla\psi)^2 = (\frac{\omega}{s})^2, \quad \Rightarrow \nabla\psi = \pm \vec{k}_a , \; n(\vec{x}) = 1 .$$

$\nabla\psi$ is the local wave vector $k = \vec{k}n$, $\vec{k}_{es}n$, $\vec{k}_a$. The curves perpendicular to $\mathrm{Re}\psi(\vec{x})$ = const. are the rays of the wave. Along such trajectories

$$\nabla \times k(\vec{x}) = \nabla \times \nabla\psi = 0$$

holds. With s being the length of a ray segment the following ray equation is obtained,

$$\frac{d}{ds}k(\vec{x}) = \frac{d}{ds}\nabla\psi = \frac{1}{|k|}(k\nabla)k = \frac{1}{|k|}\{\tfrac{1}{2}\nabla k^2 - k \times \nabla \times k\} = \nabla|k| .$$

Introducing the refractive index $n(\vec{x})$ and $k_o = \omega/c$ or $k_o = \omega/s_e$ or $k_o = k_a$ and assuming constant electron temperature for the Langmuir wave, the ray equation reads:

$$\frac{d}{ds}(\vec{k}_o n) = k_o \nabla n. \qquad (2.5)$$

Note that for s_e = const a Langmuir wave follows the same law of deflection as an electromagnetic wave.

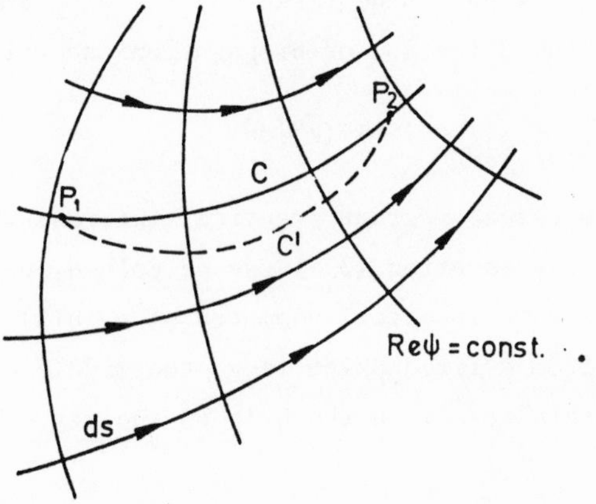

Fig. 4a- The rays form a manifold of curves which are perpendicular to the surfaces of constant phase, $Re\psi(\vec{x})$ = const.

(i) The phase difference between two points P_1, P_2 is uniquely determined by

$$\psi(P_2) - \psi(P_1) = \int_{P_1}^{P_2} |\vec{k}|ds$$

along a ray segment C. Also

$$\psi(P_2) - \psi(P_1) = \int_{C'} \vec{k}\, d\vec{s} \leqslant \int_{C'} k\, ds\ ,$$

expressing Fermat's principle[19].

(ii) As long as the optical approximation is valid, all modes decouple from each other; so do in particular the incident and the reflected waves. The general solution for one wave is

$$\vec{E}(\vec{x}) = \hat{E}_+(\vec{x})\, e^{i\int|\vec{k}|ds} + \hat{E}_-(\vec{x})\, e^{-i\int|\vec{k}|ds}$$

with the local amplitudes determined from the conservation of the energy flux included by a ray bundle. If Q(s) is its cross-section and $\vec{v}_g = \partial\omega/\partial\vec{k}$ is the group velocity ($v_g \cdot v_{Phase} = c^2$ or $s_e{}^2$ respectively)

$Q(s)|\hat{E}_\pm|^2 v_g = \text{const}$

holds (steady state!).

(iii) The presence of a finite but moderate damping coefficient ν ($\nu^2 << \omega^2$) does not alter the situation: in the ray equation the real part of $\vec{k}$ has to be taken and the wave amplitudes appear attenuated in the direction of propagation according to Beer's law(19)

$$\phi(s) = \phi_o e^{-2\int|\mathrm{Im}(\vec{k})|ds} \quad . \tag{2.6}$$

(iv) There are two cases of great practical interest in laser plasmas for which the ray equation (2.5) can be solved analytically. The first one refers to spherical geometry of n, $n(\vec{x}) = n(r)$, with r the distance from a fixed point (e.g. the middle of a laser pellet). In this case with the help of the ray equation we obtain

$$\frac{d}{ds}(\vec{r} \times n\vec{k}_o) = \frac{d\vec{r}}{ds} \times n\vec{k}_o + \vec{r} \times \frac{d}{ds}(n\vec{k}_o) = k_o \vec{r} \times \nabla n(r) = 0,$$

or in its integrated form,

$$rn(r)\sin\alpha = \text{const}, \tag{2.7}$$

where α is the angle included by the radius vector and wave vector. This implies that the angular momentum of the wave vector is conserved. A considerable amount of laser light may be deflected, according to this law, from the pellet corona. The other case of great interest relates to the so-called stratified medium in which the refractive index depends on one Cartesian coordinate only, $n(\vec{x}) = n(x)$. For this plane case the ray equation yields the solution

$$n(x)\sin\alpha = \text{const} \tag{2.8}$$

(α = angle of incidence).

(v) The optical approximation fails at critical points (1) where $n(\vec{x})$ reduces to zero and $L >> \lambda$ cannot be fulfilled, at caustics (2), i.e. envelopes of rays, and foci (3).

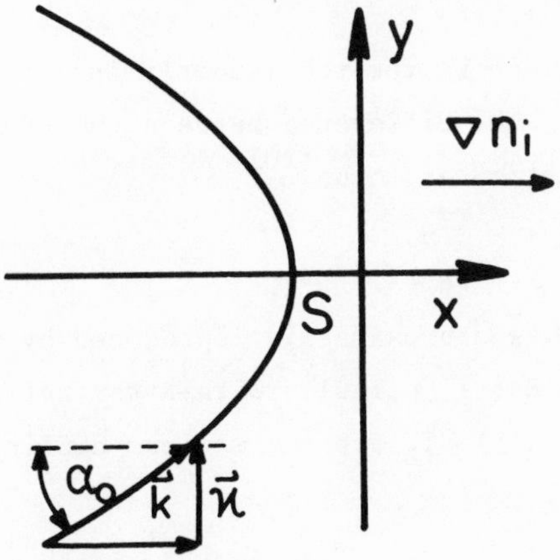

Fig. 5 - Plane z-polarized wave impinging onto a stratified plasma.

An important situation occurring in laser plasmas is the oblique incidence of a plane wave onto a stratified plasma. Let the plasma density gradient point parallel to x and the $\vec{k}$ vector lie in the (x,y) plane (Fig. 5). Further, $\vec{E}$ is assumed to be in the z-direction, $\vec{E} = E(x,y)\vec{e}_z$. Because of $\nabla\vec{E} = 0$ eq. (2.4) reduces to $\nabla^2 E + k^2 n^2(x)E = 0$. There is no y-dependence of n and consequently the Fourier ansatz

$$E(x,y) = E(x)e^{iky}$$

is appropriate. This leads to

$$E'' + k^2(n^2 - \sin^2\alpha_o)E = 0 \tag{2.9}$$

The equation is of the type

$$y'' + f(x)y = 0 \tag{2.10}$$

very often encountered in physics (light, Langmuir, sound waves, Schrödinger equation). Its WKB solution is

$$y(x) = \frac{C_+}{f^{\frac{1}{4}}} e^{i\int f^{\frac{1}{2}}dx} + \frac{C_-}{f^{\frac{1}{4}}} e^{-i\int f^{\frac{1}{2}}dx} , \tag{2.11}$$

as may be seen by direct inspection. The approximation is good as long as

$$g(x) = \frac{1}{4}\left\{\frac{f''}{f} - \frac{5}{4}\left(\frac{f'}{f}\right)^2\right\}$$

is small in comparison to f(x). At the turning point S (Fig. 5) the approximation breaks down (caustic plane perpendicular to x). If around S f(x), i.e. $n^2(x) - \sin^2\alpha_o$ can be approximated by a linear profile eq. (2.9) reduces to the Stokes equation

$$y'' + xy = 0$$

the solution of which is given by the two linearly independent Airy functions Ai(x) and Bi(x). The difference between the WKB solution and the exact one can be estimated to be[20]

$$\left|\frac{\Delta C\pm}{C_+}\right| \lesssim \frac{5}{48\, x^{\frac{3}{2}}} .$$

Even the maximum amplitude is approximately reproduced by the WKB solution (see for instance Ref.1, p.110). With R the reflection coefficient in a region of n(x) = 1, for the maximum amplitude of eq. (2.9) holds

$$|E|_{max} = 1.90(kL)^{\frac{1}{6}} R^{\frac{1}{4}} |E_{incident}|, \quad L = (n^2 - \sin^2\alpha_o)/\left|\frac{dn^2}{dx}\right| . \tag{2.12}$$

The reflected wave is shifted by the phase angle of $\phi = -\pi/4$ with respect to the incident wave. This leads to the Bohr-Sommerfeld quantization condition for the number of nodes N of a wave captured in a cavity,

$$\int_{LHS}^{RHS} f^{\frac{1}{2}} dx = (N + \tfrac{1}{2})\pi .$$

The integration extends from the left to the right turning point. An example for N = 1 is shown in Fig. 6. As long as the wave remains purely electromagnetic or electrostatic its energy density (field energy + electronic oscillatory energy) is given by

$$\rho_E = \frac{\varepsilon_o}{2} |E|^2 . \tag{2.13a}$$

At the critical point field and oscillation energy densities become equal. As long as the waves are linear and plane the corresponding intensities are

$$S_{em} = \frac{\varepsilon_o}{2} cn|\vec{E}_{em}|^2 = \varepsilon_o c^2 \vec{E} \times \vec{B}, \quad S_{es} = v_g \rho_E^{es} = \frac{\varepsilon_o}{2} s_e |\vec{E}_{ex}|^2 . \tag{2.13b}$$

Both relations can easily be deduced from Poynting's theorem, eq. (1.12). Since the Poynting vector of the Langmuir wave is zero (B = 0) we imagine it being excited by and electromagnetic wave and calculate $\nabla\vec{S}_{em} = -\vec{j}\vec{E}$. In this way S_{es} is obtained.

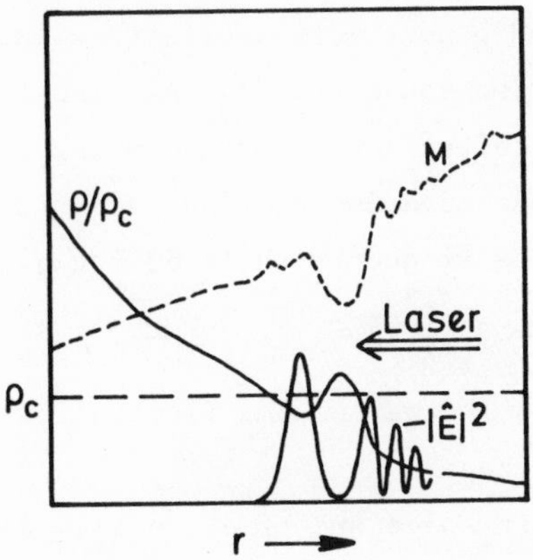

Fig. 6 - Caviton formation (three critical points) in the density can occur already due to a gentle fluctuation of laser intensity[21].

2.2 Reflection and Scattering of Laser Light

In the case of a medium such as the laser plasma which is inhomogeneous, fluctuating, and varying in time in its mean density, the question may arise of how the wave vector $\vec{k} = \vec{k}_o n(\vec{x},t)$ and the frequency ω change in time and which situations lead to appreciable reflection and diffusion of the incident laser beam.

In a medium whose optical properties are changing in time, in general $\vec{k}$ as well as ω undergo variations. The variation of $\vec{k}$ is described by the ray equation (2.5). In addition, ω must also change as can be recognized from the following situation (Fig.7).

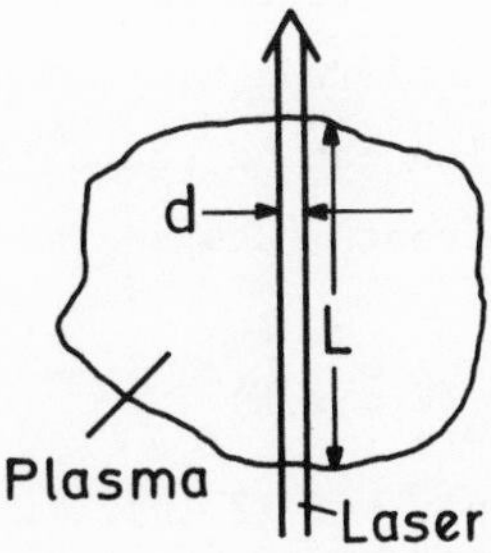

Fig. 7 - Frequency shift due to plasma rarefaction along L.

Consider a plasma of length L and averaged refractive index $\overline{n}$ at the time $t = t_1$. Assume further that along the laser beam the plasma is expelled by thermal and ponderomotive effects within the time $\tau = t_2 - t_1 \simeq d/c_s$ (c_s = ion sound speed). At time t_1 there are $L\overline{n}/\lambda$ nodes along L, whereas after τ the corresponding number is L/λ. Since ω is 2π times the number of nodes passing by a fixed point behind the plasma per unit time the frequency shift during plasma expulsion is

$$\Delta\omega = 2\pi \frac{L}{\lambda} (\overline{n} - 1) \frac{1}{\tau} = kL \frac{\partial \overline{n}}{\partial t}. \tag{2.14}$$

This relation describes the way plasma fluctuations broaden the linewidth of the laser beam.

As long as changes in space and time are slow compared to wavelength and frequency it is convenient to start from the following description of the wave

$$\vec{E}(\vec{x},t) = \hat{E}(\vec{x},t)e^{i\phi(\vec{k},\vec{x},t)}, \tag{2.15}$$

instead of using a Fourier decomposition of $\vec{E}(\vec{x},t)$. The eikonal (which now contains also the time) has to be chosen such that the amplitude $\hat{E}$ becomes a slowly varying quantity only. Then, $\vec{k}$ and ω are defined by the relations

$$\vec{k} = \nabla_x \phi \ , \quad \omega = - \frac{\partial \phi}{\partial t} . \tag{2.16}$$

For a wave of constant frequency this coincides with the earlier representation,

$$\vec{E}(\vec{x}\ t) = \hat{E}(\vec{x},t)e^{i\int \vec{k}d\vec{x} - i\omega t}.$$

If expression (2.15) is inserted in the wave equation (2.4) a linear dispersion relation results of the form

$$\omega = \omega(\vec{k},\vec{x},t). \tag{2.17}$$

If also the nonlinear currents are taken into account in the wave equation ω will depend on the amplitude $\hat{E}$, too. In addition, ω may split into several branches which can be characterized by writing ω_α where α characterizes the different modes so that the more general dispersion relation reads

$$\omega = \omega_\alpha(\hat{E},\vec{k},\vec{x},t). \tag{2.18}$$

Starting from the definitions, eq. (2.16), it can be shown that the following canonical equations of motion hold[22]

$$\frac{d\vec{x}}{dt} = \frac{\partial \omega}{\partial \vec{k}} \ , \ \frac{dk}{dt} = - \frac{\partial \omega}{\partial \vec{x}} \ , \tag{2.19}$$

and that these relations are meaningful as long as the definitions of a single $\vec{k}$ and a single ω according to eq. (2.16) make sense. A third relation follows from eq. (2.19),

$$\frac{d\omega}{dt} = \frac{\partial\omega}{\partial t} + \frac{\partial\omega}{\partial k}\frac{dk}{dt} + \frac{\partial\omega}{\partial\vec{x}}\frac{d\vec{x}}{dt} = \frac{\partial\omega}{\partial t}, \tag{2.20}$$

if d/dt means the variation an observer sees when travelling at group velocity,

$$v_g = \nabla_k\omega = \frac{\partial\omega}{\partial k}.$$

Eqs. (2.19, 2.20) are completely analogous to the Hamiltonian equations of motion as soon as ω is identified with H and $\vec{k}$ with the momentum $\vec{p}$,

$$\frac{dq_i}{dt} = \frac{\partial H}{\partial p_i}, \quad \frac{dp_i}{dt} = -\frac{\partial H}{\partial q_i}, \quad \frac{dH}{dt} = \frac{\partial H}{\partial t}.$$

Introducing the intensity I for a general radiation field of the WKB type as a function of $\omega,\vec{x},t$ and direction $\vec{\Omega}$, i.e. $I = I(\omega,\vec{x},t,\vec{\Omega})$, the following relation holds in the absence of absorption and scattering[23]

$$\frac{d}{ds}\left(\frac{I\omega}{n^2}\right) + \frac{3}{c}\left(\frac{I\omega}{n^2}\right)\frac{\partial n}{\partial t} = 0. \tag{2.21}$$

Hence, along a light path I/n^2 is conserved in the case of a steady plasma. By specializing the radiation field to a plane wave normally incident onto a plasma slab formula (2.11) is recovered,

$$I\omega/n^2 = \tfrac{1}{2}cn\hat{E}^2/n^2 = \text{const}, \Rightarrow \hat{E} = \hat{E}_o/n^{\frac{1}{2}}.$$

Another important consequence of eq. (2.21) is that Planck's radiation formula in matter has to be corrected by $n^2(\vec{x})$ (see eq. (1.34).

Remark: Eq. (2.21) may appear paradoxical if applied to an isotropic point source in a medium of constant refractive index n (say $n = 1$ for example, Fig. 8a)). First, one would be tempted to write

$$I \sim \frac{1}{r^2},$$

since the irradiated power from the point source is given by $P = 4\pi r^2\overline{S}$ for an arbitrary radius r. However, $\overline{S} = I$ does not hold in general. I is defined as the power $d\dot{E}$ which crosses the surface $d\vec{\sigma}$ in the direction $\vec{\Omega}$ and is emitted into the solid angle $d\Omega$ (Fig. 8b)),

$$d\dot{E} = I d\vec{\sigma} d\vec{\Omega} d\omega. \tag{2.22}$$

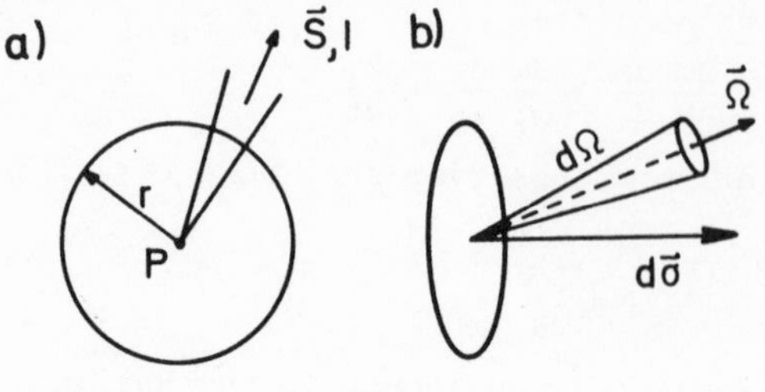

Fig. 8 - a): The Poynting vector of a point source decreases as $|\vec{S}| \sim 1/r^2$; whereas I remains constant along r.
b): Definition of I.

If the photon density in the frequency interval dω travelling into direction $\vec{\Omega}$ is $f(\omega,\vec{x},t,\vec{\Omega})$, then I can equally be expressed as

$$I = c_g \hbar\omega f(\omega,\vec{x},t,\vec{\Omega}). \tag{2.23}$$

Assume now a finite diameter ρ for P in order to illuminate the solid angle $d\Omega = \pi\rho^2/r^2$. This decreases with r^2. According to eq. (2.22), however, dΩ has to be held constant, if $d\dot{E}$ at different positions are compared and hence, I results constant, and the connection between $\overline{S}$ and I in the case of an isotropic point source is

$$I = \vec{S}\delta(\vec{\Omega} - \vec{r}/r).$$

As long as the plasma density profile varies slowly in space (WKB condition) negligible reflection occurs. To quantify this statement an Epstein transition layer of the following form is considered[24],

$$n^2(x) = 1 + P\frac{e^{kx/s}}{i + e^{kx/s}}; \tag{2.24}$$

P and s are parameters. In Fig. 9 $n^2(x)$ with $P = -\frac{3}{4}$ is sketched. The transition width is determined through s as follows,

$$\Delta = \frac{4s|P|}{k}.$$

For normal incidence the reflection is determined by

$$R = \frac{\sinh^2[\pi s(1-\sqrt{1+P})]}{\sinh^2[\pi s(1+\sqrt{1+P})]}.$$

In table 3 R normalized to the Fresnel value of the reflection coefficient $R_o = (n-1)^2/(n+1)^2$ for a refractive index step, $r = R/R_o$, is reported for $P = \pm\frac{3}{4}$ and different values of s.

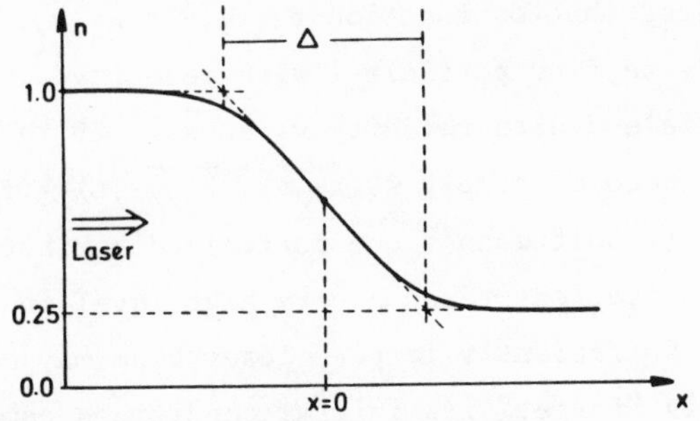

Fig. 9 - Epstein transition layer

Table 3 : Normalized reflection coefficient r for the layer of Fig. 9 as a function of the transition width Δ.

s		2	1	1/2	1/4	1/8	1/16
Δ/λ		0.95	0.48	0.24	0.12	0.06	0.03
P = + 3/4	r	6.1×10^{-10}	1.36×10^{-4}	3.92×10^{-2}	0.37	0.77	0.93
P = - 3/4	r	3.1×10^{-5}	1.5×10^{-2}	0.25	0.78	0.91	0.98

It may be surprising how fast reflection drops from its Fresnel value $r = 1$ as soon as Δ becomes larger than $\lambda/4$; at $P < -1$ total reflection occurs for all values of s.

In general reflection occurs in a laser plasma if

1. the refractive index reduces to zero (critical surface) or at caustics; in a stratified medium the latter occur at $n^2(x)-\sin^2\alpha_0$ (α_0 angle of incidence from outside);
2. there is a density jump over the depth of $\lambda/2$ or less;
3. periodic structures build up such that all small amounts of reflection from single humps interfere constructively; this is the case of back and side scattering (stimulated Brillouin and Raman scattering).

By the bulk quantities like n_e, $n(\vec{x})$ or $\sigma(\vec{x})$ reflection, diffraction, and stimulated parametric effects can be described. The phenomenon of spontaneous scattering which is due to the more or less incoherent phenomenon of single particle scattering originates from thermal or superthermal fluctuations of density. In order to describe this the two-particle distribution function $f_2(\vec{x}_1,\vec{v}_1;\ \vec{x}_2,\vec{v}_2,t)$ is needed, i.e. the probability to find particle 1 with velocity $\vec{v}_1$ at $\vec{x}_1$ and at the same time particle 2 with velocity $\vec{v}_2$ at $\vec{x}_2$. If the particles are uncorrelated f_2 becomes simply $f_2(\vec{x}_1,\vec{v}_1;\ \vec{x}_2,\vec{v}_2,t) = f(\vec{x}_1,\vec{v}_1,t)f(\vec{x}_2,\vec{v}_2,t)$ where $f(\vec{x},\vec{v},t)$ is Boltzmann's one particle distribution function. If the density fluctuations reach a very high level in the plasma and its dimensions are sufficiently large a laser beam may be scattered away and will finally "forget" its initial conditions completely.

In order to get an estimate whether such a situation may occur in a typical laser produced plasma let us assume the worst case of completely uncorrelated electrons undergoing Thomson scattering, i.e. elastic scattering from free electrons (= low energy limit of Compton scattering). The total Thomson cross-section is

$$\sigma_{Th} = \frac{8\pi}{3}\left(\frac{e^2}{4\pi\varepsilon_o m_e c^2}\right)^2 = \frac{8\pi}{3} r_o^{\,2} = 0.665 \text{ barn}; \tag{2.25}$$

hence, the beam intensity decreases according to

$$\frac{dI}{ds} = -n_e\sigma_{Th}I, \Rightarrow I(\vec{x}) = I_o e^{-\int n_e\sigma_{Th}ds} \tag{2.26}$$

For the third harmonic of Nd n_e is at most $n_e = 9 \times 10^{21}\text{cm}^{-3}$ and thus a scattering length

$$L > \frac{1}{n_e\sigma_{Th}} = \frac{1}{9 \times 10^{21} \times 0.7 \times 10^{-24}} = 1.6 \text{ m}$$

results. So, the assumption of a beam following geometrical optics in the plasma seems to be justified in general. However if there are ion species present which have a resonance at the laser frequency, resonant Rayleigh scattering occurs with a scattering cross-section increased by many orders of magnitude.

The small fraction of Thomson-scattered light was used by Baldis and Walsh to probe the plasma and to analyse for example growth and saturation of the two-plasmon decay instability[18]. In general in scattering, one has to distinguish between scattering from individual particles at $k\lambda_D \gg 1$ and scattering from collective modes at $k\lambda_D \gg 1$. The scattering vector is defined as $k = k_{inc} - k_{scatt}$ (Fig. 12).

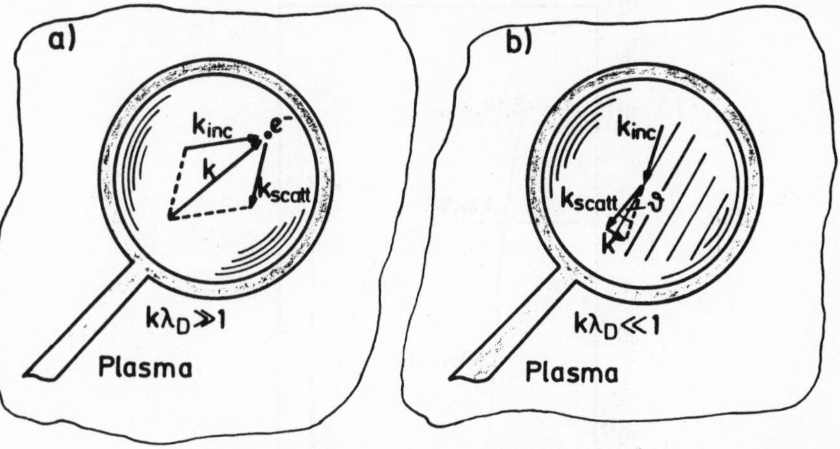

Fig. 10 - a, incoherent scattering from single electrons; Te is measured. b, scattering from collective plasma modes.

In the case of $k\lambda_D \gg 1$ the scattered frequency is broadened by $\Delta\omega \simeq kV_e$ because of the Doppler shift from the moving electrons. The line profile is a Gaussian. In the opposite case, $k\lambda_D \ll 1$, scattering reveals the collective motion of electrons. In Fig. 11 the scattered spectrum for $k\lambda_D = 0.2$ is reported. The electrostatic wave as well as the acoustic mode (broad central peak) are clearly seen. Light scattering is a powerful tool for diagnostics. Unfortunately in laser plasma experiments it is not too often applied since the experimental difficulties are considerable, mainly owing to the small size and the inhomogeneity of laser plasmas.

More dangerous than Thomson scattering for a laser beam to propagate regularly are macroscopic refractive index fluctuations in space (and time). It is well known that due to such fluctuations caustics can occur on every ray, repeatedly, on some characteristic distance scale[26]. The ray propagation becomes particularly sensitive to density fluctuations when they occur near a critical surface. In detail the problem of ray propagation in a fluctuating laser plasma was sutdied at LLE in Rochester[21]. By assuming a fluctuating electron density n_1,

$$n_e = n_o + n_1,\ \langle n_1\rangle = 0,\ \sigma = \langle n_1{}^2\rangle^{\frac{1}{2}} \ll n_c - n_o,\ \Delta n/n(x) \ll 1,$$

and a correlation length h according to

$$\langle n_1(x)n_1(x+\Delta x)\rangle = \sigma(x)\sigma(x+\Delta x)e^{-(\Delta x)^2/h^2},\ h \ll L, \qquad (2.27)$$

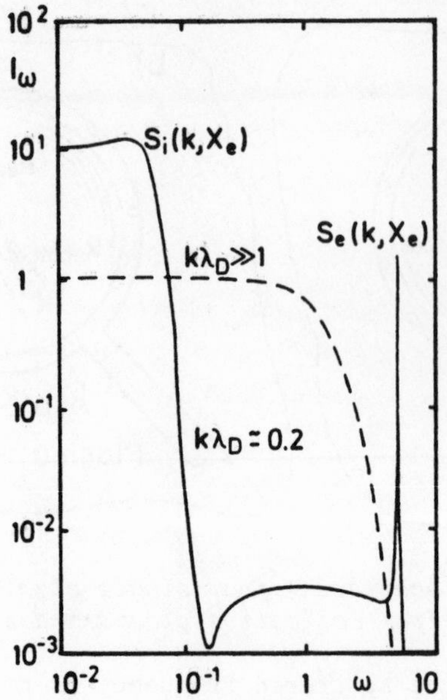

Fig. 11 - Scattered spectrum from a CO_2 laser generated plasma, $n_e = n_e/4 = 2.5 \times 10^{18} cm^{-3}$, $T_e = 100 eV$, $\nu = 22^o$, $k\lambda_D \simeq 0.2$ ($\lambda_{inc} = 0.53\mu$). $S_e(k,x_e)$, $S_i(k,x_e)$ signals from collective electron and ion fluctuations(25.

from the ray eq. (2.5) the following relation has been deduced for the ray direction vector $\vec{k}^o = \vec{k}/|\vec{k}|$,

$$\frac{d}{ds} \langle\vec{k}^o\rangle = - \frac{\nabla n_o - \vec{k}^o(\vec{k}^o\nabla n_o)}{2(n_c - n_o)} + \frac{\nabla\sigma^2 - k^o(k^o\nabla\sigma^2)}{8(n_c - n_o)^2} - \frac{\pi^{\frac{1}{2}}}{2} \frac{\sigma^2\vec{k}^o}{h(n_c - n_o)^2} \quad (2.28)$$

$\langle\vec{k}^o\rangle$ is an average taken over such a distance d that $h \ll d \ll L$ holds. The second term represents a drift term due to the fluctuations. In Fig. 12 typical density perturbations and ray deviations are shown for two different parameter sets. Solutions of eq. (2.28) show to be in good agreement with Monte Carlo calculations for the parameter sets. Although the deviations of the single rays from each other are remarkable their overall effect on absorption is nearly negligible.

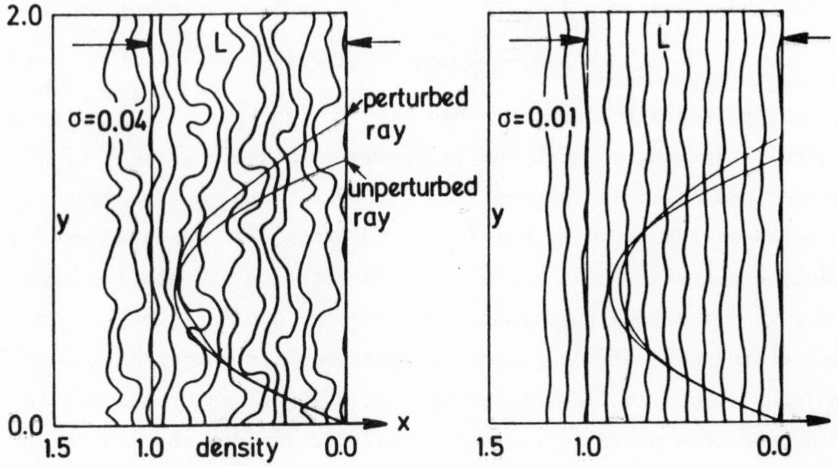

Fig. 12 - Plane-parallel uniform-gradient plasmas with superimposed density fluctuations indicated by isodensity contours. Cases with rms fluctuation amplitudes of $\sigma/n_c = 0.01$ and 0.04 are shown. The correlation length h is chosen to be $h/L = 0.1$. Each frame shows how a typical ray wanders from the unperturbed path due to the given fluctuations (after LLE Review 16, p.32).

2.3 Wave Pressure and Transient Radiation Forces

2.3.1 Definition of ponderomotive force

In contrast to the rapidly changing oscillating force which causes the quiver motion of the electrons by "ponderomotive force", "light pressure", "radiation pressure" or more generally "wave pressure (density)", we understand a secular force (density), i.e. a force which acts in the same direction over the whole duration of the wave pulse or a comparable fraction of it. It originates from the nonlinearity of the momentum equation of the electron.

It was recognised by James Clerk Maxwell in 1861 that a light beam of flux density I, when impinging normally upon a surface of reflectivit R in vacuo, exerts the pressure

$$p_r = (1 + R)\,\frac{I}{c}\,. \tag{2.29}$$

p_r is an overall force. In general an expression for the radiation force on a unit volume ($=$ force density $\vec{\pi}$), i.e. a local force, is needed in fluid theory. Such a formula was derived first for free electrons in 1958[28]. In 1967 the importance of radiation pressure was stressed for laser generated plasmas and the same formula for $\vec{\pi}$ was rederived independently[29]. It is the well-known relation

$$\vec{\pi} = -\frac{\varepsilon_o}{4}\left(\frac{\omega_p}{\omega}\right)^2 \nabla(\hat{E}\hat{E}^*). \tag{2.30}$$

Subsequently, radiation pressure has been recognised as a phenomenon of central importance. With the discovery of a whole variety of stimulated processes in laser-matter interaction in neutral gases, liquids and solids, such as Brillouin and Raman scattering as well as frequency conversion, self-focusing, filamentation and other parametric effects, it was argued that radiation pressure could be the common basis for describing some of the parametric phenomena in a unified manner[30]. The major progress in theoretical understanding of wave pressure in the last decade took place in two well separated fields: one branch of development concentrated on light pressure on single atoms, molecules and ions, the other one on all kinds of macroscopic ponderomotive effects produced in laser plasmas.

Lasers have been proposed and successfully used in optical levitation of small particles of micrometer dimensions by radiation pressure[31], in cooling of atoms for high-resolution spectroscopy[32] and for frequency standards[33,34]. For fully ionized plasmas a whole variety of derivations of $\vec{\pi}$ exist now and more general expressions allowing for relativistic effects[35] and effects due to static magnetic fields[36] were given. Among the different theoretical approaches the Hamiltonian[35,37] and Lie transform[38] treatments appeared to be particularly suitable. The common basis of all derivations of $\vec{\pi}$ is the oscillation centre approximation. For this to be valid the oscillation amplitude $\hat{\delta}_e$ must be much less than the local wavelength λ. As soon as $\hat{\delta}$ becomes comparable to λ, $\vec{\pi}$ can no longer be considered to be a local force[39]. It must also be stressed that formula (2.30) is only valid for time-independent amplitudes $\hat{E}$ and the frequency ω is that seen by an observer moving at the oscillation center speed of the electron. Hence, an additional limitation for eq. (2.30) to hold is $v_{th} \ll v_{Phase}$.

With an electric field amplitude depending on time an additional force term $\vec{\pi}_t$ of different structure appears[40]. Unfortunately, there are almost as many different expressions for $\vec{\pi}_t$ as there are papers published on this subject (see for example the sections "Discussion" of Refs. 40,41). The main reason for such a discrepancy stems from different (or not clear) definitions of the ponderomotive force. In order to obtain the correct expression for $\vec{\pi}_t$ one has to take into consideration that in the presence of an electrostatic or electromagnetic field mechanical momentum is not conserved by itself (i) and, as a consequence, the trajectory of the single fluid element is changed (ii). In accordance with Newton's second law a force density $\vec{\kappa}$ can be expressed in the following general way,

$$\vec{\kappa} = \frac{\partial}{\partial t}\vec{P} + \nabla\overleftrightarrow{T}, \tag{2.31}$$

where $\vec{P}$ and $\overleftrightarrow{T}$ represent the total momentum density and the total momentum flux density of the system under consideration. $\overleftrightarrow{T}$ is a second rank tensor. Let us assume now that in the rest frame of the fluid an electric wave of frequency ω and slowly varying amplitude in time,

$$\vec{E}(\vec{x},t) = \hat{E}(\vec{x},t)e^{-i\omega t}, \quad \nabla\times\hat{E} = i\omega B - \partial_t\hat{B} \tag{2.32}$$

is impinging on matter. Then the time-averaged ponderomotive force density, or wave pressure $\vec{\pi}$, is defined as

$$\vec{\pi} = \left\langle\frac{\partial}{\partial t}\vec{P} + \nabla\overleftrightarrow{T}\right\rangle_{\text{Wave}}, \tag{2.33}$$

where the subscript "Wave" means those components of $\vec{P}$ and $\overleftrightarrow{T}$ which arise as a consequence of the presence of $\vec{E}$ from eq. (2.32) under unchanged parameters for the rest. By the latter we mean the following: if, for instance, the fluid density profile is changed by the ponderomotive force from n_o to n_1, then eq. (2.33) contains the pressure difference $p(n_1,T,\vec{E}) - p(n\ ,T,\vec{E}=0)$ (and <u>not</u> $p-p(n_o,T,\vec{E}=0)$). The brackets indicate the time average over one cycle. From this definition of $\vec{\pi}$ follows for the motion of a fluid element

$$\rho\frac{d\vec{v}}{dt} = \vec{f} + \vec{\pi}, \tag{2.34}$$

where $\vec{f}$ comprises all force densities which are not due to the high frequency $\vec{E}$-field.

2.3.2 *Basic formulae and their physical interpretation*

All references quoted above deal with fully ionized rarefied plasmas. However, since lasers are capable of producing dense, high-Z plasmas of variable degree of ionisation, formulae of more general validity are presented in the following, as they already exist in the literature[42]. Every medium can be thought of, without loss of generality, as being composed of a classical ionic and a classical electronic fluid. The electric force density $\vec{\kappa}_{el}$ is given by the Lorentz force,

$$\vec{\kappa}_{el} = \rho_{el}\vec{E}_t + \vec{j}\times\vec{B}_t = -\frac{1}{c^2}\frac{\partial}{\partial t}\vec{S} - \nabla\overleftrightarrow{T},\quad T_{ij} = -\varepsilon_o\{E_i^tE_j^t + c^2B_i^tB_j^t - \tfrac{1}{2}\delta_{ij}(\vec{E}_t^{\,2} + c^2\vec{B}_t^{\,2}\} \tag{2.35}$$

In $\vec{E}_t$ and $\vec{B}_t$ all electric and magnetic fields are included. $\overleftrightarrow{T}$ is the Maxwellian stress tensor and $\vec{S} = \varepsilon_o c^2\vec{E}_t \times \vec{B}_t$ represents the Poynting vector. $\vec{\kappa}_{el}$ is the sink of the electromagnetic momentum and must therefore appear as a mechanical force density for the electronic and ionic fluids,

$$\vec{\kappa}_{el} = \frac{\partial}{\partial t}(\rho_e\vec{v}_e + \rho_i\vec{v}_i) + \nabla(\rho_e\vec{v}_e\vec{v}_e + \rho_i\vec{v}_i\vec{v}_i + \overleftrightarrow{p}_e + \overleftrightarrow{p}_i). \tag{2.36}$$

Bt subtracting these equations from each other the momentum conservation results in the form

$$\frac{\partial}{\partial t}(\rho_e\vec{v}_e + \rho_i\vec{v}_i + \frac{1}{c^2}\vec{S}) + \nabla(\rho_e\vec{v}_e\vec{v}_e + \rho_i\vec{v}_i\vec{v}_i + \overleftrightarrow{p}_e + \overleftrightarrow{p}_i + \overleftrightarrow{T}) = 0.$$

To obtain $\vec{\pi}$ from this relation, i.e. the secular component of the sum of forces both on electrons and ions, there is no other way than passing to a one fluid description which is accomplished by setting

$$\rho = m_en_e + m_in_i,\quad \rho_{el} = e(n_i - n_e),\quad \vec{j} = e(n_i\vec{v}_i - n_e\vec{v}_e),\quad \vec{v} = (\rho_e\vec{v}_e + \rho_i\vec{v}_i)/\rho,$$

and taking the time average. With $\vec{w} = \vec{v}_i - \vec{v}_e$ and $\overleftrightarrow{p} = \overleftrightarrow{p}_e + \overleftrightarrow{p}_i$ this leads to

$$\left\langle\frac{\partial}{\partial}\rho\vec{v} + \nabla(\rho\vec{v}\vec{v} + \overleftrightarrow{p})\right\rangle + \left\langle\frac{1}{c^2}\frac{\partial}{\partial t}\vec{S} + \nabla(\overleftrightarrow{T} + \frac{\rho_e\rho_i}{\rho}\vec{w}\vec{w})\right\rangle = 0. \tag{2.37}$$

Possible viscosity effects are included in the pressure tensors $\overleftrightarrow{p}_e$, $\overleftrightarrow{p}_i$, $\overleftrightarrow{p}$. If $\vec{E}_T$ and $\vec{B}_T$ stand for the thermoelectric fields (e.g. thermally induced magnetic fields) and p_o for $p(n_i, T_i, T_e, \vec{E}_{Wave} = 0)$, then $\vec{\pi}$ is expressed as follows,

$$\vec{\pi} = \left\langle\rho_{el}(\vec{E}_t - \vec{E}_T) + \vec{j}\times(\vec{B}_t - \vec{B}_T) - \nabla\frac{\rho_e\rho_i}{\rho}\vec{w}\vec{w} - \nabla(\overleftrightarrow{p} - \overleftrightarrow{p}_o)\right\rangle \tag{2.38}$$

This is the general non-relativistic expression of the ponderomotive force. The first part of it may alternatively be cast into the form $-\langle\partial(\vec{S} - \vec{S}_T)/\partial ct + \nabla(\overleftrightarrow{T} - \overleftrightarrow{T}_T)\rangle$. In the absence of $\vec{E}_T$ and $\vec{B}_T$ eq. (2.37) reduces to

$$\vec{\pi} = -\left\langle\frac{1}{c^2}\frac{\partial}{\partial t}\vec{S} + \nabla(\overleftrightarrow{T} + \frac{\rho_e\rho_i}{\rho}\vec{w}\vec{w} + \overleftrightarrow{p} - \overleftrightarrow{p}_o)\right\rangle \tag{2.39}$$

which clearly shows that wave pressure and Maxwellian stress tensor are not at all identical (as sometimes it is erroneously stated). By observing the mass conservation $\partial_t \rho + \nabla \rho \vec{v} = 0$ eq. (2.37) is easily transformed into Eulerian form,

$$< \rho \frac{d\vec{v}}{dt} > = - \nabla \overleftrightarrow{p}_0 + \vec{f}_T + \vec{\pi} \ . \tag{2.40}$$

For explicitly evaluating $\vec{\pi}$ it is convenient to introduce the quantity α through the electric susceptibility χ,

$$\chi = - \frac{\omega_p{}^2}{\omega^2} \alpha \ , \ \omega_p{}^2 = \frac{n_{io} Z e^2}{\varepsilon_o m_e} \ , \ Z n_{io} = n_{eo}, \ n_{eo} = <n_e>.$$

For free electrons in a dilute plasma $\alpha = 1$ holds. For reasons of simplicity the atomic charge number Z will be set equal to unity. Furthermore the following assumptions are introduced:

(1) $\hat{\delta}_e << \lambda$, $n_{eo} \simeq n_{io}$ (quasi-neutrality);

(2) ω sufficiently far from all significant electron and ion resonances;

(3) no static magnetic field present (only for simplicity; its inclusion is simple);

(4) T_e is low enough to guarantee a frequency Doppler spread $\Delta\omega << \omega$;

(5) $\vec{v}_{eo} \simeq \vec{v}_{io}$, $n_{ei} = n_e - n_{eo} << n_{eo}$.

The latter assumptions mean that in the medium neither closed loops of thermoelectric currents and accelerated or trapped electrons nor high amplitude electron plasma waves exist. With these limitations $\vec{\pi}$ can be evaluated by applying a perturbation method and it is not difficult to show that the dominant response to a harmonic electric field is the linear one of the electronic fluid for which holds[42]

$$\vec{v}_{e1} = - i \frac{e}{m_e} e^{-i\omega t} \left(1 + i \frac{\partial^2}{\partial\omega\partial t} \right) \frac{\alpha}{\omega} \hat{E}(\vec{x},t) ,$$

$$\vec{\delta}_{e1} = \frac{e}{m_e} e^{-i\omega t} \left(1 + i \frac{\partial^2}{\partial\omega\partial t} \right) \frac{\alpha}{\omega^2} \hat{E}(\vec{x},t) . \tag{2.41}$$

$\vec{x}$ is the coordinate of the oscillation centre. It also holds with good accuracy

$$e n_{e1} + \nabla e n_{io} \vec{\delta}_{e1} = -(\rho_{el,1} + \nabla \vec{P}_1) = 0, \ \vec{P} = -e n_{io} \delta_{e1},$$

$$\nabla < \frac{\rho_e \rho_i}{\rho} \vec{w}\vec{w} > = \rho_{eo} < (\vec{v}_{e1} \nabla) \vec{v}_{e1} > - < \vec{v}_{e1} \frac{\partial \rho_{e1}}{\partial t} > , \tag{2.42}$$

$$< m_e \vec{v}_{e1} \frac{\partial n_{e1}}{\partial t} - e n_{e1} \vec{E} > = m_e \frac{\partial}{\partial t} < n_{e1} \vec{v}_{e1} > + e < n_{e1} \mathrm{Re} \left[e^{-i\omega t} \left(1 + i \frac{\partial^2}{\partial\omega\partial t} \right) (\alpha - 1) \hat{E} \right] > .$$

Finally, in first order $\overline{\langle \overleftrightarrow{p} \rangle}$ equals $\overleftrightarrow{p}_o$ and the pressure disappears from eq. (2.38)[42]. With the help of approximations (2.41) and (2.42) after a somewhat lengthy calculation from eq. (2.38) one arrives at the following expression for $\vec{\pi}$:

$$\vec{\pi} = \frac{\varepsilon_o}{4} \{\frac{\chi + \chi^*}{2} \nabla \hat{E}\hat{E}^* + \nabla[(\alpha^* - 1)\chi\hat{E}^*\hat{E} + c.c.]\} + \vec{\pi}_t \tag{2.43}$$

Thereby the divergence of the second-rank tensor has to be calculated according to

$$[\nabla(\alpha^*-1)\chi\hat{E}^*\hat{E}]_i = \partial_j(\alpha^*-1)\chi\hat{E}_i{}^*\hat{E}_j.$$

For a rarefied fully ionized plasma and a transverse electric field $\alpha=1$ and $\chi = -\omega_p{}^2/\omega^2$ holds and $\vec{\pi}$ reduces to the simpler expression

$$\vec{\pi} = \frac{\varepsilon_o}{4}\frac{\omega_p{}^2}{\omega^2}\nabla \hat{E}\hat{E}^* + \vec{\pi}_t, \quad \vec{\pi}_t = i\frac{\varepsilon_o}{4\omega}\{\frac{\omega_p{}^2}{\omega^2}\nabla(\hat{E}\,\partial_t\hat{E}^*) + \partial_t[\hat{E}\,(\nabla\frac{\omega_p{}^2}{\omega^2}\hat{E}^*)] - c.c.\} \tag{2.44}$$

When $\hat{E}$ does not change with time the formula becomes identical with the expression of eq. (2.30). In dense and partially ionized plasma α differs from unity and becomes density dependent. As soon as χ is known $\vec{\pi}_t$ is also evaluated in a lengthy but straightforward way by inserting relations (2.41), (2.42) into eq. (2.38). Eq. (2.38) is valid for all kinds of periodic electromagnetic fields, transverse as well as longitudinal ones, whereas for eq. (2.43) to hold for such waves their time variation must be harmonic. In an inhomogeneous plasma the curl of $\vec{\pi}$ neither vanishes for the term $\vec{\pi}_o \sim \hat{E}\hat{E}^*$ nor for $\vec{\pi}_t$; as a consequence, a stationary wave as well as a pulse can drive dc magnetic fields. In addition, $\vec{\pi}_t$ may induce additional anisotropies in a plasma like striations[40].

The origin of $\vec{\pi}_o$ can be understood from the asymmetry of the oscillatory motion of the single electron[43]. In contrast, $\vec{\pi}_t$ from eq. (2.44) is due to global effects: the first term represents the temporal increase of the electromagnetic momentum in the unit volume element, the second one is identical with

$$\frac{\partial}{\partial t} m_e \langle n_{e1} \vec{v}_{e1} \rangle \tag{2.45}$$

which constitutes the secular increase of mechanical momentum density per unit time. According to eq. (2.31) such changes of momentum densities manifest themselves as a force $\vec{\pi}_t$ in addition to $\vec{\pi}_o$.

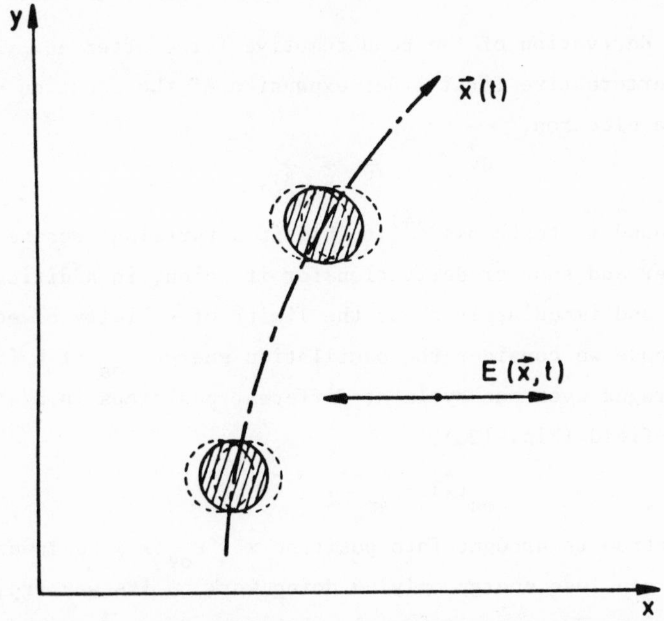

Fig. 13 - In the radiation field the electrons oscillate around the nuclei on their trajectory $\vec{x}(t)$. Owing to asymmetries in the oscillation amplitude the electrons transmit a secular force to the heavy particles. Solidly outlined circles, volume elements of the nuclei; dashed lines, electronic fluid elements at different times.

The quoted controversies about $\vec{\pi}_t$ may have their origin in the following subtlety. Newton's law

$$\frac{d}{dt}(Nm\vec{v}) = \vec{f}$$

is valid only if the number of particles N is held fixed. However, this cannot be achieved in a medium consisting of two oppositely charged fluids because the single components move into different directions (see Fig. 13). The only way to define a unique force on the fluid that fulfills Galilean or Lorentzian invariance is to pass to a one-fluid description, as done here. But then additional force terms arise, providing for the necessary corrections of particle losses during the period motion. That is the reason why radiation pressure and Maxwell's stress tensor $\overleftrightarrow{T}$ in the steady state case differ from each other by the tensor $\rho_i\rho_e\vec{w}\vec{w}/\rho$. Finally, it should be mentioned that Maxwell's expression (2.29) for p_r follows by integrating eq. (2.43) for a stationary amplitude along the space coordinate of propagation.

2.3.3 Generalized Ponderomotive Potential and Particle Acceleration

The standard derivation of the ponderomotive force after eq. (2.30) is based on a perturbative first order expansion of the equation of motion of the single electron,

$$m_e \frac{d\vec{v}}{dt} = -e\,(\vec{E} + \vec{v} \times \vec{B}),$$

and can be found in textbooks[44]. This is surprising because there is a much simpler and shorter derivation for it which, in addition, is more general and immediately shows the limits of validity of eq. (2.30). For this purpose we consider the oscillation energy ε_{os} of a free electron averaged over one cycle at different positions in a stationary sinusoidal $\vec{E}$-field (Fig. 13a),

$$\varepsilon_{os}(\vec{x}) = \frac{e^2}{4m_e\omega^2}\,\hat{E}\,\hat{E}^*. \qquad (2.46)$$

When the electron is brought into position $\vec{x}^1$, $\varepsilon_{os}(\vec{x}^1)$ is lower. But the electron can lose energy only by doing work on its way from $\vec{x}$ to $\vec{x}^1$. Since $\varepsilon_{os}(x)$ is a function of position only the force $\vec{f}$ on the electron is

$$\vec{f} = -\nabla\varepsilon_{os} = -\frac{e^2}{4m_e}\nabla\left|\frac{\hat{E}}{\omega}\right|^2.$$

Multiplying by n_e and taking ω = const yields eq. (2.30). The limitations of the latter are now evident:

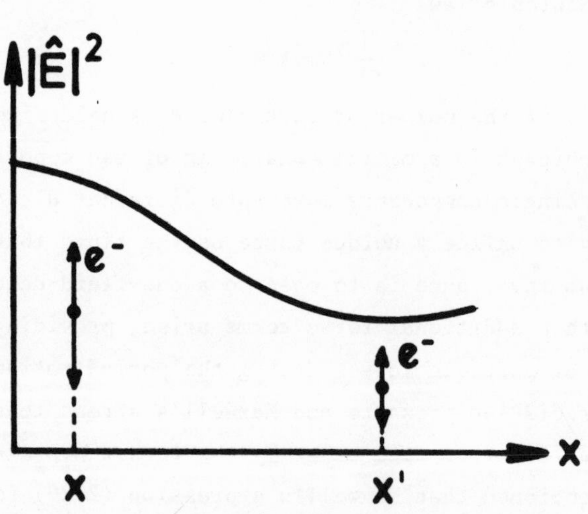

Fig. 13a - The time-averaged oscillation energy ε_{os} of a free electron is a function of position and frequency if it moves slowly through an oscillating E-field.

(1) When an electron runs into a wave pulse of amplitude $\hat{E}$ its translational motion slows down because of gradually transforming this energy into oscillation energy and eventually it is reflected once it has stopped. As a consequence, ω also changes and its variation contributes to changing the ponderomotive force (Doppler effect).

(2) In an inhomogeneous medium the frequency in the co-moving frame generally becomes space-dependent owing to such a dependence of $\vec{k}$.

(3) As soon as $e\hat{E}/k$ (longitudinal wave) or $e\hat{v}_{os}\hat{B}/k$ (transverse wave) becomes comparable to ε_{kin} the quiver motion is no longer harmonic and expression (eq. 2.43) begins to fail. However, there may still exist a ponderomotive potential since

(4) the existence of a ponderomotive potential is based only on the periodicity of motion and on the localization of ε_{os}.

With these considerations in mind one arrives at a generalized ponderomotive force of a free electron along the following line. Under very general conditions the motion of an electron in an electr(omagnet)ic field of the form of eq. (2.32) is such that its velocity splits into two components $\vec{v}_o$ and $\vec{v}_{os}(t)$ satisfying

$$\vec{v} = \vec{v}_o + \vec{v}_{os}(t), \; \langle \vec{v}^2 \rangle = \vec{v}_o^{\,2} + \langle \vec{v}_{os}^{\,2} \rangle + 2\langle \vec{v}_o \vec{v}_{os} \rangle, \; \langle \vec{v}_o \vec{v}_{os} \rangle / \langle \vec{v}_{os}^{\,2} \rangle \ll 1, \tag{2.47}$$

with $\vec{v}_o$ and $\langle \vec{v}_{os}^{\,2} \rangle$ changing slowly in time. $\vec{v}_o$ is the oscillation centre speed. This decomposition becomes unique when the additional condition of minimum curvature on $\vec{v}_o(t)$ is imposed. As a result the Hamiltonian $H_o = \langle H \rangle$ is a function of the momentum $\vec{p}_o = m_e \vec{v}_o$ and the oscillation centre coordinate $\vec{x}_o$ only,

$$H_o = \tfrac{1}{2} m_e \langle (\vec{v}_o + \vec{v}_{os})^2 \rangle \simeq \frac{\vec{p}_o^{\,2}}{2m_e} + \varepsilon_{os}(\vec{x}_o, \omega^1), \; \omega^1 = \omega - \vec{k}\,\vec{v}_o. \tag{2.48}$$

From the splitting of $\vec{v}$ clearly results that (1) ε_{os} has to be evaluated in a frame moving at speed $\vec{v}_o$ and that (2) ω^1 in H_o has to be treated as a fixed parameter (and not as a function of $\vec{p}_o$). At relativistic speeds this becomes of particular importance. $\vec{p}_o$ and $\vec{x}_o$ are canonical conjugate variables,

$$\{p_o^{\,k}, p_o^{\,l}\} = \{x_o^{\,k}, x_o^{\,l}\} = 0, \; \{p_o^{\,k}, x_o^{\,l}\} = \delta_{kl}.$$

As a consequence, with $\varepsilon_{int} = \varepsilon_{os}$

$$\frac{d}{dt}\vec{p}_o = -\frac{\partial H_o}{\partial \vec{x}_o} = -\nabla \varepsilon_{int} = \vec{f} \tag{2.49}$$

holds and $-\partial H_o/\partial \vec{x}_o$ is the secular force governing the motion of the oscillation centre $\vec{x}_o$.

The concept of ponderomotive potential is particularly useful for calculating secular forces in a fluid when these can be obtained by summing up the forces on the single particles. For fully ionized plasma this is the case. Two examples may illustrate this. The first case is that of a transverse $\vec{E}$-field after eq. (2.32) and a static magnetic field $\vec{B}(x)$ perpendicular to it. Under the assumption of weak spatial dependence of $\vec{B}$ over $\hat{\delta}_e$ the non-relativistic time-averaged oscillation energy out of resonance is given by

$$\varepsilon_{os} = \tfrac{1}{2}m\langle v_x{}^2 + v_y{}^2\rangle = \frac{e^2(\omega^2+\omega_c{}^2)}{4m_e(\omega^2-\omega_c{}^2)^2}|\hat{E}|^2, \quad \omega_c = eB(\vec{x}_o)/m_e. \tag{2.50}$$

It is valid for $\omega>\omega_c$ and $\omega<\omega_c$. The general case of a circularly polarized wave in a collisional magnetoplasma with ω near ω_c is treated in the literature[36]. The second example we consider is the relativistic motion of an electron in a linearly polarized sinusoidal wave. Here, ε_{os} is given by[45]

$$\varepsilon_{os,r} = m_e c^2\{(1 + \frac{e^2|\hat{E}|^2}{2m_e{}^2c^2\omega^2})^{\frac{1}{2}} - 1\} \tag{2.51}$$

$\vec{\pi}_o = -n_{eo}\nabla\varepsilon_{os,r}$ is the relativistic ponderomotive force density in the absence of density fluctuations. If the wave is circularly polarized the term containing $\hat{E}$ is twice as large. For values much smaller than unity the square root can be expanded and ε_{os} from eq. (2.46) is obtained.

It is useful and instructive to generalize the Hamiltonian method to the case of an electron bound by a potential $V(\vec{x}-\vec{x}_o) = ar^s$, (s = real number). Then by the virial theorem

$$\langle\varepsilon_{kin}\rangle = \frac{s}{2}\langle V\rangle$$

holds in the absence of a driving E-field. Thus, the total inner energy is $\varepsilon_i = \langle V + \varepsilon_{kin}\rangle = (1 + s/2)\langle V\rangle$, which clearly does not depend on $\vec{x}_o$. With this the time-averaged Hamiltonian becomes

$$H_o = \frac{\vec{p}_o{}^2}{2m} + \varepsilon_{int}(x_o,\omega^1) + \varepsilon_i, \quad \varepsilon_{int} = \varepsilon_{os} - \frac{s}{2}\langle V\rangle. \tag{2.52}$$

The oscillatory energy ε_{os} is the time-averaged kinetic energy relative to the oscillation centre. ε_{int} is the interaction energy representing the true ponderomotive potential from which the secular force is obtained again from eq. (2.49). As an important application the harmonic oscillator in an external electric field is considered,

$$\ddot{\vec{x}} + \omega_0{}^2(\vec{x}-\vec{x}_0) = -\frac{e}{\mu}\hat{E}\,e^{-i\omega t} \quad , \ \mu = m_e M/m_e+M) ; \ s = 2 \ . \tag{2.53}$$

If $\vec{x} = \vec{x}_0$ before $\hat{E}(\vec{x},t)$ is switched on, the solution for $\omega \neq \omega_0$ is

$$\varepsilon_{os} = \frac{\omega^2 e^2|\hat{E}|^2}{4\mu(\omega_0-\omega)^2}, \ \langle V\rangle = \frac{\omega_o{}^2 e^2|\hat{E}|^2}{4\mu(\omega_0{}^2-\omega)^2}, \ \varepsilon_{int} = \frac{e^2|\hat{E}|^2}{4\mu(\omega^2-\omega_0{}^2)} \ . \tag{2.54}$$

Thus, the ponderomotive force on an induced electric dipole out of resonance is

$$\vec{f} = -\frac{e^2}{4\mu}\nabla\frac{|\hat{E}|^2}{\omega^2-\omega_0{}^2} \ , \tag{2.55}$$

a result which is correct also quantum mechanically. For $\omega > \omega_0$ the dipole behaves like a free electron; for $\omega < \omega_0$ it is driven into the direction of increasing E-field; at resonance $\varepsilon_{int} = \varepsilon_{os} - \langle V\rangle = 0$ holds and $\vec{f}$ is zero (again in agreement with quantum mechanics). For $\omega_0 = 0$ eq. (2.46) results. This example also shows that in using Hamiltonian formulation it is important to take the right canonical conjugate variables at the beginning; in eq. (2.54) ε_{os} also depends on $\vec{x}_0$ through $\hat{E}(\vec{x}_0,t)$, but $\nabla\varepsilon_{os}$ would yield the wrong force.

At resonance or very close to it there is an additional ponderomotive force on a system which is due to spontaneous absorption and emission of quanta[46] (photons, plasmons, phonons, magnons, etc.). Therefore, in order to be precise, only induced wave pressure is treated here.

By using arguments similar to those exposed in this section it was possible to determine the spectrum of non-interacting trapped particles in resonance absorption or stimulated Raman scattering[47]. The excited electron plasma wave shows an amplitude distribution which may be idealized as sketched in Fig. 13b (small picture). Particles entering the wave from the RHS (A) may cross the wave or, at lower energies, get reflected in the ponderomotive potential. Particles from LHS (B,C) may be reflected or get trapped and accelerated by the potential of the electrostatic wave. On the assumption that the process of trapping and detrapping is adiabatic (in reality adiabaticity is slightly violated) the contours of possible particle energies in

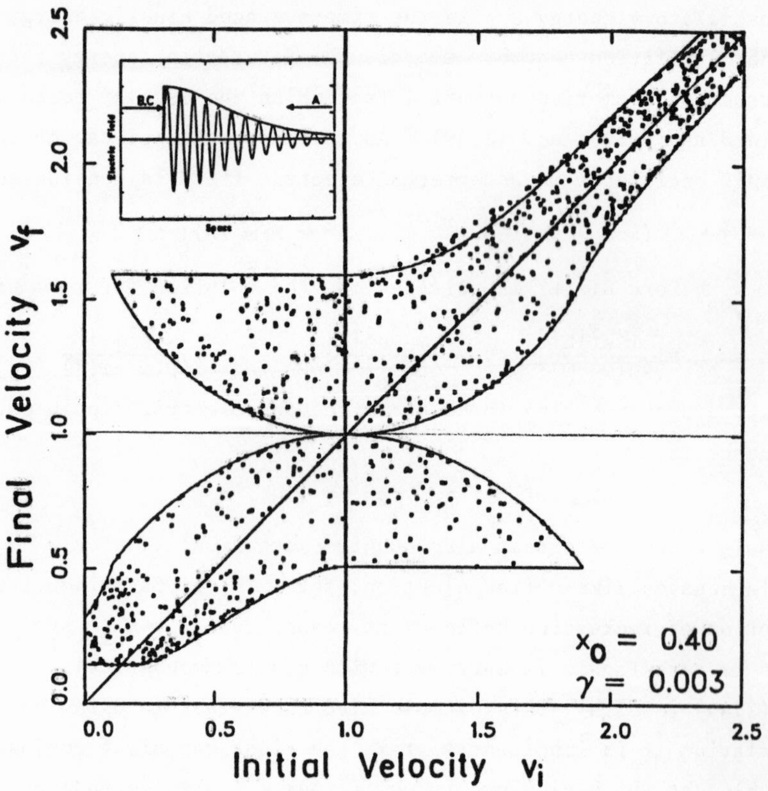

Fig. 13b - Velocity spectrum of accelerated electrons after detrapping. The spoon-shaped contours are analytical curves; the dots represent numerical simulations. Velocities are measured in units of phase speed of the Langmuir wave. The amplitude of the resonantly excited Langmuir wave (small picture) may be approximated by $\hat{E} = 4x_0 \exp(\gamma kx)/(1+\exp(\gamma kx))^2$ *The wave is moving to the RHS.*

Fig. 13b could be calculated analytically. The dots were obtained by numerically solving 2500 individual trajectories. The precision of the analytical prediction is impressive.

The examples reported may illustrate the usefulness of the ponderomotive potential. In general it is easier to calculate ε_{int} than $\vec{\pi}$ directly. However, in cases for which no Hamiltonian exists (e.g. friction) or when collective effects play a decisive role $\vec{\pi}$ must be determined directly according to the preceding section.

3.
RESONANCE ABSORPTION

3.1 Linear Resonance Absorption

In contrast to the s-polarised wave of eq. (2.9) a p-polarised electromagnetic field with its $\vec{E}$-vector in the plane of incidence does not decouple from the electron plasma wave. Considering again a layered medium as in Fig. 5, $\vec{E}(x,y) = \vec{E}(x)\exp(i\kappa y)$ can be set. Then, with $\beta = s_e/c$ and $\nabla n_i = 0$ at high frequency, eq. (2.4) transforms into

$$\frac{\partial^2}{\partial x^2}\vec{E} + k^2(n^2(x) - \kappa^2/k^2)\vec{E} = (1-\beta^2)(\vec{e}_x\frac{\partial}{\partial x} + i\kappa\vec{e}_y)(\frac{\partial E_x}{\partial x} + i\kappa E_y) \; . \qquad (3.1)$$

For the validity of this equation it must be assumed that the thermal velocity is small compared with phase and light velocities and that the density scale length L is large in comparison to the Debye length, i.e.

$$|n^2(x)| = |\varepsilon| << 1, \beta^2 << 1, (\beta/k_o L)^{\frac{2}{3}} << 1.$$

By splitting eq. (3.1) into the single components and suppressing small terms in β one obtains

$$E_x'' + \frac{k^2}{\beta^2}(n^2 - \sin^2\alpha_o)E_x = i\frac{k}{\beta^2}\sin\alpha_o\frac{\partial E_y}{\partial x}, \; E_y'' + k^2n^2E_y = ik\sin\alpha_o\frac{\partial E_x}{\partial x} \; . \; (3.2)$$

The interpretation of these equations is as follows. At the turning point S of the obliquely incident laser wave the electric vector oscillates in the direction of the density gradient thereby causing periodic accumulations of charges which propagate as a longitudinal plasma wave down the plasma density profile. In order to be effective the excitation must occur resonantly with the driving laser field. This imposes the conditions on the electromagnetic (em) and electrostatic (es) waves:

$$\omega^2 = \omega_p^{\;2} + c^2k_{em}^{\;2} \; , \; \omega^2 = \omega_p^{\;2} + s_e^{\;2}k_{es}^{\;2} \; , \; \vec{k}_{es} = \vec{k}_{em}$$

Because of the different speeds of light c and electron sound s_e, resonance is fulfilled for $k_{es} = k_{em} = 0$, i.e. in a small region around the critical point only. In the overdense region E_x^{es} as well as E_y^{es} become evanescent similarly to E_z^{em} of eq. (2.9), the main difference being the more rapid decay in correspondence to their wave number ~$k/\beta >> k$. From the second of eqs. (3.2) follows

$$i\kappa \int_{x>x_c}^{x\simeq x_c} E_x'\,dx = i\kappa E_x(x) = \int_{x>x_c}^{x\simeq x_c} (E_y'' + k^2 n^2 E_y)\,dx = E_y'(x) + \varepsilon \simeq E_y'(x),$$

which inserted in the first equation leads to

$$E_x'' + \frac{k^2}{\beta^2} n^2 E_x = -\,i\,\frac{k}{\beta^2}\sin\alpha_o \varepsilon. \tag{3.3}$$

Because of the $e^{-i\omega t}$ time dependence of E_x, E_y and ε this is equivalent to the equation of a resonantly driven electron plasma wave,

$$\frac{\partial^2}{\partial t^2} E_x - s_e^{\,2} \frac{\partial^2}{\partial x^2} E_x + \omega_p^{\,2} E_x = ikc^2 \sin\alpha_o \varepsilon. \tag{3.4}$$

From $E_y'(x) \simeq i\kappa E_x(x)$ and the fact that $|E_x|^2$ looks similar to E_z^{em} of eq. (2.9) around $x_c(=0)$, with its typical scaling factor now being of the order of magnitude of k/β, the estimate follows

$$|E_y| \simeq |E_x| \beta \sin\alpha_o \ll |E_x| \,. \tag{3.5}$$

That is the main result of a more quantitative evaluation (48). As a consequence, $\nabla\vec{E} \simeq \partial E_x/\partial x$ is a good approximation and the displacement of an electronic fluid element $\vec{\delta}$ is nearly δ_x. It is the content of the so-called capacitor model.

As an effective absorption mechanism for laser light, resonance absorption was proposed for the first time by R.P. Godwin in 1971 (49). This mechanism of linear mode conversion, however, had already been treated before in two classical papers (50,24;51), the first one (Denisov/Ginzburg) considering the phenomenon in the cold plasma limit, i.e. $\beta = 0$, and the second one (Piliya) allowing for finite electron temperature (51). An important step towards a general treatment is made by two other papers (52,53) which show that the cold plasma model leads to essentially the same conversion rate as the warm plasma. The most complete treatment of this subject was provided by Kull (54). This author also dealt with the cases of relativistic temperature and relativistic flow (55). In the following, results and figures are taken from Ref.54.

For a more detailed investigation of resonance absorption it is convenient to eliminate E_y through $i\omega\vec{B} = \nabla x\vec{E}$, i.e. $i\omega B = i\omega B_z = E_y' - i\kappa E_x$, and to introduce

$$k_1 = k/\beta,\ \gamma^{-2} = 1 - \beta^2.$$

Then eqs. (3.2) read

$$E_x'' + k_1^2(\varepsilon-\beta^2\sin^2\alpha_o)E_x = -\frac{k_1^2}{c\gamma^2}\sin\alpha_o B,$$
$$B'' - \frac{\varepsilon'}{\varepsilon-\beta^2\sin^2\alpha_o}B' + k^2(\varepsilon-\sin^2\alpha_o) = -\frac{c\beta^2k^2\varepsilon'\sin\alpha_o}{\varepsilon-\beta^2\sin^2\alpha_o}E_x. \tag{3.6}$$

In general this system has to be solved numerically. There are no simple analytical solutions available. Fig. 14 shows the time evolution of E_x and B over the dimensionless space coordinate $\rho = (kL)^{\frac{2}{3}}x/L$ for a linear density profile. The strength of excitation of the electron plasma wave at fixed frequency ω and electron temperature T_e are determined by L and α_o. For a cold plasma Ginzburg introduced

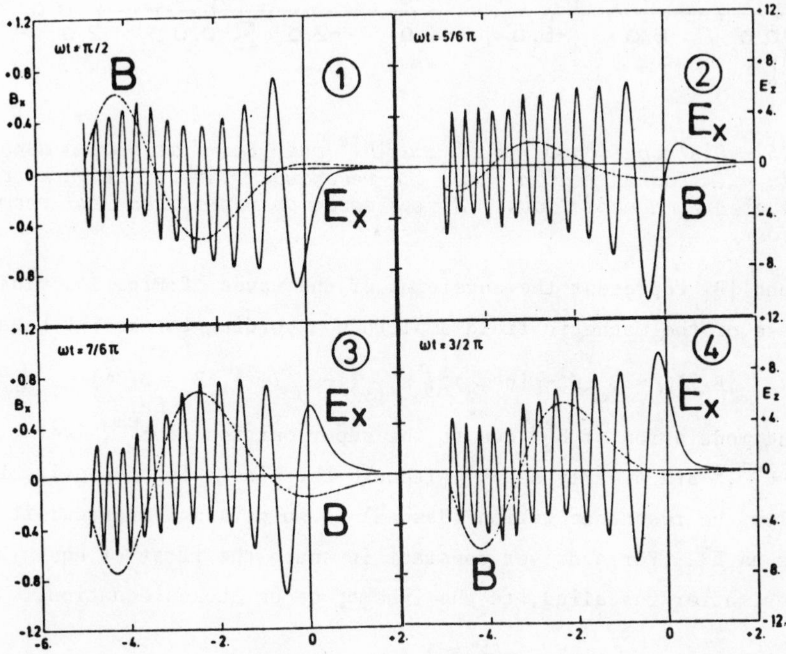

Fig. 14 - E_x and B evolving in time. E_x is the superposition of E_x^{es} (small wavelength) and E_x^{em} (the same wavelength as B!). Linear density profile. Note that B is almost constant in the resonance region (ρ≃0); q=0.5. $\rho=(kL)^{\frac{2}{3}}x/L$.

the dimensionless parameter $q = \sin^2\alpha_o(kL)^{\frac{2}{3}}$. Conversion is best characterized by plotting $|E_x|^2$, $|B|^2$. For q = 0.5 and $\beta^2 = 0.01$ (corresponding to $T_e = 1.8$ keV) these quantities are shown in Fig. 15.

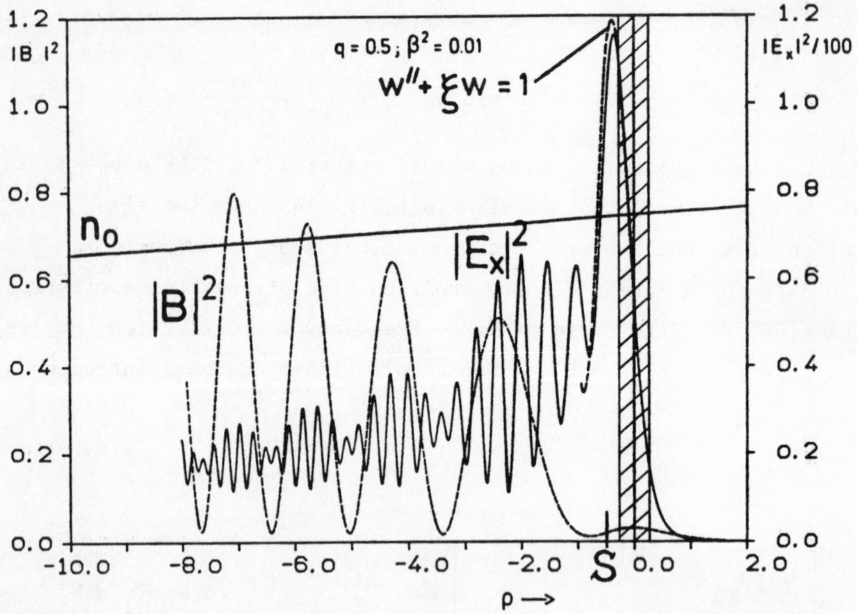

Fig. 15 - Distribution of $|E_x|^2$ and $|B|^2$ over the flat plasma density profile n_0. Dashed region shows the resonance width. S turning point of the electromagnetic wave. Stokes equation reproduces well resonance behaviour.

$|E_x|$ and $|B|$ represent the envelopes of the waves of Fig. 14. The decrease of the magnetic field amplitude according to the WKB behaviour,

$$|B(x)| = B_0(n^2 - \sin^2\alpha_0)^{\frac{1}{4}}, \; n^2 = (1-\omega_p{}^2/\omega^2), \; B_0 = B(-\infty) \tag{3.7}$$

and the modulation of E_x^{es} due to the superposition of E_x^{em}, i.e. $|E_x^{es} + E_e^{em}|^2$ are clearly seen. Although the plasma inhomogeneity is very low the resonance region (dashed) is surprisingly narrow; it varies as $L^{\frac{1}{3}}$. For a driver constant in space the first of eqs. (3.6) reduces, after rescaling, to the inhomogeneous Stokes equation,

$$W'' + \xi W = 1 \; ; \tag{3.8}$$

$$\xi = -(k_1/L^2)^{\frac{1}{3}}(x + x_0), \; x_0 = L\beta^2\sin^2\alpha_0, \; W = -\gamma^2 E_x/c(k_1L)^{\frac{2}{3}}\sin\alpha_0 B(o).$$

As can be seen from the figure the solution $W(\xi)$ reproduces the real solution in the resonance region well indeed. $|W|^2$ reaches its maximum of $|W|^2 = 3.5$ at $\xi = 1,8$, its halfwidth being $\Delta\xi = 3$. The maximum of $|W|$ is already out of the resonance region, the electrostatic wave propagate as a free wave following its own dispersion, eqs. (2.3). At the latest from $\xi = 2.8$ on the Langmuir wave can be described by the WKB approximation. According to the considerations of sec. 2.1 the magnitude of its wave vector is k/β. Then, owing to the constancy of $\kappa = k\sin\alpha_o$, for the angle α_1 which the Langmuir wave forms with the x-axis must hold

$$\kappa^2 + (k/\beta)^2(1-\sin^2\alpha_1) = (k/\beta)^2, \Rightarrow \sin\alpha_1 = \beta\sin\alpha_o, \tag{3.9}$$

i.e. the electron plasma wave is emitted nearly parallel to the x-axis (for $\beta = 0.1$ the maximum possible α_1 is less than 6 degrees). The electric field E_x^{es} varies as

$$E_x^{es}(x) = \frac{C\cos\alpha_1}{(1-\sin^2\alpha_1)^{\frac{1}{4}}}, \quad C = \text{const.} \tag{3.10}$$

The smallness of α_1 can also be interpreted in the following way: since the wave is excited very near to the critical point its angle of "incidence" α_1 must be low in order to have its turning point S_1 at a position x_1 that is very close to $x = x_c$; in fact,

$$\sin^2\alpha_1 = 1 - \frac{n_e(x_1)}{n_c} .$$

Resonance absorption, i.e. the strength of plasma wave excitation, clearly depends on the angle α_o of incidence. Graphs analogous to that of Fig. 15 showing such a dependence are depicted in Fig. 16. They clearly show that it is the product

$$B(o)\sin\alpha_o$$

which acts as the driver.

Let us now determine the conversion rate $A = |S_x^{es}/S_x^{em}|$. From eqs. (2.13b),(3.6),(3.8) one calculates ($\gamma = 1$)

$$S_x^{em} = \tfrac{1}{2}(1-\sin^2\alpha_o)^{\frac{1}{2}}c\varepsilon_o|E_o|^2, \quad S_x^{es} = \frac{\pi}{2}c\varepsilon_o kL\sin^2\alpha_o|B(o)|^2,$$

$$A = \pi kL\frac{\sin^2\alpha_o}{(1-\sin^2\alpha_o)^{\frac{1}{2}}}\left|\frac{B(o)}{E_o}\right|^2 . \tag{3.11}$$

For warm ($\beta^2 > 0$) and cold plasma the conversion factor =(absorption) is shown in Fig. 17. There is almost no temperature dependence for $\beta^2 \lesssim 10^{-2}$. Maximum absorption of 49% occurs at $q = 0.5$. For relativistic temperatures ($\beta^2 = 1$) total conversion occurs at $q \simeq 0.7 - 0.9$.

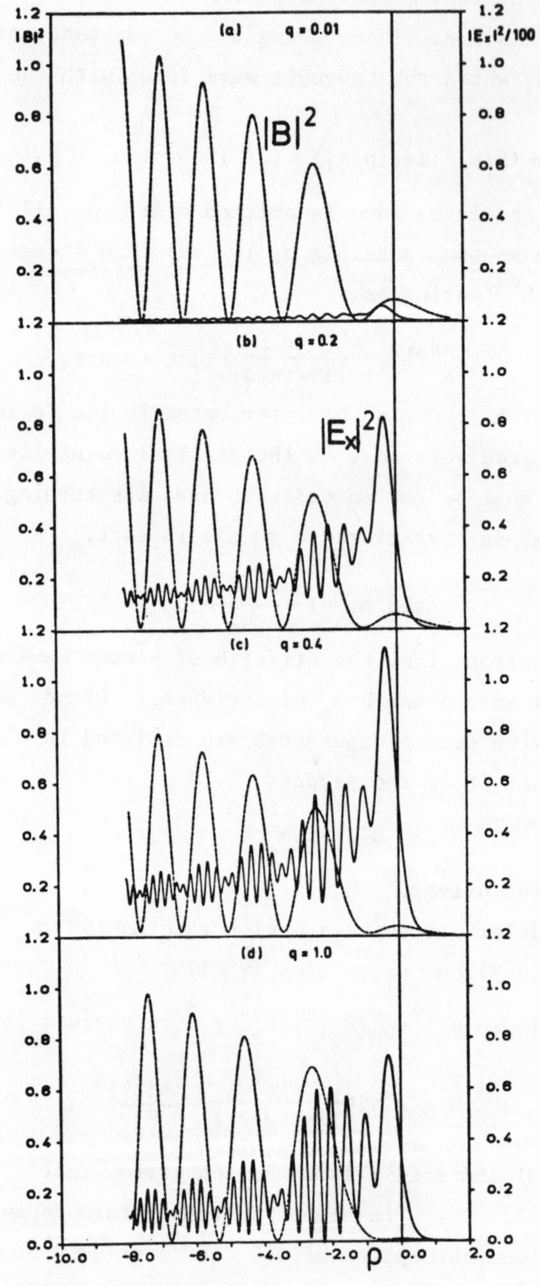

Fig. 16 – Field distributions as in Fig. 15 for different angles of incidence.

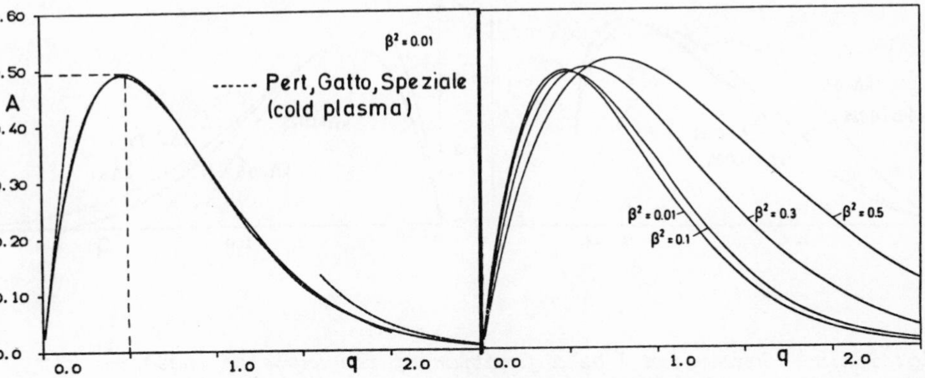

Fig. 17 - Conversion rates A in cold and warm plasmas.

From eq. (3.8) the shape of the resonance field is determined analytically for $\xi \lesssim 2$:

$$|E_x/E_o|^2 = (1-\sin^2\alpha_o)^{\frac{1}{2}}(kL)^{\frac{1}{3}}\frac{A}{\pi\beta^{\frac{4}{3}}}|W|^2 \ ; \tag{3.12}$$

specialising for the maximum value $|E_x^{max}|$ yields

$$|E_x^{max}|^2 = 1.1 \times (1-\sin^2\alpha_o)^{\frac{1}{2}}(kL)^{\frac{1}{3}}\frac{A}{\beta^{\frac{4}{3}}} \ . \tag{3.13}$$

Resonance absorption can also occur in thin transition layers of the type of Fig. 9. More precisely, if an Epstein layer of the form

$$n^2(x) = n_1^2 - \frac{2n_1^2}{1+\exp(-x/\Delta)} \qquad n_2^2(\infty) = -n_1^2 \ , \tag{3.14}$$

is chosen the following characteristic absorption curves are obtained as functions of the angle of incidence

$$\alpha_1 = \frac{\sin\alpha_o}{n_1}$$

for different transition widths $\overline{k}\Delta = n_1 k\Delta$ (Fig. 18a) (54).

If such curves are drawn as a function of the Ginzburg parameter q instead of α_1 Fig. 18b is obtained: the maximum of A increases slightly and shifts towards smaller values of $q = (kL)^{\frac{2}{3}}\sin^2\alpha_o$. For $\overline{k}\Delta \geqslant 1$ the Ginzburg curve of the cold plasma represents a good approximation. Deviations from that curve start at higher angles of incidence. As it

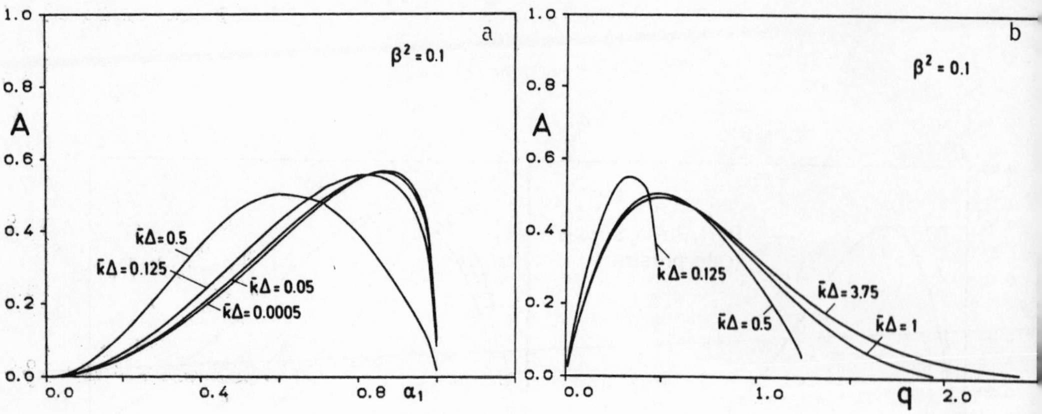

Fig. 18a,b - Conversion A as a function of the angle of incidence for various widths of the Epstein transition layer, eq. (3.14). With decreasing Δ the angle of maxA increases. $\alpha_1 = \sin\alpha_o/n_1$, $q = \sin^2\alpha_o (kL)^{\frac{2}{3}}$

was first found by H.J. Kull total conversion at all temperatures may occur in very steep density profiles if the incident wave forms an angle of $\pi/2$ with the asymptotic direction of the outgoing Langmuir wave.

Resonance absorption was shown to occur experimentally(56-59) and to be in agreement with linear theory at laser intensities of $I = 10^{14}$ W/cm^2 Maaswinkel et al.(56) used a 30 ps Nd laser of 0.7 J energy output and measured the s- and p-polarized reflection from solid target as a function of the angle of incidence at $\lambda = 1.06\mu m$ and $\lambda = 0.53\mu m$. A clear difference between s- and p-absorption was found at all angles of incidence $\alpha_o \leqslant 60^o$.

The most accurate and complete measurements in the intensity range from 10^{13} to 10^{14} W/cm^2 were performed by Dinger et al.(59) on plexiglass targets with a Nd-YAG laser of 35 ps pulse length. The absorption was detected simultaneously in five parallel channels of observation: reflection of the fundamental wave, second harmonic generation, charge and velocity distribution of the expanding plasma cloud, and relative ablation pressure on the target. All five channels gave equivalent results and the absorption of p-polarization exceeded that for s-polarisation by about a factor of 2 on the average (Fig. 19). The excited Langmuir wave can interact non-linearly with the incident laser light,

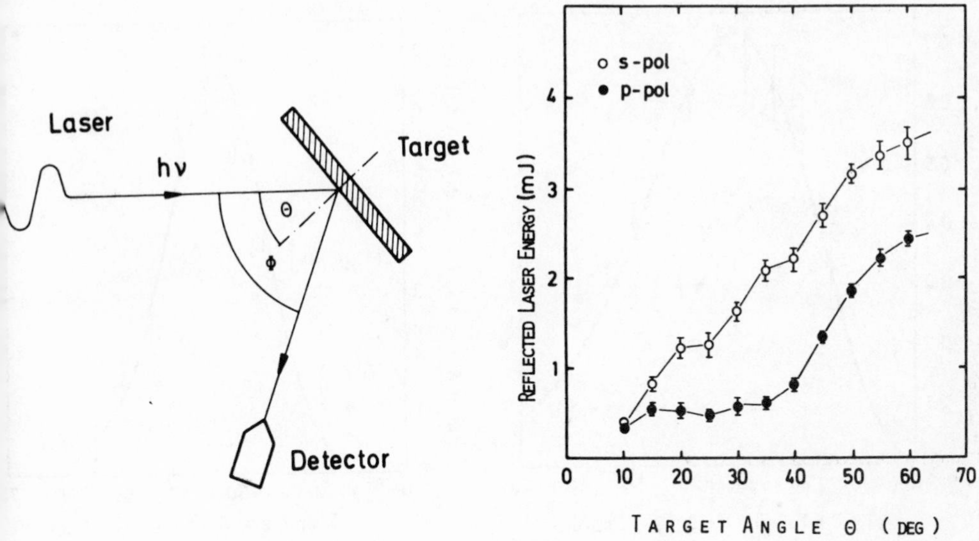

Fig. 19 - Resonance absorption vs. angle of incidence of Nd laser pulse of 35 ps time duration in plexiglass (after Dinger et al.).

thus creating the second harmonic, $2\omega_{Nd}$. If this effect occurs $I_{2\omega_{Nd}}$ has to be proportional to A. Fig. 20 clearly shows that this is indeed the case[59]. Some characteristic deviations between experiments and theory of resonance absorption may have to be attributed to fluctuations of the plasma density profile as indicated in Fig. 12. Theoretically the influence of stochastic density oscillations was studied by several authors[60,61]. It was found that the excitation level of the Langmuir wave reduces monotonically with increasing size of the density fluctuations.

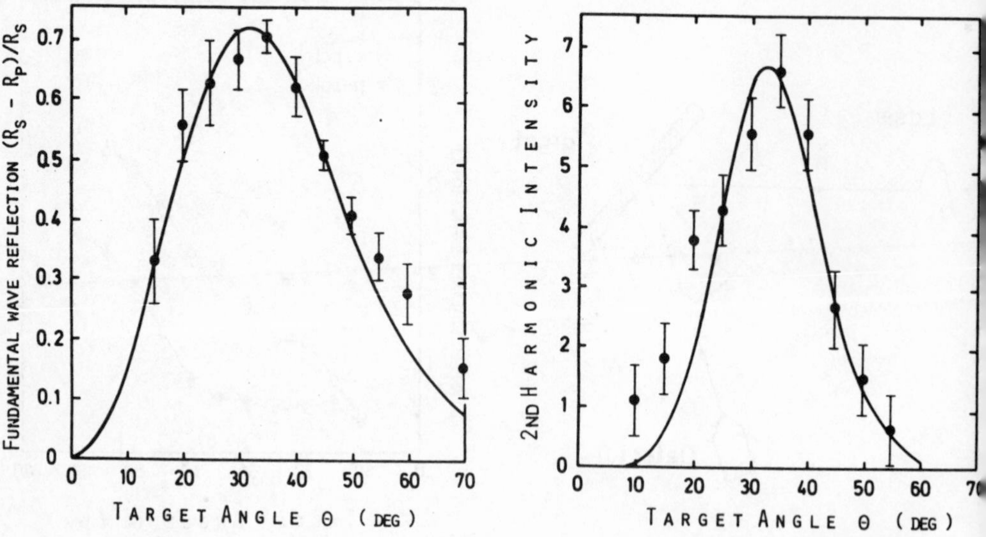

Fig. 20 - The intensity of 2nd harmonic light follows the intensity of the Langmuir wave resonantly excited by the Nd laser pulse (after Dinger et al.).

3.2 Nonlinear Resonance Absorption

Under the influence of a strong driver field the equation of motion of the electrons can no longer be linearized and the nonlinear terms of the current density in eq. (2.1) have to be taken into consideration Strong experimental evidence for the excitation of a highly nonlinear plasma wave is the appearance of 47 harmonics of CO_2 laser light from a solid target plasma[(62)]. It may be estimated that in plasmas produced by a Nd laser at its fundamental frequency a linear treatment is satisfactory for intensities not higher than 10^{14}W/cm^2 whereas the same limit for a CO_2 laser is reached at a 100 times lower intensity.

The basic aspects of high amplitude Langmuir wave conversion can be studied in one dimension by making the simplifications that all space dependencies occur in x-direction only. The justification for this assumption is inequality, eq. (3.5) (capacitor model). For the electromagnetic fields this implies

$$\vec{E} = (E,0,0),\ \mathrm{div}\vec{E} = \frac{\partial E}{\partial x}\ ,\ \nabla \times \vec{B} = (i\kappa B,0,0). \qquad (3.15)$$

With the electron density $n_e(x,t)$ and the electric current density $j = e(n_o v_o - n_e v)$ Maxwell's equations and the equation of motion for the electrons read in this approximation:

$$\frac{\partial E}{\partial t} = i\kappa c^2 B - j$$

$$\frac{\partial E}{\partial x} = \frac{e}{\varepsilon_o}(n_o - n_e) \qquad (3.16)$$

$$\frac{dv}{dt} = -\frac{s_e^2}{n_e}\frac{\partial n_e}{\partial x} - \frac{e}{m_e}E$$

From these equations follows

$$\frac{d^2 v}{dt^2} + \frac{d}{dt}\left(s_e^{\,2}\frac{1}{n_e}\frac{\partial n_e}{\partial x}\right) + \omega_p^{\,2}(v - v_o) = -i\frac{ec^2}{m_e}\kappa B \qquad (3.17)$$

The usefulness of this equation lies in the fact that $\omega_p^{\,2}$ contains n_o and not the rapidly varying electron density n_e, $\omega_p^{\,2} = e^2 n_o(x,t)/\varepsilon_o m_e$. The ion flow velocity v_o has been introduced to avoid the usual singularity in the cold plasma case[63,64].

Replacing x and t by the Lagrangian coordinates a and t, in which a designates the initial position of the electron and ion fluid elements, the actual position of them at time t is

$$x_o(a,t) = a + \int_0^t v_o dt, \quad x_e(a,t) = a + \int_0^t v dt.$$

It is assumed throughout this section that for $t < 0$ no driving field is present. In the new coordinates eq. (3.17) reads for the oscillatory velocity $v_e = v - v_o$ of the electrons

$$\frac{\partial^2}{\partial t^2}v_e + \frac{s_o^2}{\gamma n_o^{\gamma}}\frac{\partial^2}{\partial t\partial a}\left(\frac{n_o}{1+\partial/\partial a\int v_e dt}\right)^{\gamma} + \omega_p^{\,2}(a,t).v_e = \frac{1}{2}\omega^2 v_d e^{-i\omega t} + cc. \qquad (3.18)$$

s_o represents the electron sound speed at $n_e = n_o$. The validity of the oscillation centre approximation used in deriving eq. (3.18) is well confirmed[64]. The electronic fluid is assumed to obey an adiabatic law, $p_e = \text{const} \times n_e^{\gamma}$. v_d is

$$v_d = -i\frac{ec^2}{m_e\omega^2}\kappa B .$$

In the cold plasma $s_o = 0$ holds and eq. (3.18) reduces to that of an oscillator with time varying eigenfrequency $\omega_p(a,t)$. Therefore, even in the nonlinear oscillation regime the oscillatory velocity remains sinusoidal. The density n_e is related to $\delta_e = \int v_e dt$ by

$$n_e = \frac{1}{1+\partial\delta_e/\partial a} \simeq \frac{1}{1+\varepsilon \sin\omega_p t} . \qquad (3.19)$$

Thus, at larger values of ε (e.g. $\varepsilon > 0.5$) n_e becomes peaked and, for $\varepsilon \to 1$, it breaks.

In a warm plasma energy transport out of the resonance region mainly occurs by the Langmuir wave. Thus, v_o may be set equal to zero. As a consequence, in a steady state density profile $\omega_p(a)$ does no longer depend on time and, according to eq. (2.20), ω of the Langmuir wave does not change.

At moderate amplitudes the denominator in eq. (3.19) can be expanded in a Taylor series. By summing up the dominant terms the amplitude dependent Bohm-Gross dispersion relation results,

$$\omega^2 = \omega_p^{\ 2}\{1+g(\frac{s_o k_{es}}{\omega_p})^2\}, \ g = 1+\frac{1}{4}(\frac{\gamma+2}{2})k_{es}^{\ 2}|\delta_1|^2 + (\frac{\gamma+4}{4})k_{es}^{\ 4}|\delta_1|^4 + \ldots \qquad (3.20)$$

$|\delta_1|$ is the amplitude of the fundamental mode(65). The phase velocity increases with the amplitude $|\delta_1|$ because of self-interaction of the fundamental mode. This is a characteristic phenomenon of nonlinear waves.

With the help of eq. (3.20) the relative contributions of the second and third harmonics are determined as follows:

$$\hat{\delta}_2/\hat{\delta}_1 = \frac{\gamma+1}{2} \frac{n^{\frac{3}{2}} g^{-\frac{1}{2}} \omega^2}{3\omega_p^{\ 2} g + 4(g-1)n^2\omega^2} \frac{\hat{v}_1}{s_o} , \ n^2 = 1 - \omega_p^{\ 2}/\omega^2 ,$$

$$\hat{\delta}_3/\hat{\delta}_1 = \frac{\gamma+1}{4} \frac{n^3\omega^2}{g^2} (\frac{\hat{v}_1}{s_o})^2 \left[3(\gamma+1) \frac{\omega^2}{3\omega_p^{\ 2} g + 4(g-1)n^2\omega^2} + \frac{\gamma+2}{2} n\right] /$$

$$\left[8\omega_p^{\ 2} + 9(g-1)n^2\omega^2\right] .$$

In the resonance region $n^2 \ll 1$, $g \simeq 1$, $\omega^{\ 2} \simeq \omega^2$ holds and these relations simplify to

$$\hat{\delta}_2/\hat{\delta}_1 \simeq \frac{\gamma+1}{6} n^{\frac{3}{2}} \frac{v_1}{s_o} , \ \hat{\delta}_3/\hat{\delta}_1 \simeq \frac{1}{2}(\frac{\gamma+1}{4})^2 n^4 (\frac{v_1}{s_o})^2 ,$$

showing thereby the interaction of modes: as the excited wave moves away from the resonance region towards lower densities the refractive index n increases from zero to one. Due to the excitation of higher harmonics at finite electron temperature v_e or δ no longer remain sinusoidal as can clearly be seen from Fig. 21. The maxima of v_e are peaked now and $|v_e|_{max} > |v_e|_{min}$ holds. No exact analytical expression exists for the breaking limit in the resonance region up to now. An approximate limit may be obtained by keeping in mind that the thermal particle motion acts as an effective flow velocity v_c for the transport of the oscillatory energy out of the resonance region. By projecting a Maxwellian distribution at T_e on the x-direction

$$v_e = (k_B T_e/2\pi m_e)^{\frac{1}{2}} = 0.4 v_{th} \ , \ v_{th} = (k_B T_e/m_e)^{\frac{1}{2}} \qquad (3.21)$$

is obtained. For breaking, in order to occur at the position of maximum amplitude, $v_d/v_c = 2.8$ must be fulfilled[64].

As long as breaking is considered in regions where n is still small and higher harmonics are weakly excited only, a more accurate breaking criterion is deduced by identifying v_c with a group velocity v_g of the electron plasma wave. In this case k_{es} can be obtained from the linear theory of resonance absorption[54] if the breaking point is chosen in a region where the WKB-approximation applies. With the dimensionless variable $\xi = (\omega L_c/s_o)^{\frac{2}{3}} n^2$ results

$$v_g = s_o^{\,2}/v_{ph} = \xi^{\frac{1}{2}} s_o^{\frac{4}{3}}/(\omega L_c)^{\frac{1}{3}} \ . \qquad (3.22)$$

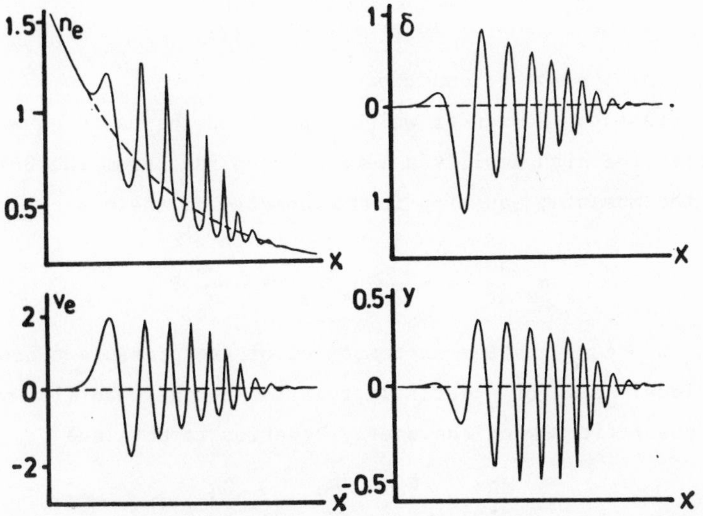

Fig. 21 - Normalized electron density n_e, electron velocity v_e, excursion $\delta = \int v_e dt$ and $y = \partial\delta/\partial a$ as functions of z. Nearly the same picture results if these quantities are plotted as functions of a.

At the position of maximum amplitude WKB is fulfilled and ξ becomes 1.8. Interesting enough, with $T_e = 1$ keV, $\gamma = 3$, $\omega = 1.7 \times 10^{15}$, $L_c = 10^{-4}$cm from eq. (3.22) $v_g = 7 \times 10^8$cm/s results whereas eq. (3.21) yields $v_c = 5 \times 10^8$cm/s, so that in both cases the limiting driver intensities differ by a factor of 2 only.

Out of the resonance region a criterion for wave breaking is obtained in the following way. By assuming a steady state situation eqs. (3.16) reduce in the wave frame to

$$m_e \frac{s_o^2}{\gamma-1}\left[(n_e/n_o)^{\gamma-1} - 1\right] - e(V - V_o) = m_e(v_{ph}^2 - u^2) \ , \ u = v - v_{ph} . \quad (3.23)$$

V is the electrostatic potential of the wave, $V = \int E dz$. It is convenient to eliminate V and u and to integrate eq. (3.23) once. Then, with $\mu = n_e/n_o$ the relation follows:

$$|(s_o^2/v_{ph}^2)\beta^{\gamma-2} - \mu^{-3}|\mu^1 = 2(\omega_p/v_{ph})^2 \int(\mu - 1)dx,$$

if the integration is taken from a position where μ is a minimum. In order to obtain periodic solutions μ must obey the inequality

$$\mu \lessgtr \mu_o = (v_{ph}/s_o)^{2/(\gamma+1)} . \quad (3.24)$$

This is the desired criterion for wave breaking out of resonance. It also expressed the tendency for an electrostatic wave to break at lower amplitudes in the region of lower ion density.

Limitations of the moment equations

It is important to show under which conditions the basic equations used to describe high amplitude waves are valid. From the Boltzmann equation the momentum equation in the geometry of the capacitor model follows:

$$n_e \frac{dv}{dt} = - \frac{\partial}{\partial x} n_e \langle w_x^2 \rangle - e \frac{n_e}{m_e} E \ ,$$

where $m_e n_e \langle w_x^2 \rangle = p_{xx}$ is the xx-component of the pressure tensor. $\vec{w}$ is the individual speed of a particle relative to the mean flow velocity v From the quadratic moment the energy equation is obtained

$$m_e \frac{d\varepsilon}{dt} = \frac{p_{xx}}{n_e^2} \frac{dn_e}{dt} - \frac{1}{n_e} \frac{\partial q}{\partial x} \ , \quad (3.25)$$

with $\varepsilon = \frac{1}{2}\langle \vec{w}^2 \rangle$ the internal energy per unit mass and $q = m_e n_e \langle \vec{w}^2 w_x \rangle/2$ the electron heat flow density. p_{xx} is a sum of terms which contain thermal pressure and volume viscosity. In order to get an adiabatic law as used in this paper it is sufficient that the divergence of q vanishes.

It can be assumed that during one oscillation almost no coupling between w_x and the other two components w_z, w_y occurs and $2d\varepsilon/dt = d(p_{xx}/m_e n_e)/dt$ holds. Then eq. (3.25) reads with $q = 0$

$$(1/p_{xx})\ dp_{xx}/dt - (3/n_e)\ dn_e/dt = 0\ ,$$

the solution of which is $p_{xx}n_e^{-3}$ = const. Under the further assumption of

$$\partial p_{xx}/\partial x = (n_e/n_o)^3\ \partial p_{xx}^o/\partial x + p_{xx}^o\ \partial(n_e/n_o)^3/\partial x \simeq p_{xx}^o\ \partial(n_e/n_o)^3/\partial x,$$

which is reasonable, eq. (3.16_3) is deduced if $s_o^2 = \langle 3\ w_x^2\rangle_o$ is taken.

3.3 Considerations on Wave Breaking

Usually wave breaking is mentioned in the literature in connection with trapping of particles, or the appearance of trapped and successively accelerated fast electrons is made responsible for wave breaking. However, breaking of a Langmuir wave in itself is a purely hydrodynamic phenomenon: it can occur also in the absence of fast particles. Wavebreaking is defined as the loss of the periodic character of the wave. In a cold plasma this occurs at $n_{e,max} = \infty$; in a warm plasma the limiting electron density amplitude is given by eq. (3.24). In order to understand breaking in physical terms the mechanical model of Fig. 22 should be considered. The analogy of a warm plasma wave is the motion of a chain of spheres in a periodic potential structure V which are connected by springs. The springs react only to compression and do not transmit any cohesive force. In the wave frame V is static and the chain moves through it with phase velocity. Consider first a series of disconnected balls (Fig. 22, LHS): they all climb over the maxima of V or none of them. Observe also the different distances at different positions. Now, if the balls are connected by the springs mentioned (Fig. 22, RHS) it can happen that the first spheres are pushed over a hill with $V_{max} > \frac{1}{2} m_e v_{max}^2$ whereas the last ball falls back because of having lost too much energy due to pushing. Thus, one may say: the nonlinear wave generates such a high V by self-interaction of modes that a part of the fluid cannot superate it any more and gets

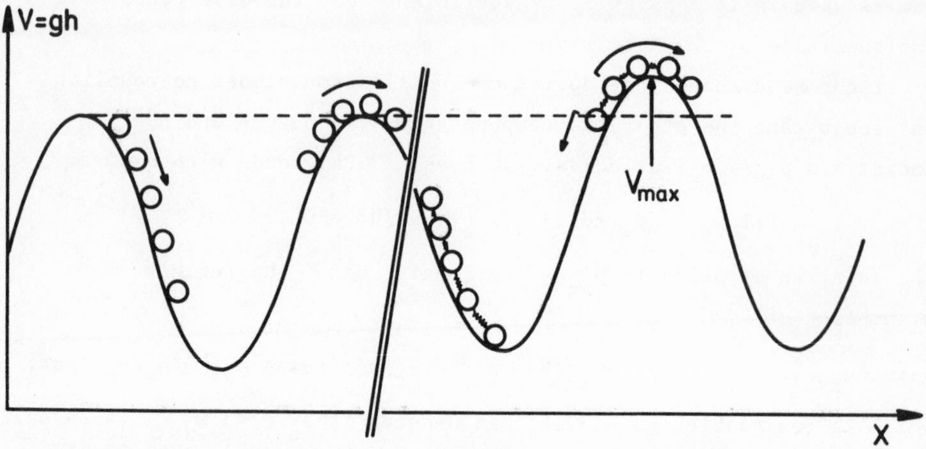

Fig. 22 - Mechanical model of wavebreaking. Cold plasma: independent spheres; warm plasma: spheres connected by springs which transmit compression but not cohesion.

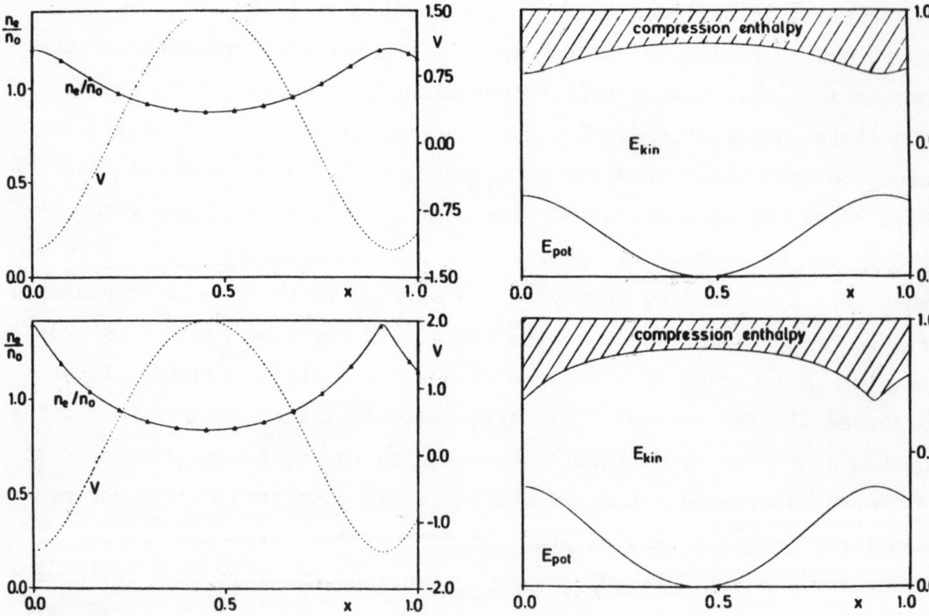

Fig. 23 - Distribution of n_e/n_o, potential V, density of kinetic and potential energy and compressional enthalpy in space, far from breaking (upper case) and near breaking (lower case); $V_{phase}/V_{sound} = 2$. No trapped particles.

reflected because of the work done by the thermal pressure. The numerical calculations seem to confirm this explanation. In Fig. 23 the partition of the total enthalpy in compressional enthalpy and kinetic and potential energy densities in space is reported[66]. In the lower diagrams the wave is near to the breaking limit and its density is also more peaked.

It has to be kept in mind that the steady state picture does not hold in the resonance region where the wave is driven in phase with the electromagnetic field of the laser.

Wavebreaking was also studied in the presence of trapped particles[66,67]. In the model used the free electrons were treated as a fluid whereas for the trapped electrons the model of noninteracting particles with a constant velocity distribution function was assumed[67]. When there are no trapped particles present the dispersion function follows the corrected Bohm-Gross dispersion relation after eq. (3.20) (see Fig. 24, LHS; the lower curve is for vanishing amplitude, the upper one for the amplitude at the breaking limit). With increasing number of trapped particles the dispersion bands between zero and maximum amplitude broaden and move towards lower frequencies (note that $\omega<\omega_p$ is now possible). It was also found that the tendency for the Langmuir wave to break at fixed amplitude increases with rising number of trapped particles.

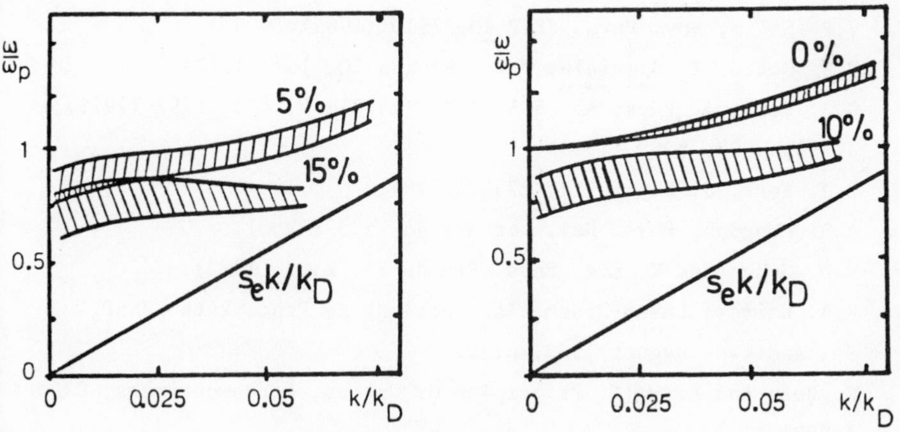

Fig. 24 - Dispersion bands of a nonlinear electron plasma wave with 0, 5, 10, and 15% of trapped particles.

REFERENCES

1. R.A. Cairns and J.J. Sanderson, Laser-Plasma Interactions 1, Proc. SUSSP, St. Andrews, August 1979.
2. I.P. Shkarofsky, T.W. Johnston, M.P. Bachynski, The Particle Kinetics of Plasmas, Addison-Wesley Publ. Co., Reading, Mass. (1966); chap.8. For Bremsstrahlung Radiation see chap.6.
3. J.M. Dawson and C. Oberman, Phys. Fluids 5, 517 (1962).
4. L. Spitzer, Physics of Fully Ionized Gases, Intersci. Publ., New York (1962), p.139.
5. S.I. Braginskii, Transport Processes in Plasma, in Review of Plasma Physics, ed. M.A. Leontovich, Consultants Bureau, New York (1965), §4.
6. T.W. Johnston and J.M. Dawson, Phys. Fluids 16, 722 (1973).
7. J.D. Jackson, Classical Electrodynamics, John Wiley & Sons, New York (1975).
8. Ya. B. Zel'dovich, Yu. P. Raizer, Physics of Shock Waves and High-Temperature Hydrodynamic Phenomena, Acad. Press, New York (1966), Vol.I, pp.120, 259.
9. K.R. Lang, Astrophysical Formulae, Springer-Verlag, New York (1974) pp.46, 47.
10. R.L. Shoemaker, in Laser and Coherence Spectroscopy, ed. J.I. Steinfeld, Plenum Press, New York (1978).
11. S. Rand, Phys. Rev. 136B, 231 (1964).
12. V.P. Silin, Sov. Phys. JETP 20, 1510 (1965).
13. P.J. Catto, Th Speziale, Phys. Fluids 20, 167 (1977).
14. G.J. Pert, J. Phys. A5, 506, 1221 (1972); 9, 463, 1797 (1976); J. Phys. B8, 3069 (1975).
15. G.J. Pert, J. Phys. B12, 2755 (1979).
16. A.B. Langdon, Phys. Rev. Letters 44, 575 (1980).
17. R.D. Jones and K. Lee, Phys. Fluids 25, 2307 (1982).
18. R.A. Cairns, Laser-Plasma Interactions 2, Proc. 24th SUSSP, St. Andrews, August 1982, p.32.
19. M. Born and E. Wolf, Principles of Optics, Pergamon Press, Oxford (1980).
20. Jon Mathews and R.L. Walker, Mathematical Methods of Physics, Benjamin, Menlo Park, Ca. (1970), p.32.

21. P. Mulser and G. Spindler, Radiation Pressure Dominated Plasma Flow, Garching Rep., PLF 12 (Dec. 1978).

22. G.B. Whitham, Linear and Nonlinear Waves, John Wiley, New York (1974), chap.11.

23. G.C. Pomraning, Radiation Hydrodynamics, Pergamon Press, Oxford (1973), chap.5.

24. V.L. Ginzburg, The Propagation of Electromagnetic Waves in Plasma, Pergamon Press, Oxford (1964), p.204.

25. H.A. Baldis and C.J. Walsh, Phys. Fluids 26, 1364 (1983).

26. V.A. Kulkarny and B.S. White, Phys. Fluids 25, 1770 (1982).

27. LLE Review 16, 26-38 (1983).

28. H.A.H. Boot, S.A. Self and R.B.R. Shersby-Harvie, J. Electron. Control 4, 434 (1958). A.V. Gapunov and M.A. Miller, Sov. Phys. JETP 7, 168 (1958).

29. H. Hora, D. Pfirsch and A. Schlüter, Z. Naturforsch. 22a, 278 (1967).

30. F.F. Chen, in Laser Interaction and Related Plasma Phenomena, eds. H.J. Schwarz and H. Hora, Plenum Press, New York (1974), p.294.

31. R. Thurn and W. Kiefer, Appl. Spectrosc. 38, 78 (1984).

32. J. Prodan, A. Migdall, W.D. Phillips, I. So, H. Metcalf and J. Dalibard, Phys. Rev. Letters 54, 992 (1985). W. Ertmer, R. Blatt, J.L. Hall and M. Zhu, Phys. Rev. Letters 54, 996 (1985). D.E. Pritchard, Phys. Rev. Letters 51, 1336 (1983). A.L. Migdall, J.V. Prodan, W.D. Phillips, Th. H. Bergeman and H.J. Metcalf, Phys. Rev. Letters 54, 2596 (1985).

33. J.J. Bollinger, J.D. Prestage, W.M. Itano and D.J. Wineland, Phys. Rev. Letters 54, 1000 (1985).

34. Reviews on "Mechanical Effects of Light", J. Opt. Soc. Am. B 2, No.11 (Nov. 1985), pp.1707-1860.

35. C. Grebogi and R.G. Littlejohn, Phys. Fluids 27, 1996 (1984). W.M. Manheimer, Phys. Fluids 28, 1569 (1985).

36. M.L. Sawley, J. Plasma Phys. 32, 487 (1984). Ph. L. Similon and A.N. Kaufman, Phys. Rev. Letters 53, 1061 (1984).

37. M.M. Skorić, D. ter Haar, Phys. Fluids 27, 2375 (1984).

38. J.R. Cary and A.N. Kaufman, Phys. Fluids 24, 1238 (1981).

39. T. Hatori, Phys. Fluids 28, 219 (1985).

40. A. Zeidler, H. Schnabl and P. Mulser, Phys. Fluids 28, 372 (1985).

41. N.C. Lee and G.K. Parks, Phys. Fluids 26, 724 (1983).

42. P. Mulser, J. Opt. Soc. Am. B 2, 1814 (1985).
43. R.A. Cairns, Laser-Plasma Interactions 2, Proc. SUSSP, St. Andrews August 1982, p.46.
44. G. Schmidt, Physics of High Temperature Plasmas, Acad. Press, New York (1979), pp.47-49.
45. L.D. Landau and E.M. Lifshitz, The Classical Theory of Fields, Pergamon, Oxford (1980), p.118.
46. V.S. Letokhov and V.G. Minogin, Laser Radiation Pressure on Free Atoms, Phys. Reports 73, 1-65 (1981).
47. W. Schneider, Elektronen beschlennigung durch inhomogene Langmuirwellen hoher Amplitude (electron acceleration by high amplitude Langmuir waves), Thesis T.H. Darmstadt (1984).
48. Th. Speziale and P.J. Catto, Phys. Fluids 22, 681 (1979).
49. R.P. Godwin, Phys. Rev. Letters 28, 85 (1972).
50. N.G. Denisov, Sov. Phys. JETP 4, 544 (1957).
51. A.D. Piliya, Sov. Phys. - Tech. Phys. 11, 609 (1966).
52. T. Speziale, P.J. Catto, Phys. Fluids 20, 990 (1977).
53. G.J. Pert, Plasma Phys. 20, 175 (1978).
54. H.J. Kull, Phys. Fluids 26, 1881 (1983). See also Report MPQ 50 (July 1981), 115 pages.
55. H.J. Kull, Linear Mode Conversion in Laser Plasmas, Thesis at Tech. Univ. Munich (1981).
56. A.G.M. Maaswinkel, K. Eidmann and R. Sigel, Phys. Rev. Letters 42, 1625 (1979).
57. J.E. Balmer and T.P. Donaldson, Phys. Rev. Letters 39, 1084 (1977)
58. A.G.M. Maaswinkel, Opt. Comm. 35, 236 (1980).
59. R. Dinger, K. Rohr and H. Weber, Polarisation Dependent Light Absorption in Laser Produced Plasmas, to be published.
60. J.C. Adam, A. Gourdin Serveniere and G. Laval, Phys. Fluids 25, 376 (1982).
61. R. Dragila, Phys. Fluids 26, 1682 (1983).
62. R.L. Carmen, D.W. Forslund and J.M. Kindel, Phys. Rev. Letters 46, 29 (1981).
63. P. Koch and J. Albritton, Phys. Rev. Letters 32, 1420 (1974).
64. P. Mulser, K. Takabe and K. Mima, Z. Naturforsch. 37a, 208 (1982).
65. P. Mulser and H. Schnabl, Laser and Particle Beams 1, 379 (1983).
66. A. Bergmann, Excitation of Nonlinear Langmuir Waves in Laser Plasmas, IAP Report, T.H. Darmstadt, Nov. 1985.
67. M. Kono and P. Mulser, Phys. Fluids 26, 3004 (1983).

LASER-DRIVEN INSTABILITIES IN LONG SCALELENGTH PLASMAS*

W. L. Kruer
Lawrence Livermore National Laboratory

ABSTRACT

In this update lecture we focus on laser-driven instabilities in long scalelength underdense plasmas. Particular attention is given to some recent experiments on Raman scattering of intense laser light. Many important features are in accord with theoretical expectations as discussed in lectures at previous summer schools. These features include a correlation of hot electron generation with Raman scattering, an increase in this scattering as the density scale length increases, and collisional suppression of the instability. Some challenging aspects of the growing data base as well as various deficiencies in the understanding are discussed. The role of the $2\omega_{pe}$, Brillouin, and filamentation instabilities is also briefly considered.

1. INTRODUCTION

Long scalelength plasmas are expected in laser fusion applications, since high gain capsules will be irradiated with long shaped pulses.(1) Effective pulse widths are of order 10 ns, leading to underdense plasma with density scale lengths of order 1 mm. As an illustrative example,(2) consider an Au disk irradiated with a generic shaped reactor pulse of 0.26 μ light. Near the end of the pulse , the underdense plasma consists of two parts. The first is an ablatively steepened portion where inverse bremsstrahlung is efficiently depositing energy. In this high Z plasma, inverse bremsstrahlung begins to be quite efficient at densities of about a tenth of the critical density (n_{cr}), so the steepened portion of the profile extends to densities below 0.1 n_{cr}. The density scale length here is set by a competition between the deposition and the energy transport and is ∿ 400 μ for this example. At still lower densities, the plasma becomes isothermal with an electron temperature of ∿ 5 keV and a density scale length of ∿ 2 mm determined by the plasma expansion. The peak intensity is about 3×10^{15} W/cm^2 in this example, although lower peak intensities could be used. Note that the density scale length in the underdense plasma is $\gtrsim 2000\ \lambda_0$, where λ_0 is the free space wavelength of the light. Underdense plasmas with similar scale lengths are expected in low Z targets for direct drive applications.

As discussed in lectures at previous summer schools,(3) intense laser light can excite a variety of instabilities in long scalelength underdense plasmas. Most of these instabilities can be simply represented as the resonant decay of the incident light wave into two other waves. For example, if the unstable waves are a scattered light wave and an electron plasma wave (ion acoustic wave), we have the Raman (Brillouin) instability. If the unstable waves are both electron plasma waves, we have the two plasmon decay ($2\omega_{pe}$) instability. In addition, the incident light beam can filament and self-focus. These instabilities can lead to hot electron generation, scattering of the incident light, as well as significant nonuniformity in the irradiation. Hence, it is of considerable interest to understand the efficiency of these processes in long scalelength plasmas.

In these update lectures, we will consider some of the progress made in understanding and characterizing laser-driven instabilities in long scalelength plasmas. Our emphasis will be on Raman scattering, which to date has received the most attention because of its role in generating hot electrons which preheat the fuel. Then the role of the $2\omega_{pe}$, Brillouin and filamentation instabilities will be briefly considered.

2.
RAMAN SCATTERING

Stimulated Raman scattering is a well-known process. The frequency and wave number matching conditions are

$$\omega_0 = \omega_s + \omega_{ek}$$
$$\underline{k}_0 = \underline{k}_s + \underline{k} \quad , \tag{2.1}$$

were ω_0 (ω_s) and $\underline{k}_0$ ($\underline{k}_s$) are the frequency and wavenumber of the incident (scattered) light wave and ω_{ek} and $\underline{k}$ are the frequency and wavenumber of the electron plasma wave. This process leads to an instability since there's a feedback loop. Laser light with electric field $\underline{E}_L$ oscillating electrons in the presence of a small density fluctuation δn produces a transverse current ($\alpha\delta n\ \underline{E}_L$) which generates a small scattered light wave with electric field $\underline{E}_s$. This scattered light wave in turn beats with the incident field to reinforce the density fluctuation via the ponderomotive force ($\propto \underline{E}_L \cdot \underline{E}_S$). Hence the plasma wave and the scattered light wave grow at the expense of the incident light.

Since $\omega_s \geq \omega_{pe}$ (the electron plasma frequency), the process is limited to densities $\leq 1/4 n_{cr}$, where n_{cr} is the critical density. The maximum growth rate γ occurs for backscatter and is[3]

$$\gamma = \frac{k\ v_{os}}{4} \sqrt{\frac{\omega_{ek}}{\omega_s}} \ , \tag{2.2}$$

where v_{os} is the oscillatory velocity of an electron in the field of the laser light. As an example, for .35 μ light with an intensity of 10^{15} W/cm^2, $\gamma \simeq 2 \times 10^{-3}\ \omega_0$ at $n = 0.1\ n_{cr}$. The growth time is about .1 ps, quite short. Of course, the wave coupling occurs even if it's not stimulated (i.e., if the instability is below its threshold). If the plasma wave is thermal level or enhanced by other processes, we simply have ordinary Raman scatter.

The threshold intensity for the Raman instability is determined by either damping of the unstable waves or by inhomogeneity in the plasma. A density gradient limits the region of resonant interaction. Propagation of the waves out of this interaction region represents a dissipation. As discussed in previous lectures, the gradient threshold intensities for backscatter (I_{TB}) and for sidescatter (I_{TS}) are

$$I_{TB} \simeq \frac{4 \times 10^{17}}{L_\mu \lambda_\mu} \frac{v_{gs}}{c} \frac{W}{cm^2}, \tag{2.3}$$

$$I_{TS} \simeq \frac{5 \times 10^{16}}{L_\mu^{4/3} \lambda_\mu^{2/3}} \frac{W}{cm^2} \quad .$$

Here L_μ is the density scale length ($L^{-1} = 1/n\ \partial n/\partial x$) and $\lambda\mu$ the wavelength of the light in microns, and v_{gs} is the group velocity of the scattered light wave. The backscatter threshold corresponds to a convective amplification of $e^{2\pi}$, and the sidescatter threshold(4) is estimated for $n = 0.1\ n_{cr}$. Sidescatter has the lower gradient threshold, since the scattered light wave spends a longer time in the interaction region. For $n \sim 1/4\ n_{cr}$, $v_{gs}/c \sim (k_0 L)^{-1/3}$ and the thresholds for back and sidescatter become comparable.

In practice, gradients usually determine the threshold intensities. To illustrate the magnitudes, consider 0.35 μ laser light and a plasma with a scale length $L/\lambda_0 = 10^3$. Then $I_{TB} \sim 10^{15} W/cm^2$ and $I_{TS} \sim 3 \times 10^{13}\ W/cm^2$. The gradient thresholds can be significantly lower in a plasma with a density maximum.(5)

Threshold intensities are useful indicators but are, of course, only estimates since the calculations are ideal and the plasma gradients are usually not well known. For example, threshold calculations typically assume a plane coherent light wave and a planar plasma slab. Experiments are characterized by a finite beam spot, intensity structure in the beam, and nonplanar plasma expansion. Hence, the scale length in the transverse direction can be comparable to that in the direction of expansion, an effect which can increase the threshold for sidescatter. In addition, temporal(6) and spatial incoherence(7) in the laser beam can raise the threshold. Significant affects are expected when $\Delta\omega \gtrsim \gamma$ or when $(\Delta\theta)^2 \gtrsim \gamma/\omega_0$. Here $\Delta\omega$ is the bandwidth, $(\Delta\theta)^2$ is the mean angular spread in the wave vectors of the pump, and γ is the growth rate of the instability.

In addition to the sizeable frequency shift of the scattered light, there are several other features of Raman scattering to be noted. First, sidescattering occurs preferentially out of the plane of polarization. The ponderomotive force due to the beat between the incident and scattered light waves then maximizes. In addition, we expect hot electron generation concomitant with Raman scattering. Part of the laser light energy is coupled into an electron plasma wave. When this plasma wave damps in a collisionless plasma, the faster, resonant particles are preferentially heated, resulting in suprathermal tails on the electron velocity distribution. As shown by the Manley-Rowe relations, $f_H \simeq \omega_{ek}/\omega_s$ fs, where f_H (f_s) is the fraction of the laser energy absorbed into suprathermal electrons (scattered).

3.
SOME EXPERIMENTS ON RAMAN SCATTERING

With these general features in mind, let's now consider some recent experiments on Raman scattering in long scalelength plasmas. Many of the earlier observations(8-15) were already referred to in the lectures(3) at the previous summer school. These experiments included measurements(14) by Offenberger et al. of a Raman reflectivity of ∿.7% using 10.6 μ light as well as measurements(15) by Phillion et al. of a reflectivity of ∿10%

using 1.06 μ light. Since then many more experiments have been reported.[5,16-23]

A review is beyond the scope of these lectures. Instead, some recent experiments at the Lawrence Livermore National Laboratory on Raman scattering in long scalelength plasmas will be briefly discussed. These experiments serve to indicate some important trends as well as some areas in which the understanding is deficient. These observations include a correlation of hot electron generation with Raman scattering, an increase in this scattering as the density scale length increases, and collisional suppression of the instability. Poorly understood aspects of the data include the detailed frequency spectrum of the scattered light and the intensity thresholds for onset of the scattering. Indeed a low level signal is observed for low intensity irradiation, which might be simply ordinary Raman scattering from an enhanced level of plasma waves produced by another process.

Raman scattering is usually identified in experiments by the frequency (or wavelength) spectrum of the scattered light. Since the maximum allowed density is about $n_{cr}/4$, the frequency matching conditions give

$$\omega_0/2 \lesssim \omega_s \lesssim \omega_0 \quad , \tag{3.1}$$

$$\lambda_0 \lesssim \lambda_s \lesssim 2\,\lambda_0 \quad ,$$

where λ denotes the free space wavelength of the light. Figure 1 shows a typical measurement[22] of the scattered light signal as a function of wavelength in an experiment in which an Au disk is irradiated with 0.53 μ laser light with a peak intensity of a few times 10^{15}W/cm^2. The signal is strongest at wavelengths which correspond to Raman scattering from densities $0.1 \lesssim n/n_{cr} \lesssim 0.2$. There is a small signal with wavelength near $2\lambda_0$, which may be Raman scattering near 1/4 n_{cr} or perhaps mode conversion of plasma waves generated by the $2\omega_{pe}$ instability. Note the "gap" in the signal which would correspond to Raman scattering for densities $0.2 \lesssim n/n_{cr} \lesssim 0.24$. Note also the strong decrease in the signal at short wavelengths. Similar spectra have been observed in experiments at other laboratories.

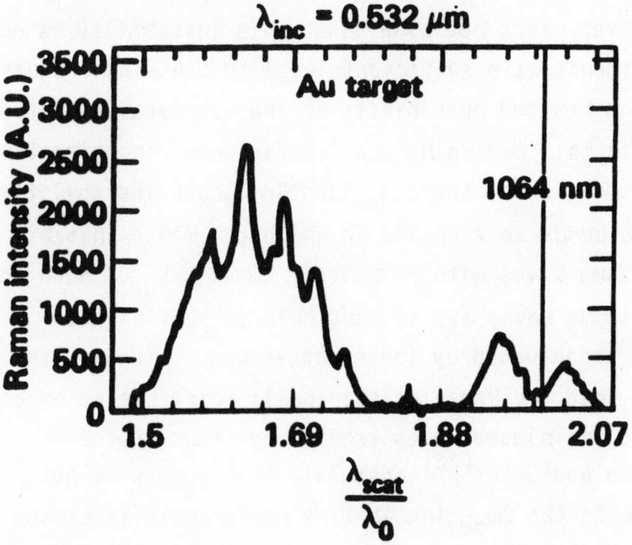

Fig. 1 The wavelength spectrum of light scattered by electron plasma waves in a Au disk(22) irradiated with a 1 ns pulse of .53μ light. The peak intensity was about 2×10^{15} W/cm^2.

The short wavelength cut-off is typically attributed to the suppression of the Raman instability at low plasma density due to strong Landau damping of the plasma wave. Indeed, the observed cut-off can be used to estimate(16) the electron temperature in the low density plasma. If we estimate that strong Landau damping onsets when $k\,\lambda_{De} \simeq 0.3$, a cut-off at $\lambda_s \simeq 1.5\,\lambda_0$ indicates that $\theta_e \simeq 3$ keV. Such estimates have been found to agree(22) with design code calculations and x-ray spectroscopy in thin CH foils. However, code calculations are typically somewhat higher (1.5-2x) than these estimates when thick disks are irradiated. This trend could provide a clue for modeling transport, since the thick disk targets are more sensitive to heat flow to high density.

The gap in the spectrum corresponding to Raman scattering from $0.2 \lesssim n/n_{cr} \lesssim 0.24$ presents a challenging puzzle. A variety of explanations have been offerred. Local steepening of the density profile near 1/4 n_{cr} by the 2 ω_{pe} instability might cause this effect. However, it's not clear that this instability is operative in the experiments at a sufficient level to cause the required steepening. A related possibility is the suppression of the Raman instability in this region by ion fluctuations concomitant with the nonlinear saturation of the 2 ω_{pe} and Brillouin instabilities. The gap corresponds to a regime in which the Raman instability generates plasma waves with relatively small $k \lambda_{De}$. Such long wavelength plasma waves may be particularly sensitive to collapse(25) accentuated by ion fluctuations. Finally, the gap may indicate that the Raman scattering is being seeded by an enhanced level of plasma waves excited by other processes. For example, Simon and Short(26) postulate that bursts of hot electrons due to the $2\omega_{pe}$ instability preferentially excite the plasma waves in the lower density region. Below the Raman instability threshold, we would then have ordinary Raman scattering from enhanced fluctuations. Above threshold, the instability grows from the enhanced levels. Here more calculations are needed to illustrate the plausible level of this enhancement.

Well above threshold, the Raman sidescattered light is observed to be preferentially out of the plane of polarization. Figure 2 shows the Raman scattered light (integrated over frequency) as a function of angle both in and out of the plane of polarization.(22) In the experiment a thin (2 μ) CH foil was irradiated with a 1 ns pulse of 0.53 μ light with a peak intensity of about 10^{15} W/cm^2. Note the strong asymmetry at significant angles between the measurements in and out of the plasma of polarization, as qualitatively expected.

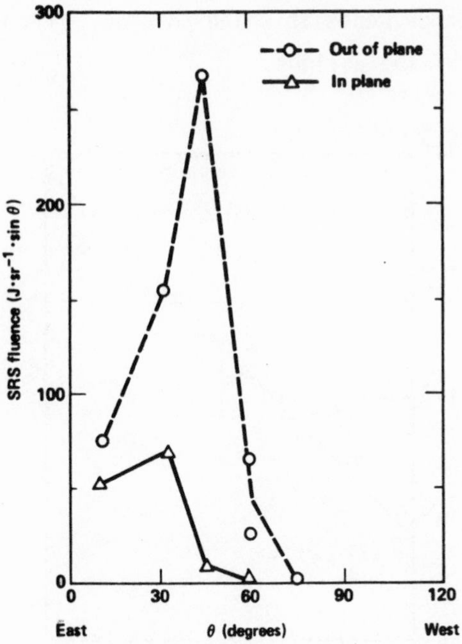

Fig. 2 An angular distribution of the light which was Raman-scattered from a CH foil(22) irradiated by a 1 ns pulse of .53μ light with a peak intensity of about 10^{15}W/cm^2. The dashed (solid) line denotes measurements out of (in) the plane of polarization.

Raman scattering has also been observed to correlate with hot electron generation in experiments in which Au disk were irradiated with 1 ns pulses of 0.53 μ light. In these experiments,(21) the laser energy varied from .5 - 4 kJ and the nominal intensity from about 10^{14} - 2 X 10^{16} W/cm^2. Figure 3 shows the fraction of the laser energy deposited into hot electrons as inferred from the level of the hard x-rays versus the measured fraction of the laser energy which was in Raman-scattered light. Note the impressive correlation. The solid line represents the expected correlation using the Manley-Rowe relations with the measured mean value of the frequency of the scattered light. Because of the error bars, it's quite possible that other processes such as the $2\omega_{pe}$ instability are also contributing to hot electron generation. A correlation between the Raman scattering and the hot electron generation (the high energy x-rays) has also been observed in time-resolved measurements in experiments in which thin Au foils were irradiated with 1 ns pulses of .53 μ light.

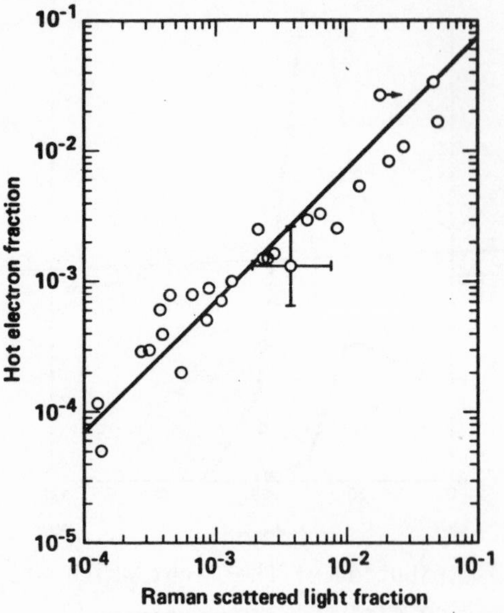

Fig. 3 The fraction of the laser energy absorbed into hot electrons versus the fraction in Raman-scattered light in Au disks irradiated by 1 ns pulses of .53μ light. See ref. 21.

The density scalelength near 0.1 n_{cr} in the thick disk experiments is estimated to be $L/\lambda_0 \sim 400$, varying somewhat as the focal spot was decreased to achieve the higher intensities. Hence the threshold for backscatter is about 3×10^{15} W/cm^2, and the threshold for sidescatter is about 10^{14} W/cm^2. Although there was significant scatter in the Raman signal as a function of intensity, the general trend was for the Raman reflectivity to increase from about 10^{-4} to several percent as the nominal intensity was increased from 10^{14} to 10^{16} W/cm^2. The reflectivity of 10^{-4} at intensities near 10^{14} W/cm^2 could be partially due to hot spots in the laser beam. Or this reflectivity could indicate an enhanced level of plasma waves generated in some other way. Another puzzling feature of the data is the growth of the Raman signal as a function of angle. Backscatter is observed[22] to onset at about the same intensity as sidescatter does, rather than at the expected higher threshold intensity. This may indicate some difficulty in the backscatter theory or possibly microstructure in the plasma.

Finally, Raman scattering has been observed[23] both to increase when the density scalelength is further increased and to fall off dramatically near the collisional threshold of the instability. Underdense plasmas with larger density scale lengths were accessed by irradiating thin foil targets which expand to about 0.1 - 0.2 n_{cr} near the peak of the pulse. Some measurements of the fraction of the laser energy which is Raman scattered are shown in Fig. 4 for both CH and Au foils irradiated with either 0.53 μ or 0.26 μ laser light. In these experiments, the nominal intensity was $\gtrsim 10^{15}$ W/cm^2 and the pulse lengths were about 1 ns. The density scalelengths accessed are estimated to be $L/\lambda_o \gtrsim 10^3$. Note that an average Raman reflectivity of about 10% is found in the CH targets irradiated with 0.53 μ light (the peak reflectivity is even larger). Such a level is quite comparable to the predictions of computer simulations[27] and simple models.

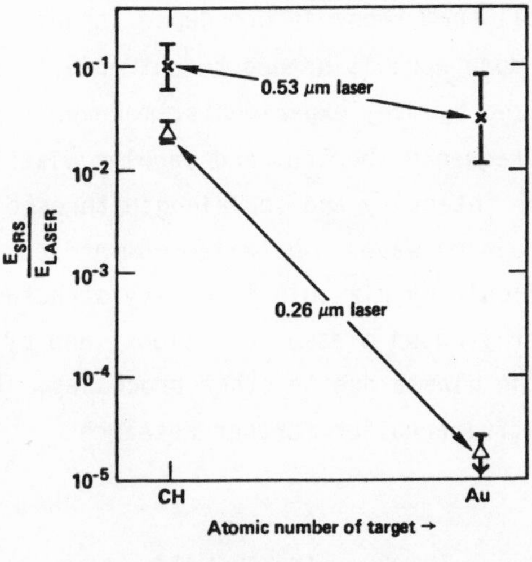

Fig. 4. Energy in Raman-scattered light from CH and Au thin-foil targets. All cases have intensities $\gtrsim 10^{15}$ W/cm^2. The gold data point for 0.26 μm is an upper limit. See ref. 23.

The results in Fig. 4 also illustrate the effect of collisionality on Raman scattering. When CH foils are irradiated with 0.26 μ light, the plasma is denser and more collisional. The reflectivity dropped a factor of about three. Alternatively, the collisionality can be increased by using an Au foil. When Au foils were irradiated with 0.53 μ light, the Raman reflectivity was several times less than that observed with CH foils. Finally, for Au foils irradiated with 0.26 μ light, the plasma is estimated to be sufficiently collisional that the instability is stabilized. Indeed, the Raman reflectivity is observed(23) to drop to a very low level.

This brief discussion illustrates some of the progress in characterizing and controlling Raman scattering in long scalelength plasmas. Many significant trends well above threshold are in accord with expectations, but the detailed understanding is far from complete. Thin foil experiments are beginning to access plasmas with density scale lengths comparable to those expected in high gain targets. The results clearly indicate that sizeable Raman scattering can occur when the gradient and collisional thresholds are far exceeded. Of course, the plasma conditions in thin foils will differ in detail from those in overdense targets driven by long shaped pulses, and more work is needed to extrapolate to this regime. As emphasized by many experiments, more work is also needed to understand the frequency spectrum and angular distribution of the scattered light, the intensity and scalelength thresholds, and the noise level of the plasma waves. Detailed comparisons with theory are often made difficult by sizeable intensity structure in the laser beams, by poorly known plasma conditions, and by microstructure in the plasma due to other processes. This is a rich and no doubt a fruitful area for further research.

4.
THE $2\omega_{pe}$ INSTABILITY

Let's now briefly consider a related instability which can be operative in long scalelength plasmas. The $2\omega_{pe}$ instability corresponds to decay of the laser light into two electron plasma waves. As apparent from frequency matching, the instability is limited to a narrow region near 1/4 n_{cr}. The maximum growth rate is equal to that of the Raman instability near 1/4 n_{cr}. The intensity threshold(28) I_T due to a density gradient is

$$I_T \sim \frac{4 \times 10^{15} \Theta_{keV}}{\lambda_\mu L_\mu} \; \frac{W}{cm^2} \; , \qquad (4.1)$$

where Θ_{keV} is the electron temperature in keV. This threshold is typically lower than that for Raman sidescatter, except in fairly long scale length plasmas. For example, for $\Theta_{keV} \simeq 2$ the sidescatter threshold becomes lower when $L/\lambda_o \gtrsim 400$.

Since this instability generates high phase velocity electron plasma waves, hot electron production is a characteristic feature of the nonlinear state. The nonlinear state(28) is also characterized by local steepening of the density profile and by ion fluctuations nonlinearly generated by the plasma waves. The growth rate of the plasma waves, the density profile steepening, and the generation of ion waves have been measured in some detail in experiments(29) with 10.6 μ light. Although the results are qualitatively consistent with simulations, there are also discrepancies (for example, in the wavelengths(30) of the nonlinearly generated ion fluctuations). A useful signature of the instability is emission near $3/2\omega_o$, which arises from the coupling of the incident and reflected light with a plasma wave near 1/4 n_{cr}. Unfortunately the level of the instability is difficult to estimate from this signal since the emission only indirectly indicates the level of part of the spectrum of the plasma waves.

The 3/2 ω_o emission is frequently diagnosed in laser plasma experiments.(31-33) Some experiments(33) at the University of Rochester provide a recent example. In these experiments, CH spheres were irradiated with a 600-700 ps pulse of 0.35 μ light. The 3/2 ω_o emission was observed to onset at an intensity of about 2×10^{14} W/cm^2, the estimated threshold intensity of the

$2\omega_{pe}$ instability. The level of the emission increased with the intensity of irradiation but then saturated for an intensity of about 6 X 10^{14} W/cm^2 at a rather low level (10^{-9} of the incident energy). Hard x-rays indicating suprathermal electrons with a temperature of about 35 keV were observed to be correlated with the 3/2 ω_0 emission. The inferred fraction of the laser energy in these suprathermal electron also saturated at a low value of about 10^{-4} of the incident energy. The density scalelength in these experiments was estimated to be $L/\lambda_0 \lesssim 150$, decreasing as the target radius was reduced to achieve higher intensity. In experiments at Lawrence Livermore National Laboratory with longer scalelength plasmas and .53μ light, higher levels of $3/2\omega_0$ emission have been observed (up to about 10^{-4} of the incident energy). As already discussed, the hot electrons in these experiments correlated best with the Raman signal, but within the error bars there may also be some contribution from the $2\omega_{pe}$ instability.

Although difficult to quantify in experiments, the level of the 2 ω_{pe} instability seems puzzling. Simulations(28) without self-generated magnetic fields typically suggest an instability absorption $\gtrsim$ 5%, which is significantly greater than that generally attributed to this instability in experiments. Since the heat transport is not well understood, the profile steepening due to energy depositon near the critical density could extend down to densities < 1/4 n_{cr}. Alternatively, the rather low level might indicate additional choking of the instability by suprathermal tails if the fast electron transport is inhibited by magnetic fields(34). Another possibilty is additional suppression due to ion fluctuations excited by the Brillouin instability. More work in this area is clearly needed.

5.
BRILLOUIN SCATTERING AND FILAMENTATION

Before concluding these update lectures, let's very briefly consider several instabilities which involve the growth of perturbations in ion density. It's well-known that stimulated Brillouin scattering can occur in long scalelength plasmas. The feedback mechanism is similar to that for the Raman instability, except now the density fluctuation is associated with a low frequency ion acoustic wave. The growing ion fluctuations scatter the light, modifying the absorption and/or its location. In addition, these ion waves can indirectly affect other processes as mentioned in the previous sections.

For Brillouin scatter, gradients in the expansion velocity of the plasma typically limit the region of resonant interaction. The gradient threshold intensity I_{TB} for Brillouin backscatter then is

$$I_{TB} \sim 6 \times 10^{15} \frac{\Theta_{keV}}{L_{\nabla v}\lambda_{\mu}} \frac{n_{cr}}{n} \frac{W}{cm^2} , \tag{5.1}$$

where $L_{\nabla v}$ is the scalelength in microns for variation of the expansion velocity v_{exp}. In particular, $L_{\nabla v}^{-1} = 1/c_s$ $\partial v_{exp}/\partial x$, where c_s is the sound speed. As is the case for the Raman instability, the threshold for sidescatter is lower by a factor of order $(\omega_0 L/c)^{1/3}$. As an example, for 0.35 μ light in a plasma with $n = 0.1\, n_{cr}$, $\Theta_e = 2$ keV and $L_{\nabla v} = 10^3 \lambda_0$, $I_{TB} \sim 10^{15} W/cm^2$. The threshold for sidescatter is about a factor of 10 lower. A brief discussion of some of the nonlinear effects can be found in Ref 35.

Brillouin scattering remains a significant issue for long scalelength plasmas. Identification of the scattering by its frequency spectrum is often uncertain because of Doppler shifts in the expanding plasma. In some experiments(24) with large focal spots, much more light with frquency near ω_0 is observed to be scattered out of the plane of polarization than into the plane. In a number of experiments(36-39) with 1.06 μ and 10.6 μ light, a Brillouin reflectivity up to about 50% has been observed. However, in current experiments with shorter wavelength light, the reflectivety attributed to Brillouin scattering is typically $\lesssim$ 20%. Important questions include the scaling of the reflectivity to longer scalelength plasmas and its angular distribution.

Filamentation is another potentially important process involving perturbations in ion density. In this instability, perturbations in the intensity profile of an incident light beam grow in amplitude, causing the beam to break up into intense filaments. The feedback mechanism leading to instability is easy to understand. A local increase in the light intensity creates a depression in plasma density either directly via the ponderomotive force or indirectly via enhanced collisional absorption and subsequent plasma expansion. The density depression refracts the light wave into the lower density region, enhancing the intensity perturbations. The instability is termed either ponderomotive(42) or thermal(43-45) filamentation, depending on which mechanism generates the density depression. Self-focusing is the analogous process involving the entire beam.

Filamentation can significantly impact laser plasma coupling. Enhancements in intensity can introduce or modify other instabilities, change the location of the energy deposition, and possibly aggravate deleterious collective effects such as hot electron generation. Spatial structure in the irradiation pattern can enhance magnetic field generation, modify energy transport, and even degrade the symmetry and stability of implosions if the scalelength of the structure is sufficiently long. Filamentation can also complicate the interpretation of coupling and transport experiments, since the intensity in the underdense plasma is no longer a controlled variable and may be particularly sensitive to details of the beam profile.

A very simple treatment(46) suffices to illustrate the essential features of filamentation. Let's consider a plane light wave with intensity I propagating in the z direction in a plasma with uniform density n_0. We first calculate in the quasi-static limit the density perturbation, δn_e, induced by a perturbation in the intensity profile, where $I = I_0(1+\alpha \cos ky)$. In the case of ponderomotive filamentation,

$$n_e = n_0 e^{-\frac{I}{2 n_{cr} \theta_e c}} , \tag{5.2}$$

where n_e is the electron density, n_{cr} is the critical density, θ_e is the electron temperature, and c is the velocity of light. This equation is easily obtained by balancing the variation in electron pressure with the variation in the light wave pressure.

Hence Eq. 5.2 gives

$$\frac{\delta n_e}{n_0} \simeq \frac{-\alpha\, I_o \cos ky}{2 n_{cr} \theta_e c} \quad , \tag{5.3}$$

where we have assumed that $I_0 << 2n_{cr}\theta_e c$.

In the case of thermal filamentation, we simply balance the electron heat flow with the absorbed intensity:

$$\nabla \cdot (\kappa^T \nabla \theta_e) = -\kappa\, I \; , \tag{5.4}$$

where κ^T is the classical electron thermal conductivity and κ is the spatial decay rate due to collisional absorption. The perturbation in intensity drives a perturbation in electron temperature which leads to a variation in electron density. In particular, we postulate

$$\begin{aligned} I &= I_0\, [1 + \alpha \cos ky], \\ \theta_e &= \theta_0\, [1 + \beta \cos ky], \\ n_e &= n_0\, [1 - \beta \cos ky], \end{aligned} \tag{5.5}$$

where we have invoked pressure balance transverse to the beam; i.e., the quasi-static limit. Only first order corrections are retained. With these variations in θ_e and n_e, we also have

$$\kappa^T = \kappa_0^T\, [1 + \tfrac{5}{2}\beta \cos ky], \tag{5.6}$$

$$\kappa = \kappa_0\, [1 - \tfrac{7}{2}\beta \cos ky], \tag{5.7}$$

where the subscript zero denotes the thermal conductivity and absorption coefficient evaluated at θ_0 and n_0, the electron temperature and density in the absence of an intensity perturbation.

Substituting Equations (5.6) and (5.7) into Equation (5.4) gives to lowest order

$$\frac{\partial}{\partial z}\, \kappa_0^T\, \frac{\partial \theta_0}{\partial z} = -\kappa_0 I, \tag{5.8}$$

and to next order

$$\beta = \frac{\kappa_0 I_0 \alpha}{6\kappa_0 I_0 + k^2 \kappa_0^T \theta_0} \quad . \tag{5.9}$$

Equation (5.8) determines θ_0, the electron temperature generated by the unmodulated beam. Equations (5.5) and (5.9) yield

$$\frac{\delta n_e}{n_0} \simeq \frac{-\kappa_0 I_0 \alpha \cos ky}{k^2 \kappa_0^T \theta_0} \quad . \tag{5.10}$$

Here we have assumed that $k^2 \kappa_0^T \theta_0 >> 6\kappa_0 I_0$, i.e., that the scalelength for temperature variation in the direction of propagation of the light wave is much longer than the wavelength of the intensity modulation in the transverse direction.

Comparing Equations (5.3) and (5.10) shows that

$$\left(\frac{\delta n_e}{n_0}\right)_t = \left(\frac{\delta n_e}{n_0}\right)_p \frac{2 n_{cr} \kappa_0 c}{k^2 \kappa_0^T} \quad , \tag{5.11}$$

where the subscripts t and p denote thermal and ponderomotive, respectively. For a high Z plasma,

$$\frac{2\, n_{cr} \kappa_0 c}{k^2 \kappa_0^T} \simeq \frac{1}{7} \frac{\nu_{ei}^2}{k^2 v^2} \quad , \tag{5.12}$$

where ν_{ei} is the collision frequency appropriate to the high frequency resistivity and v_e is the electron thermal velocity. Hence thermal filamentation dominates for wavelengths $\lambda >> 10\lambda_{ei}$, where $\lambda_{ei} = v_e/\nu_{ei}$. To include the implicit Z dependence of the thermal conductivity, the right hand side of Eq. (5.12) is multiplied by a factor of about $(1 + \frac{3.3}{Z})$.

The growth of these zero-frequency intensity modulations along with their self-consistent density fluctuations can easily be obtained from the standard dispersion relation(3) familiar from discussions of Brillouin scatter:

$$\omega^2-k^2c_s^2 = \frac{k^2v_{os}^2\omega_{pi}^2}{4}\left\{\frac{1}{D(\omega-\omega_0,\underline{k}-\underline{k}_0)} + \frac{1}{D(\omega+\omega_0,\underline{k}+\underline{k}_0)}\right\}. \qquad (5.13)$$

Here $D = \omega^2-k^2c^2-\omega_{pe}^2$, c_s is the sound speed, ω_{pi} (ω_{pe}) is the ion (electron) plasma frequency, ω_0 and $\underline{k}_0$ are the frequency and wavenumber of the incident light wave, and v_{os} is the oscillatory velocity of an electron in the incident light wave. The derivation assumes that there are no variations along the direction of the electric field vector of the light wave. To obtain ponderomotive filamentation, we look for zero frequency fluctuations with a wave vector orthogonal to the direction of propagation $\underline{k}_0$. In particular, we set $\omega = 0$ and take $\underline{k} = \underline{k}_r + i\underline{k}_i$, where $\underline{k}_r\cdot\underline{k}_o = 0$ and $\underline{k}_i$ is parallel to $\underline{k}_0$. Equation (5.13) then gives the dispersion relation

$$4k_i^2k_0^2 + k_r^4 - k_r^2\frac{\omega_{pe}^2}{c^2}\frac{v_{os}^2}{2v_e^2} = 0 . \qquad (5.14)$$

Growth maximizes ($\partial k_i/\partial k_r = 0$) when

$$k_r = \frac{1}{2}\left(\frac{v_{os}}{v_e}\right)\frac{\omega_{pe}}{c} ,$$

$$k_i = \frac{1}{8}\left(\frac{v_{os}}{v_e}\right)^2\frac{\omega_{pe}^2}{k_0c^2} , \qquad (5.15)$$

where we have also assumed that $Z\theta_e >> \theta_i$.

As shown in Eq. (5.11) the dispersion relation for thermal filamentation can be obtained from Eq. (5.13) by simply multiplying the intensity term by $\sim (7k_r^2\lambda_{ei}^2)^{-1}$. Then

$$4k_i^2k_0^2 + k_r^4 - \frac{v_{os}^2}{14v_e^2}\frac{\omega_{pe}^2}{c^2\,\lambda_{ei}^2} = 0 , \qquad (5.16)$$

where we have again assumed that $Z >> 1$. The growth now maximizes for long wavelength:

$$k_r \ll \left(\frac{\omega_{pe}}{3.6c} \frac{v_{os}}{v_e} \frac{1}{\lambda_{ei}} \right)^{1/2} ,$$

$$k_i \simeq \frac{\omega_{pe}}{7.5\, k_o c} \left(\frac{v_{os}}{v_e} \right) \frac{1}{\lambda_{ei}} . \tag{5.17}$$

We note that the maximum spatial gain coefficient for thermal filamentation exceeds that for ponderomotive filamentation when $\nu_{ei}/\omega_{pe} > v_{os}/c$. Hence the thermal mechanism is especially competitive in the denser, more collisional plasmas produced by short wavelength laser light.

To illustrate the numbers, let's consider some conditions typical of recent experiments[21] in which Au disks were irradiated with 1 ns pulses of .53 μ laser light. We take $I \simeq 2 \times 10^{15}$ W/cm^2, $L_\mu \simeq 200$ at $n_o/n_{cr} = .1$, $\theta_{kev} \simeq 2$, and $Z \simeq 50$. Then the minimum growth length ($\ell g = 1/k_i$) for ponderomotive filamentation is about 60 μ for a filament with a wavelength of about 10 μ. There are several growth lengths for ponderomotive filamentation. The minimum growth length for thermal filamentation at $n = .1\, n_{cr}$ is about 300 μ for a filament with a wavelength much longer than about 20 μ. Of course, filamentation can be operative in longer scalelength plasmas at smaller intensities.

Unfortunately, filamentation in laser plasmas is perhaps the least understood and characterized of the processes we have discussed. An introductory discussion of possible nonlinear consequences is given in Ref. 46. Filamentation has been difficult to quantify in laser plasmas. Much of the evidence is rather indirect: inferences from structure[47-49] in x-ray pictures of the heated plasma or in images of the back-reflected light as well as inferences from the angular distribution[16] of half-harmonic light or from frequency shifts[40] in the reflected light. Recently, filamentary structures have been directly observed by using optical shadowgraphy,[50] by imaging the second harmonic emission,[(51)] and by Thomson scattering[(52)] from electron plasma waves generated in the walls of the filament. Some experiments[(50)] with short wavelength light have also indicated that implosions are degraded when signs of filamentation are present. A better understanding of filamentation is clearly needed.

6.
SUMMARY

In summary, a variety of instabilities can play a role in the coupling of intense laser light with long scalelength plasmas. Improved understanding of these instabilities is important for the optimum use of large lasers in many applications. Experiments indeed show that at least some of the instabilities can be significant in laser fusion applications, although more work is needed to quantitatively extrapolate to targets irradiated with long shaped pulses. For example, some recent experiments on Raman scattering show many expected features, such as an increase in the scattering with density scalelength, a correlation with hot electron generation, and collisional suppression of the Raman instability in high Z plasmas irradiated with short wavelength light. As briefly discussed, there are also many aspects of the data which point to deficiencies in our understanding. Important topics for further study also include the competition of the instabilities and the role of filamentation in laser plasma experiments.

ACKNOWLEDGMENTS

I acknowledge valuable discussions with M. Campbell, P. Drake, K. Estabrook, R. Kauffman, B. Langdon, B. Lasinski, D. Phillion, R. Turner and E. Williams.

*Work performed under the auspices of the U. S. Department of Energy by the Lawrence Livermore National Laboratory under contract number W-7405-ENG-48.

References

1. J. Nuckolls, L. Wood, A. Thiessen, and G. Zimmerman, Nature 239, 139 (1972).
2. Barbara Lasinski (private communication, 1984).
3. W. L. Kruer, in Laser-Plasma Interactions (SUSSP Publications, Edinburgh, 1980), edited by R. A. Cairns and J. J. Sanderson, p. 388-430; also Laser-Plasma Interactions II (SUSSP Publications, Edinburgh, 1983), edited by R. A. Cairns, p. 185-204; and many references therein.
4. B. B. Afeyan and E. A. Williams, Lawrence Livermore National Laboratory UCRL-91595 (1984).
5. H. Figueroa, C. Joshi, H. Azechi, N. A. Ebrahim, and K. Estabrook, Phys. Fluids 27, 1887 (1984).
6. J. J. Thomson, Phys. Fluids 21, 2082 (1978); and references therein.
7. A. M. Rubenchik (private communication, 1983); A. B. Langdon, Lawrence Livermore National Laboratory, UCRL-50021-83 (1984), p. 3-35
8. J. L. Bobin, M. Decroisette, B. Meyer and Y. Vittel, Phys. Rev. Letters 30, 594 (1973).
9. R. G. Watt, R. D. Brooks and Z. A. Pietrzyk, Phys. Rev. Letters 41, 170 (1978).
10. D. W. Phillion and D. L. Banner, Lawrence Livermore National Laboratory, UCRL-84854 (1980).
11. C. Joshi, T. Tajima, J. M. Dawson, H. A. Baldis and N. A. Ebrahim, Phys. Rev. Letters 47, 1285 (1981).
12. J. Elazar, W. Toner and E. R. Wooding, Plasma Phys. 23, 813 (1981).
13. K. Tanaka, L. M. Goldman, W. Seka, M. C. Richardson, J. M. Soures and E. A. Williams, Phys. Rev. Letters 48, 1179 (1982).
14. A. A. Offenberger, R. Fedosejevs, W. Tighe, and W. Rozmus, Phys. Rev. Letters 49, 371 (1982).
15. D. W. Phillion, E. M. Campbell, K. G. Estabrook, G. E. Phillips and F. Ze, Phys. Rev. Letters 49, 1405 (1982).
16. W. Seka, E. A. Williams, R. S. Craxton, L. M. Goldman, R. W. Short, and K. Tanaka, Phys. Fluids 27, 2181 (1984).
17. C. L. Shephard, J. A. Tarvin, R. L. Berger, Gar. E. Busch, R. R. Johnson, and R. J. Schroeder, KMSF U1562 (1985).

18. C. J. Walsh, D. M. Villeneuve, and H. A. Baldis, Phys. Rev. Letters 53, 1445 (1984).
19. R. G. Berger, R. D. Brooks, and Z. A. Pietrzyk, Phys. Fluids 26, 354 (1983).
20. R. E. Turner, D. W. Phillion, E. M. Campbell, and K. G. Estabrook, Phys. Fluids 26, 579 (1983).
21. R. P. Drake, R. E. Turner, B. F. Lasinski, K. G. Estabrook, E. M. Campbell, C. L. Wang, D. W. Phillion, E. A. Williams and W. L. Kruer; Phys. Rev. Letters 53, 1739 (1984).
22. R. P. Drake, Bull. Am. Phys. Soc. 29, 1229 (1984), E. M. Campbell et al., 14th Anomalous Absorption Conference, Charlottesville, Va. (1984), paper A4; R. P. Drake et al, ibid, paper A9; R. E. Turner et al, ibid, paper A10; D. W. Phillion et al, ibid, paper A11; R. L. Kaufmann et al, ibid, paper A12.
23. R. E. Turner, K. G. Estabrook, R. L. Kauffman, D. R. Bach, R. P. Drake, D. W. Phillion, B. F. Lasinski, E. M. Campbell, W. L. Kruer and E. A. Wiliams, Phys. Rev. Letters 54, 189 (1985).
24. E. M. Campbell, in Radiation in Plasmas (World Scientific Publishing Co., Singapore, 1984) Vol. I, edited by B. McNamara, P. 579-622.
25. D. Dubois, H. A. Rose, and D. R. Nicholson, Phys. Fluids 28, 202 (1985); and references therein.
26. A. Simon and R. L. Short, Phys. Rev. Letters 53, 1912 (1984).
27. K. Estabrook and W. L. Kruer, Phys. Rev. Letters 53, 465 (1984), and many references therein.
28. A. B. Langdon, B. F. Lasinski and W. L. Kruer, Phys. Rev. Letters 43, 133 (1979); B. F. Lasinski, A. B. Langdon, and W. L. Kruer, Lawrence Livermore National Laboratory, UCRL-50021-81 (1982), p. 3-30 and references therein.
29. H. A. Baldis and C. Walsh, Phys. Rev. Letters 47, 1658 (1981); also Phys. Fluids 26, 1364 (1981); and references therein.
30. D. M. Villeneuve and H. A. Baldis, paper D2, 15th Anomalous Absorption Conference, Banff, Canada, 1985.
31. R. E. Turner, D. W. Phillion, B. F. Lasinski and E. M. Campbell, Phys. Fluids 27, 511 (1983).
32. P. D. Carter, S. M. L. Sim, H. C. Barr and R. G. Evans, Phys. Rev. Letters 44, 1407 (1980)

33. R. L. Keck, R. L. McCrory, W. Seka and J. M. Soures, Phys. Rev. Letters 54, 1656 (1985).
34. J. M. Kindel, D. W. Forslund, W. B. Mori, C. Joshi, and J. M. Dawson paper D3, 15th Anomalous Absorption Conference, Banff, Canada (1985)
35. W. L. Kruer and K. G. Estabrook, in Laser Interaction and Related Plasma Phenomena, Vol. 5, Edited by H. Schwarz, H. Hora, M. Lubin ar B. Yaakobi (Plenum Press, New York, 1981), p. 783-800.
36. B. H. Ripin, F. C. Young, J. A. Stamper, C. M. Armstrong, R. Decoste E. A. McLean and S. E. Bodner, Phys. Rev. Letters 39, 611 (1977).
37. D. W. Phillion, W. L. Kruer and V. C. Rupert, Phys. Rev. Letters 39, 1529 (1977).
38. A. Ng, L. Pitt, D. Salzmann and A. A. Offenberger, Phys. Rev. Letter 42, 307 (1979).
39. F. J. Mayer, G. E. Busch, C. M. Kinzer and K. G. Estabrook, Phys. Rev. Letters 44, 1498 (1980).
40. K. Tanaka, L. M. Goldman, W. Seka, R. W. Short and E. A. Williams, Phys. Fluids 27, 2960 (1984).
41. W. C. Mead, E. M. Campbell et al., Phys. Fluids 26, 2316 (1983).
42. P. K. Kaw, G. Schmidt and T. Wilcox, Phys. Fluids 16, 1522 (1973); B. I. Cohen and C. E. Max, Phys. Fluids 22, 1115 (1979); and references therein.
43. M. S. Sodha, A. K. Ghatak, and V. K. Tripathi, in Progress in Optics Vol. 13, edited by E. Wolf (North Holland, Amsterdam, 1976); V. K. Tripath, and L. A. Pitale, J. Appl. Phys. 48, 3288 (1977); and references therein.
44. M. J. Herbst, J. A. Stamper, R. H. Lehmberg, et.al., Naval Research Laboratory Report 4983 (1981), and Proceedings of the 1981 Topical Conference on Symmetry Aspects of Inertial Fusion Implosions, Ed. S. Bodner, Naval Research Laboratory, Washington, D.C.
45. R. S. Craxton and R. L. McCrory, J. Appl. Phys. 56, 108 (1984); Kent Estabrook, W. L. Kruer and D. S. Bailey, Phys. Fluids 28, 19 (1985).
46. W. L. Kruer, Comments Plasma Phys. Controlled Fusion 9, 63 (1985).
47. R. A. Haas, M. J. Boyle, K. Manes and J. E. Swain, J. Applied Phys. 47, 1318 (1976); H. Shay et al., Phys. Fluids 21, 1634 (1978).

48. C. E. Max, et al., in Laser Interaction and Related Plasma Phenomena, Vol. 6, edited by H. Hora and G. Miley (Plenum, N.Y., 1984); W. C. Mead, et al., Phys. Fluids 27, 1301 (1984).
49. A. Ng, D. Saltzmann, and A. A. Offenberger, Phys. Rev. Letters 43, 1502 (1979).
50. O. Willi, in Laser Interaction and Related Plasma Phenomena, Vol. 6, edited by H. Hora and G. Miley (Plenum, NY, 1984).
51. J. Stamper, et al., Naval Research Laboratory Report 5173 (1984); M. J. Herbst, J. A. Stamper, R. R. Whitlock, R. H. Lehmberg, and B. H. Ripin, Phys. Rev. Letters 46, 328 (1981).
52. H. A. Baldis and P. B. Corkum, Phys. Rev. Letters 45, 1260 (1980).

STUDIES OF THERMAL ELECTRON TRANSPORT IN LASER FUSION PLASMAS

D. Shvarts

Nuclear Research Center Negev
P.O. Box 9001, Beer-Sheva, Israel

1. INTRODUCTION

Thermal conduction of energy by electrons plays a dominant role in the behaviour of ablatively accelerated laser fusion targets[1]. The laser light is absorbed in the low density corona up to the critical density where the remaining light is fully reflected. The energy absorbed is then conducted inward, by electron thermal transport, from the hot corona to the cold ablation surface. This inward heat flow is then balanced by an outward flow of plasma kinetic energy and PdV work from the ablative surface. The newly heated material at the surface of the cold dense part of the target then expands outward, carrying with it kinetic energy and momentum. Momentum conservation then implies that the rest of the target will be accelerated inwards, as in the rocket action.

Since the efficiency of the ablation process is determined by the efficiency of the heat transport from the critical surface to the ablation surface, any change in the thermal transport process will effect the efficiency of the ablation process.

Understanding the mechanisms of the thermal conduction process is therefore an essential ingredient in a proper description of the ablative acceleration process required to achieve efficient successful laser-driven implosion of thermonuclear targets.

The commonly used description of thermal conduction by electrons in a plasma was derived by Spitzer and Härm[2] and will be described in Section 3.1. However, it is useful to review the fundamentals of the kinetic theory of gases in order to understand the key problem in applying Spitzer's derivation to plasmas produced by high power lasers.

The heat flow in a uniform density plasma can be written as :

$$q = -\frac{\bar{\lambda}_c \cdot \bar{v}}{3} \cdot n_e \cdot k \cdot \nabla T \tag{1.1}$$

where : n_e - is the electron density

$\bar{\lambda}_c$ - is the average collision mean free path of electrons with plasma ions and electrons

$\bar{v}$ - is the average velocity of the electrons

k - is the Boltzmann constant

and T - is the electron temperature.

Defining the temperature gradient scale length, $L \equiv T/|\nabla T|$, we can write the heat flow as :

$$q = \frac{\bar{\lambda}_c \cdot \bar{v} \cdot n_e \cdot kT}{3 \cdot L} \tag{1.2}$$

The kinetic theory, from which (1.1) is derived, assumes that $\bar{\lambda}_c \ll L$. However, when the gradient is so steep that L becomes less than the average mean free path, $\bar{\lambda}_c$, the heat flux q might exceed its physical upper limit, usually called the "free streaming" value:

$$q_f = f \cdot n_e \cdot \bar{v} \cdot kT \tag{1.3}$$

where f is of the order of unity.

Under conditions relevant to plasmas produced by high power lasers the temperature gradient scale lengths are often smaller than the average collision mean free path, leading to the violation of the basic assumption of (1.1). To avoid such non physical behaviour a common remedy has been to postulate q_f as an upper limit to the heat flux in the region where q is larger than q_f. Heuristic arguments, using free streaming distribution functions, suggest that the flux limiter coefficient f is of the order of 0.5. However, the analysis of many experimental results[3] suggests that f should be smaller by about an order of magnitude, typically $0.03 \lesssim f \lesssim 0.1$. The use of such a small value of f, without a physical basis is unsatisfactory and has led to large uncertainties in target design and the simulation of experiments.

The small value of f has been attributed to a variety of anomalous processes, including ion acoustic instabilities[4] and electric fields generated by suprathermal electrons[5], but the importance of the above processes has not yet been demonstrated. In a review of the ion acoustic turbulence model, Mead[6] has shown that the fluctuation levels, required to reduce the flux limiter to the small values needed to model experiments, are much too large to be plausible. The fact that almost the same flux limiter is needed to interpret short wavelength experiments, where the generation of suprathermal electrons is greatly reduced, as for long wavelength experiments, indicates that electric fields generated by suprathermal electrons are not likely to cause the required thermal transport inhibition. The effect of a magnetic field that may exist in the region inside the critical surface on thermal transport, will be discussed shortly in Section 3.5.

An alternative explanation for the inhibition of thermal transport has been that the Spitzer-Härm (S-H) description should not be applied to steep temperature gradients, and that a correct treatment of classical transport based on classical Coulomb collisions would result in lower values for the thermal conductivity than previously suggested. It is the main aim of the present lectures to present some of the work done in the last 5 years in the area of classical kinetic theory, leading to a better understanding of the thermal transport process under conditions relevant to laser produced plasmas. A similar review has been written recently by J. Delettrez from the University of Rochester[7] and parts of the present lectures are based on that review.

2
EXPERIMENTAL STUDY OF THERMAL TRANSPORT

Electron thermal flux cannot be measured directly but must be inferred from other measurable quantities. The most relevant experiment is of burnthrough type which measures the depth and time of the heat front penetration. The target usually consists of a radiating material substrate coated with a layer of low-z material. The heat front propagates through the poorly radiating low-z material, and when it reaches the substrate, the target emits x-ray radiation, characteristic of the substrate material.

The transport characteristics can be deduced in two experimental ways:

(a) By measuring the intensity of appropriate x-ray lines from the substrate layer as a function of the overcoating layer thickness[8]. Fig.1 shows, for example, the relative intensity of the 2P-1S line of $A\ell^{+12}$ at 1.73 kev as a function of the parylene (CH) coating thickness over the $A\ell$ substrate[9]. The experiment was conducted at LLE using the 24-beam OMEGA facility at 1.06μm. From the fall-off position of the curve one can infer the "penetration depth" of the heat front in the coating material.

(b) By measuring the x-ray emission as a function of time. A sharp rise in the emission intensity indicates the penetration of the heat front into the substrate layer. Fig.2 shows the x-ray streak traces for a few CH coating thicknesses which were taken in the same experiment as above[9].

Knowing the penetration depth, d, and the burnthrough time, τ, one can deduce the mass ablation rate:

$$\dot{m} = \rho \cdot \frac{d}{\tau} \qquad (2.1)$$

where for burnthrough depth measurements, τ is usually taken to be the laser pulse width, since no time dependent measurements are taken.

Another method to determine the mass ablation rate is through ion velocity spectra. Ions of the blowing-off plasma are relevant, since

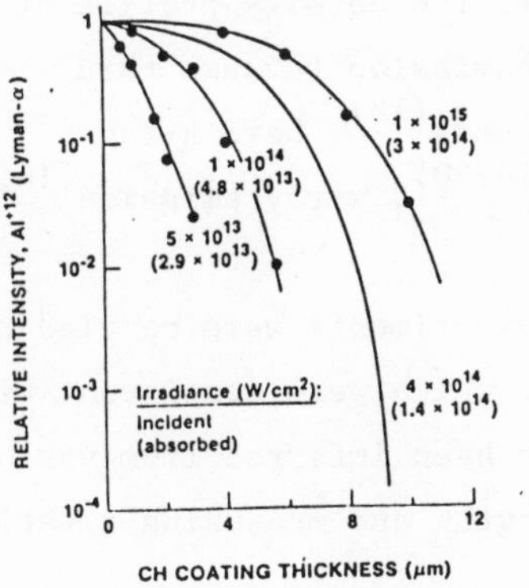

Fig.1: *The burnthrough of CH into Aℓ in spherical irradiation*(9).

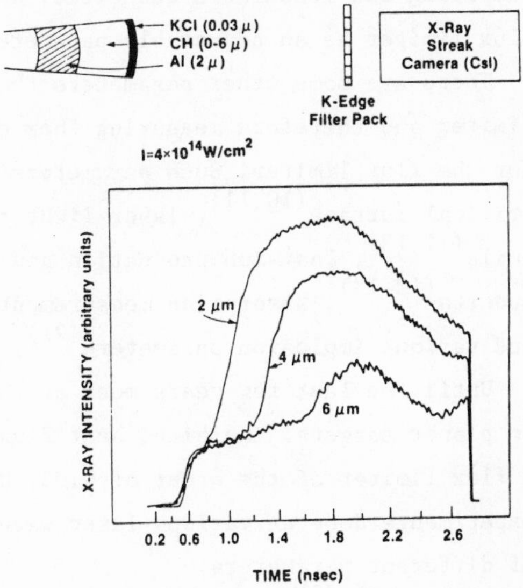

Fig.2: X-ray streak traces as a function of CH coating thickness. The first rise in intensity (at 0.5nsec) marks the heating of the KCℓ layer. The second rise is due to the penetration into the Aℓ layer(9).

inhibition of the inward heat flux leaves more energy to heat up the corona, accelerating the blowoff plasma to higher velocities. The ion velocity spectrum of the expanded plasma can be detected by charge collectors. Knowing the average velocity of the expanded plasma the mass ablation rate can be inferred:

$$\dot{m} = \frac{2 \cdot I_{ABS}}{\langle v^2 \rangle} \tag{2.2}$$

where I_{ABS} is the average absorbed intensity.

Simple scaling laws predict that $\dot{m}$ should scale with $f^{2/3}$ and therefore the mass ablation rate is a sensitive measure to the flux limiter. The flux limiter is inferred from the experimental data by comparing the measurable quantities with code predictions using the flux limiter as an adjustable parameter.

There are some other parameters that are sensitive to the flux limiter and therefore measuring them can point out a preferred value for the flux limiter. Such parameters are: the density profile at the critical surface(10,11), laser light transmission through thin foils(12,13), fast ion production and spectra(13), hard x-ray spectra(13-15), absorption measurements(16-20), x-ray emission(16,17) and various implosion parameters(21).

Until the last few years most of the experiments were carried out on planar targets. Inhibited heat fluxes, which were consistent with a flux limiter of the order of 0.03, have been inferred from various experiments done at various laser wavelengths and measuring a variety of different parameters.

Fig.3 shows(8) the burnthrough depth measurements conducted at LLE using the 0.35μm single-beam GDL laser. The inferred flux limiter from

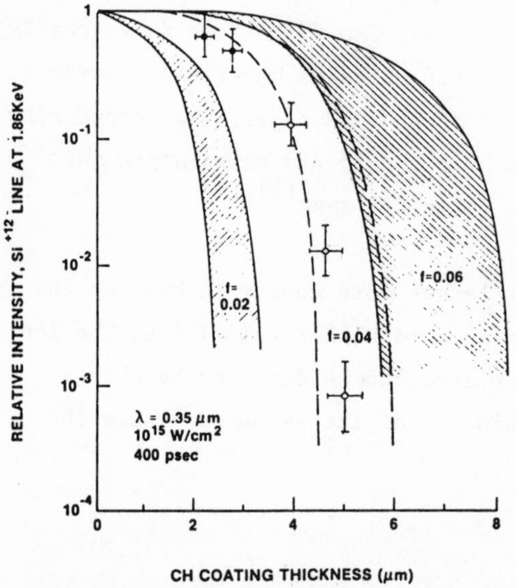

Fig.3: Relative intensity of the 1.86keV line of Si^{+12} from a glass target as a function of the plastic coating thickness. f is the flux limiter value used in the calculations(8).

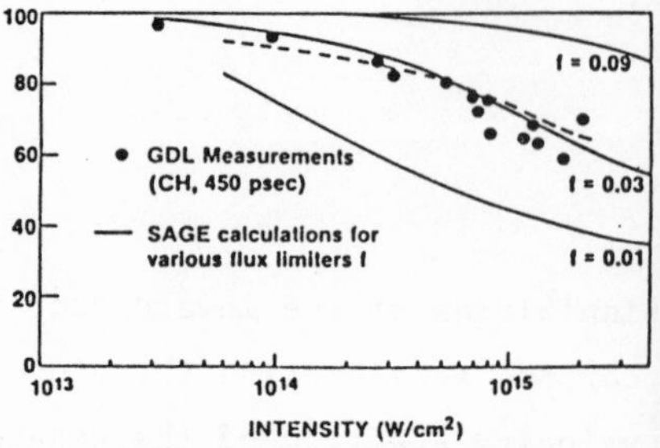

Fig.4: Absorption fraction for λ=0.35μm for CH targets(20).

these measurements was f = 0.04±0.01 (the band in Fig.3 represents calculations at intensities between $I_{nom}/\sqrt{2}$ and $I_{nom} \cdot \sqrt{2}$ in order to account for the intensity modulation in the laser focal spot). A similar flux limiter has been inferred from absorption measurements as is shown in Fig.4 (20). It is worth notice that these last experiments used a short wavelength laser (0.35μm) where the fast electron fraction is small and does not effect the plasma behaviour as was the case in the longer wavelength experiments (22,23).

Recently the availability of multi-beam facilities has permitted the conduction of transport experiments in spherical geometry. Since the goal of the ICF program is to drive a spherically symmetric target to high density, these experiments are of direct interest to the program, rather than the planar geometry experiments which are of a slightly different nature. Furthermore, the spherical geometry greatly simplifies interpretation of the data, which for planar geometry experiments is complicated by lateral transport (24), magnetic field generation (25-27) and other edge effects. The spherical nature of these experiments greatly assists in the computer modeling, since 1-D modeling is adequate, as opposed to the less accurate 2-D modeling required for simulating finite spot planar targets. An important effect, in planar target experiments, might be the decoupling from the target of the almost collisionless electrons of the hot corona, which are responsible for the axial heat conduction to the ablation surface (energies of about 4-10kT). This decoupling can arise from the azimuthal magnetic field, which will force the electrons to orbit in the corona and deposit their energy outside the laser spot (26,27), and in the expanding corona.

However, it is still questionable how all those many planar target experiments done at various wavelengths and spot sizes, have arrived almost at the same value of f (about 0.03). In some of those experiments the focal spot was large enough, so that no significant lateral conduction or spot size dependence has been expected or detected (28), and still a severe flux inhibition has been inferred. Further work on interpreting the planar target experiments is needed.

Spherical geometry transport experiments were conducted at three laboratories: The Rutherford Laboratory in England, the Laboratory for Laser Energetics at the University of Rochester and KMS Fusion Inc.

in Michigan. At Rutherford a six-beam laser system at two laser wavelengths, 1.05μm and 0.53μm was used to measure mass ablation rates[29]. The targets were glass spheres or empty glass shells with a coating of plastic overcoated with a thin layer of Aℓ. The mass ablation rate was determined by both x-ray streak spectroscopy and ion calorimetry. The burnthrough time in the spectroscopy measurement was defined as the time between the fiducial emission of the $A\ell^{+12}$ 1S-2p line and the emission of the Si^{+13} 1S-2p line in the underlying glass. The absorbed laser intensity in these experiments was in the range $5x10^{12}$-$5x10^{13}$ W/cm^2. The agreement between the x-ray and the ion derived mass ablation rates indicates that fast electrons do not have a significant effect in that intensity range. The mass ablation rate was found to scale strongly with absorbed intensity, $\dot{m} \sim I_a^{0.8}$, in contrast to $I_a^{0.5}$ in most plane targets, as is shown in fig.5. The heat flux inferred from those measurements was greater than 10% of the free streaming value, i.e. f > 0.1 (above that value $\dot{m}$ is insensitive to f). It should be noticed that the burnthrough time in those experiments was smaller than the laser pulse length, and therefore the

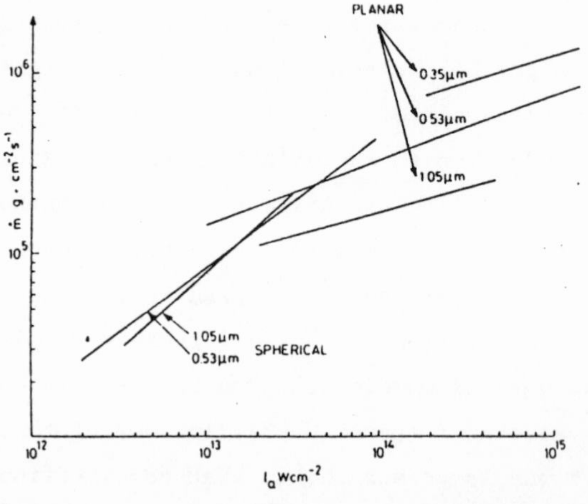

Fig.5: The specific mass ablation rate $\dot{m}$ as a function of absorbed irradiance I_a in spherical geometry and in plane geometry. Note the difference in the scaling[29].

mass ablation rate measured is actually the time-averaged mass ablation rate during the rising part of the laser pulse. This introduces a correction factor to the data and may lead to some uncertainty in the results.

At the University of Rochester, parylene coated targets over different substrate materials were illuminated with the 24-beam OMEGA laser system[9]. The mass ablation rate was obtained both from the charge collector traces and from the burnthrough depth into the plastic as measured by the emission of particular lines of the substrate materials. The substrate materials were Aℓ, Ni and Ti which gave signatures for different isotherm penetration: Aℓ for temperatures of about 300-500 eV and Ti and Ni for temperatures near approximately 1 keV and 1.5 keV respectively. The temporal progress of the heat front was also observed with an x-ray streak camera. Results show a much larger penetration to the Aℓ substrate with an uniform illumination than to the Ti or Ni substrate, or with single beam illumination, which is a planar geometry experiment done using 24 isolated laser spots (See Fig.6).

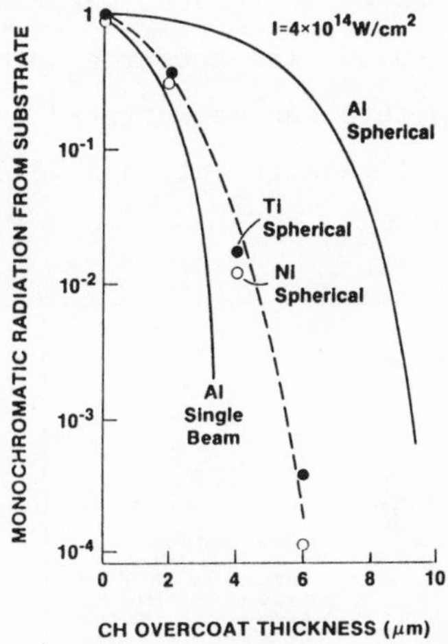

Fig.6: Burnthrough of parylene coating into various substrate materials. $A\ell^{+12}$, Ti^{+20} and Ni^{+26} lines at 1.75, 4.75 and 7.8 keV probe temperatures of about 0.4, 1 and 1.5 respectively[9].

Code simulations, using a flux limiter of 0.04, match the burn-through depths in the high Z substrate and the single beam experiments, as in the planar experiments. However, the uninhibited classical conductivity, or even a larger one, was needed to fit the Aℓ penetration depth, which was about three times larger. This large penetration depth to the Aℓ substrate, cannot be explained by the presence of hot spots in the illumination pattern because it would require that the hot spot intensity be an order of magnitude larger than the average irradiance; such hot spots have not been observed in nonuniformity calculations of the illumination or in the x-ray microscope images of the target. One interpretation of these results is that the heat front is preceeded by a low-temperature (less than 500 eV) "foot" in the uniform irradiation experiment, and that this foot does not exist in the single beam experiments because the long mean-free-path electrons responsible for the larger penetration are lost to the cold material surrounding the spot. Fig.7 compares the mass ablation rates obtained from this experiment using the various diagnostics and those measured by the Rutherford Laboratory. At the higher intensities the mass ablation rate scales as $I_a^{0.5}$ as compared to the $I_a^{0.86}$ scaling obtained at lower intensities. The Rochester charge collector data are consistent with a flux limiter of 0.06-0.08. The spectroscopic data are very spread and various flux limiters can be inferred from the various substrate used, 0.03 from the Ti measurements and uninhibited heat flux from the Aℓ data.

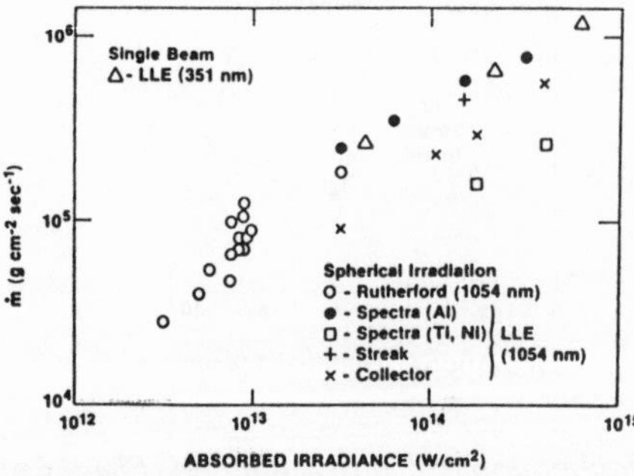

Fig.7: Comparison of mass ablation rates as a function of absorbed intensity from different measurements[9].

A recent experiment done at the Institute of Laser Engineering at Osaka University, Japan(30), using multilayered targets, has indicated a non-local behaviour of the heat front, similar to that in the LLE experiment, where a low-temperature "foot" was observed(9).

Another series of experiments was conducted at KMS with a two-beam clamshell focusing laser system at 1.05μm. The first series(31) was similar to the experiments at Rutherford. The mass ablation rate was obtained by measuring the burnthrough times through a plastic layer from x-ray signals generated by signature layers. Conditions are somewhat different from the other experiments in that the laser pulse had a fast rise-time and ramins almost constant for about 800 ps and the targets were smaller giving rise to steeper gradients. Comparison with the model of Max, McKee, and Mead(1) gives a best fit for f=0.06±0.02 (See Fig.8), which was confirmed by numerical simulations(32).

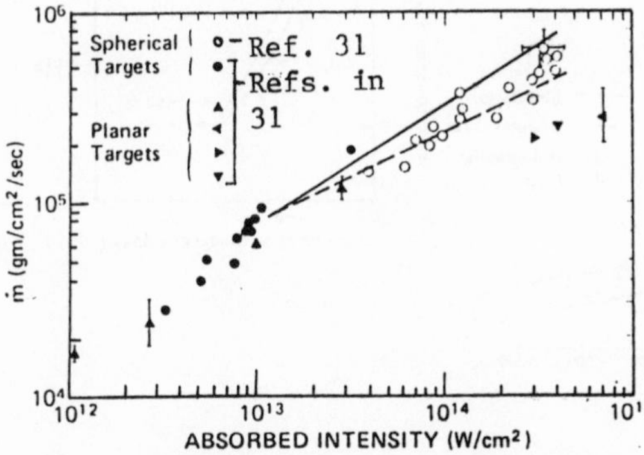

Fig.8: Mass-ablation rate vs. absorbed intensity for λ=1.05μm. Curves are from the model of Ref.1. Solid line, f=0.25; dashed line, f=0.03(31).

The second series of experiments obtained the mass ablation rate from measurements of the electron density profile in the corona and the excursion of the critical surfaces using interferometry(11). The density scalelength and the excursion of the critical density are both sensitive to the mass ablation rate. Comparison with code simulations showed that the data are consistent with a flux limiter of 0.03–0.06 (See Fig.9 and Fig.10).

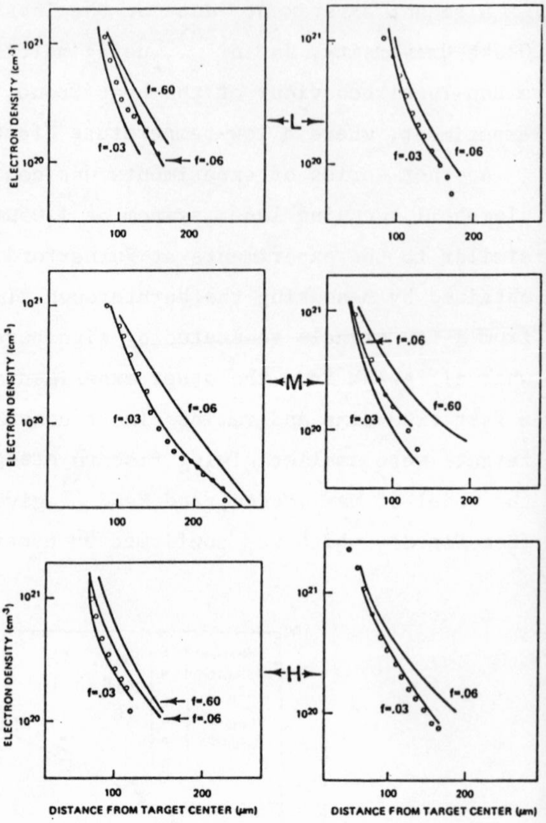

Fig.9: Plasma density profiles derived from experimental holographic data (open circles) are compared to simulated plasma density profiles (solid curves) for three values of the flux-limit parameter. L,M and H identify the incident laser intensity as low ($I<3\times10^{14}$ W/cm^2), medium (3×10^{14} $<I<6\times10^{14}$ W/cm^2), or high ($I>6\times10^{14}$ W/cm^2)[11].

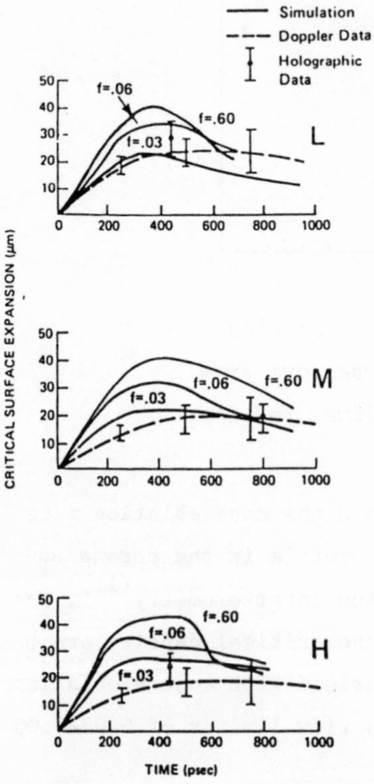

Fig.10: Trajectories for the critical density surface obtained from simulations (solid curves) are compared to experimental measurements of critical surface location for three individual experiments. L,M and H refer to the laser intensity (see Fig.9)[11].

In a review of these spherical experiments, M. Rosen[32] suggested that from a combination of all the spherical targets mass ablation data, except the Rochester spectroscopy data with Aℓ, an intriguing picture seems to emerge (see Fig.11). Below some threshold corresponding to an absorbed irradiance of mid 10^{13} W/cm^2 the data lies on the f=0.5 curve, implying non anomalous flux inhibition for those conditions. Above that threshold, however, the usual picture of a need for f=0.03 emerges. Rosen has noticed that the intensity at which the LASNEX code predicts no difference between the two values of f (0.5 and 0.03), due to large scale lengths and small mean free paths, is an order of magnitude lower than the mid 10^{13} W/cm^2 intensity, at which there are still differences in the code's predictions.

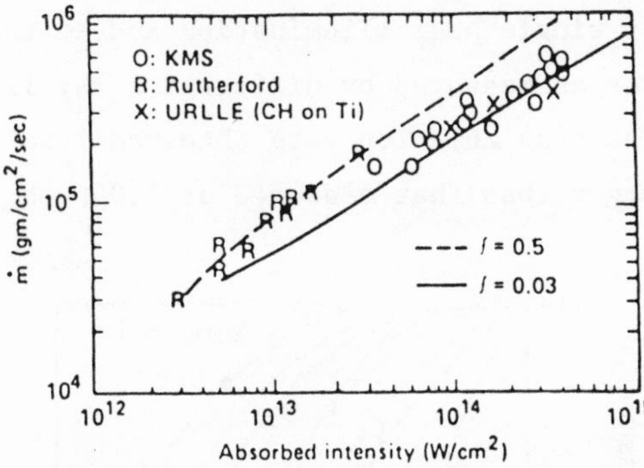

Fig.11: Compilation of many laboratories data on mass ablation rate of CH in spherical systems, illuminated by 1.06μm light[32].

A more recent experiment at the Rutherford Laboratory[33] used targets which consisted of solid glass spheres, containing Si and Ca, coated with three layers: CH, a thin layer of Aℓ, and CH again. The mass ablation rate was measured by taking the delay between the Aℓ and Si emissions as the time needed to ablate the plastic layer between them. This experiment was simulated with the hydrocode LASNEX and good agreement was obtained for f=0.08±0.02. The mass ablation rates deduced overlap the KMS data and the LLE charge collector data. Temporal observation of the Si and Ca lines indicate that the heat front must be steep, a substantial difference when compared with the

Rochester and Osaka experiments.

A first series of experiments at the u.v. laser wavelength of 0.35μm has been carried out at LLE[34] with six beams. The experiment was similar to their previous one at 1.05μm. Results proved to be even more puzzling than those at 1.05μm. Fig.12 shows the measured and calculated $A\ell^{+12}$ line emission for several laser intensities as a function of the CH-overcoat thickness. The burnthrough depths are very large and cannot be modeled by code simulation. At 1×10^{14} W/cm^2, where flux limiting has no effect, there is a factor of 2.5 between the observed and calculated burnthrough depths. Such large discrepancie can be caused by hot spots in the beam which may lead to self-focussing and large penetration depths. Another difference relative to the 1.05μm data, is that there are now no differences in the burnthough of CH between uniform or single-beam illumination and in the burnthrough of different isotherms as measured by different x-ray lines from a titanium substrate. The mass ablation rate obtained from the charge collector data is larger than that measured at 1.05μm by about a factor

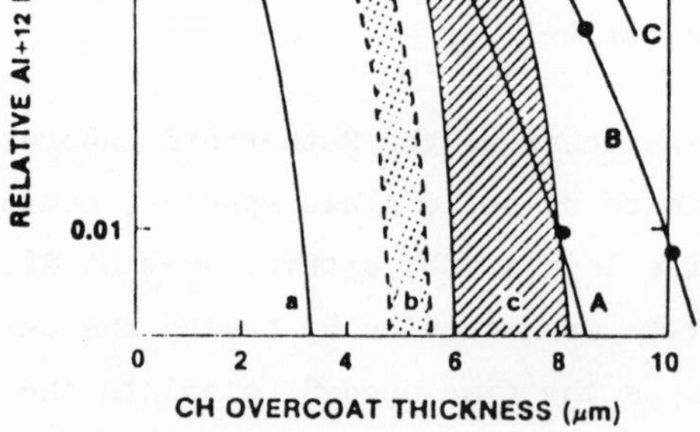

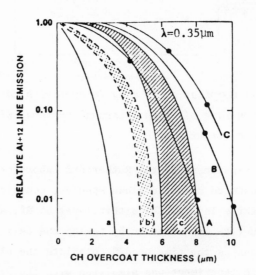

Fig.12: Experimental burnthrough curves: intensity of the $A\ell^{+12}$ line, as a function of the CH-overcoat-layer thickness. Irradiance (in W/cm^2): A-10^{14}, B-5×10^{14}, C-10^{15}. Theoretical curves, for flux limiters between 0.04 and 0.1, are shown in bands a, b, c for irradiance A, B, C, respectively[30].

of two for the same absorbed intensity but is lower than that obtained from spectroscopy. The mass ablation rate for 0.35μm irradiation deduced from charge collectors follows the scaling law $\dot{m} \sim 3.8\times10^{5}$ $(I_a/10^{14})^{0.53}$. As in the case of the spectroscopic data, code simulations at $f \geqslant 0.1$ could not match the larger values obtained from the charge collector data. Experiments are being conducted at LLE with 12 and 24 beams of OMEGA at 0.35μm and preliminarely results support the nonuniformity argument. The experimental measured mass ablation rate with 12-beam irradiation, and even more so with 24-beam irradiation, is in much closer agreement with the code prediction using $f = 0.1$ [35].

In summary, even though no conclusive result can be made on the value of the flux-limiter deduced from spherical target experiments, it seems that a higher value of the flux limiter, $f \sim 0.06$–0.1, is needed in the simulations in order to match most of the spherical geometry experiments. This value is greater than the one inferred from the planar geometry experiments, $f \sim 0.03$–0.06. The difference between the two types of experiments may be due to the difference in the plasma expansion behaviour or to the presence of magnetic fields, which may cause lateral energy transport by convection of the fast collisionless electrons, in the planar experiments.

3.
THEORY OF THERMAL ELECTRON TRANSPORT

3.1 The Fokker-Planck Equation and Spitzer-Härm Thermal Conductivity

The transport of electrons in a fully ionized plasma is described by the Boltzmann equation:

$$\frac{\partial f}{\partial t} + \overline{v} \cdot \frac{\partial f}{\partial \overline{r}} + \frac{e\overline{E}}{m_e} \cdot \frac{\partial f}{\partial \overline{v}} = \left(\frac{\partial f}{\partial t}\right)_c \tag{3.1}$$

where $f(\overline{r},\overline{v},t)$ is the electron distribution function in space, $\overline{r}$, velocity, $\overline{v}$, and time, t. $\overline{E}$ is the electrical field in the plasma, e and m_e are the charge and mass of the electron, respectively. Magnetic fields are not included in (3.1) and their effects on electron transport will be discussed later. $(\frac{\partial f}{\partial t})_c$ in (3.1) is the collision operator which accounts for collisions between the electrons and the plasma's ions and electrons. Since the Coulomb interaction is mainly a long range interaction, a common description of these collisions is to use the small angle approximation, known as the Fokker-Planck approximation to the collision operator[36]:

$$(\frac{\partial f}{\partial t})_c = \frac{1}{v}\cdot\frac{\partial}{\partial v}\left(v^2\cdot(\frac{D_{\parallel}}{2}\cdot\frac{\partial f}{\partial v} + C\cdot f) \right) + \frac{D_{\perp}}{2\cdot v^2}\cdot\frac{\partial}{\partial \mu}\left((1-\mu^2)\cdot\frac{\partial f}{\partial \mu} \right) \qquad (3.2)$$

where, for simplicity, we have limited the discussion to one dimensional geometry, and μ is the cosine of the angle between the velocity vector and the spatial direction. C, $D_{\parallel}$ and $D_{\perp}$ are respectively the coefficients for slowing down, diffusion in velocity space and angular space, due to electron-electron and electron-ion collisions. The plasma's collective effects are included by introducing a screened Coulomb potential for the interaction, resulting in the Coulomb logarithm, $\ell n\Lambda$, in all three coefficients. These coefficients depend on the distribution function of the plasma's electrons and ions.

Electron-electron collisions are characterized by changes in the electron velocity, slowing down and velocity diffusion, while electron-ion collisions are characterized by changes in the electron direction. Therefore the angular diffusion term in (3.2) is about Z times larger than the velocity diffusion term, where Z is the ionic charge.

In the presence of small gradients, and since the angular scattering term is the dominant one in (3.2), it is common to assume that the distribution function has a weak angular dependence and that it can be expressed by the first two terms in a perturbation expansion:

$$f\,(x,v,\mu,t) = f\,(x,v,t) + \mu\cdot f_1(x,v,t) \qquad (3.3)$$

where f_0 and f_1 are the isotropic and anisotropic components of the distribution function, respectively. Higher terms in the expansion can be shown to be small compared to f_0 and f_1 since the large electron-ion collision frequency tends to maintain the distribution function near isotropic. When (3.3) is substituted into (3.1) and (3.2) and the first two angular (μ) moments of the equation are taken, the diffusion approximation to the Fokker-Planck equation results:

$$\frac{\partial f_0}{\partial t} + \frac{1}{3}\cdot\left(v\cdot\frac{\partial}{\partial x} + \frac{eE}{m_e}\cdot\frac{\partial}{\partial v} \right) f_1 = \frac{1}{v^2}\cdot\frac{\partial}{\partial v}\left(v^2\cdot\left(\frac{D_\parallel}{2}\cdot\frac{\partial f_0}{\partial v} + C\cdot f_0 \right) \right) \quad (3.4)$$

$$\frac{\partial f_1}{\partial t} + \left(v\cdot\frac{\partial}{\partial x} + \frac{eE}{m_e}\cdot\frac{\partial}{\partial v} \right) f_0 = \frac{1}{v^2}\cdot\frac{\partial}{\partial v}\left(v^2\cdot\left(\frac{D_\parallel}{2}\cdot\frac{\partial f_1}{\partial v} + C\cdot f_1 \right) \right) - \frac{D_\perp}{v^3}\cdot f_1 \quad (3.5)$$

In general, equations (3.4) and (3.5) should be solved for f_0 and f_1 in order to calculate the heat flux as a function of time and space. This can be done only numerically and such solutions will be discussed in 3.4. However, using some simplifications, useful results can be derived without solving the full Fokker-Planck equation.

For small temperature and density gradients Spitzer and Härm[(2)] have solved (3.5) assuming that the isotropic component of the distribution function, f_0, is the local maxwellian, $f_{M.B.}$. In order to derive f_1 from (3.5) they assumed a steady state situation, i.e. $\frac{\partial f_1}{\partial t} = 0$, and charge neutrality in the plasma, which is equivalent to the zero current condition given by

$$J = \frac{4\pi e}{3} \int v^3 \cdot f_1 \, dv = 0 \quad (3.6)$$

yielding an expression for the self consistent electric field.

An analytical expression for f_1 is obtained for a Lorentz gas, in which the electron-electron collision terms in (3.5) are neglected and the ions are assumed at rest. The collisional mean free path is then $\lambda(v) = \lambda_0\cdot(v/v_{th})^4$, where v_{th} is the thermal velocity $(2kT/m_e)^{1/2}$ and λ_0 is the mean free path for 90^0 scattering by multiple collisions at kT, $\lambda_0 = (kT)^2/(\pi n_e Z e^4 \ell n\Lambda)$. Using these assumptions one finds the ratio f_1/f_0 to be:

$$\frac{f_1}{f_0} = \frac{\lambda_0}{L} \cdot \left(\frac{v}{v_{th}}\right)^4 \cdot \left\{ \left(\frac{v}{v_{th}}\right)^2 - 4 \right\} \quad (3.7)$$

where the gradient length L is defined by $L \equiv T/|dT/dx|$. The net heat flux Q is then defined by $Q \equiv (4\pi m_e/6) \int v^5 \cdot f_1 \, dv$, which upon substitution of (3.7), yields the Fourier's law for heat conductivity, $Q = -\kappa \cdot dT/dx$, where κ is the Spitzer-Härm (S-H) electron thermal conductivity for high Z plasmas:

$$\kappa = \frac{\varepsilon \cdot 20 \cdot \left(\frac{2}{\pi}\right)^{3/2} \cdot (kT)^{5/2} \cdot k}{m_e^{1/2} \cdot e^4 \cdot Z \cdot \ell n\Lambda} \tag{3.8}$$

Here ε is a reduction factor due to the neutralizing electric field, and is equal to 0.4 for the case of a Lorentz gas. If the electron-electron collision term in (3.5) cannot be neglected, such as is the case for low Z plasmas, (3.5) together with (3.6) must be solved numerically for f_1. The results are usually expressed in terms of the Lorentz gas thermal conductivity by multiplying it by a coefficient δ. The values of ε and δ depend on Z and have been calculated by Spitzer and Härm[(2)]. A useful approximation for their product is given by the expression[(1)], $0.095\cdot(Z+0.24)/(1+0.24\cdot Z)$.

The thermal transport problem is best illustrated by plotting the ratio f_1/f_0 and the differential heat flux, $Q(v) = \frac{4\pi}{6}\cdot m_e \cdot v^5 \cdot f_1$, see Fig.13. The heat flux is negative for velocities up to $2\cdot v_{th}$. Thus the heat is carried primarily by electrons with velocities between $2\cdot v_{th}$ to $4\cdot v_{th}$, and the peak heat flux occurs for electrons with

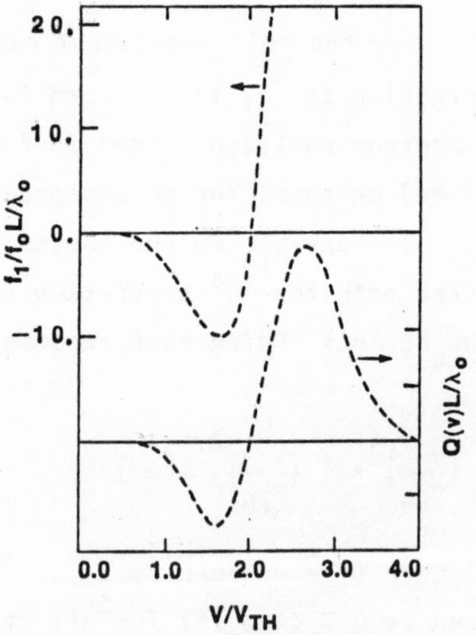

Fig.13: Particle (f_1/f_0) and heat ($Q(v)$) fluxes from Spitzer-Härm theory (normalized by L/λ_0).

velocities of about $2.5 \cdot v_{th}$. The negative f_1 at low velocities results from the negative electron density gradient at these velocities and the self-consistent electric field needed to preserve neutrality. This electric field, commonly called the thermoelectric field, is of the order of kT/L, and acts to cancel the electron flux down the temperature gradient, thus reducing the heat flux.

From (3.7), it can be seen that f_1/f_0 increases with λ_0/L, and at some velocity, depending on λ_0/L, it becomes greater than unity. However, the S-H description, which is a perturbation approach, is not valid when $f_1 \gtrsim f_0$. When f_1 exceeds f_0 the diffusion approximation for the distribution function, $f(\mu)$ becomes negative for some μ [37]. Furthermore, for any transport description the particle flux, $v \int d\mu \cdot \mu \cdot f(\mu) \equiv f_1 v/3$, cannot exceed the free-streaming value $\mu_{max} \cdot f_0 v$, where μ_{max} is the maximal average μ of the distribution function, and f_0 is the particle density $f_0 \equiv \int d\mu\, f(\mu)$. Therefore, at those velocities for which f_1 exceeds f_0, the S-H heat flux, $Q(v)$, becomes unphysically large, independent of the assumed transport treatment.

For $\lambda_0/L \lesssim 10^{-3}$, the above violation of the diffusion approximation occurs at velocities above the range where significant heat is carried, $2\text{-}3.5 \cdot v_{th}$, and the S-H theory is expected to be accurate. However, as the gradient becomes steeper, λ_0/L increases, and the violation occurs at velocities at which a significant part of the heat is carried (for $\lambda_0/L = 0.1$, f_1/f_0 exceeds unity at about $2.1 v_{th}$). The S-H theory will fail to predict the heat flow in such gradients.

When the gradients are steep, another assumption of S-H theory fails. In the theory it is assumed that f_0 is the local maxwellian distribution. However, when λ_0/L increases, there will not be as many electron-electron collisions over a gradient length, as are needed to prevent the distribution function from deviating from the local maxwellian. Non-local transport of the long mean-free-path electrons will tend to modify the distribution function, and energy will be transported over a few gradient lengths, causing a modification of the temperature profile.

3.2 Local Extension of Spitzer-Härm Theory

In order to prevent the unphysical situation where $f_1 > f_0$, which implies that more particles are streaming in the gradient direction than the number of particles that exist, a common remedy in transport theory (see Section 1) is to limit the flux, f_1, to the number density, f_0. By applying this limitation Shvarts et al[38] have extended the Spitzer-Härm local model to the region of steep temperature gradients. Since no attempt has been made to solve the transport equation to get the actual distribution function, f_0 was taken as the local maxwellian distribution, as was assumed by Spitzer and Härm.

Even though a modified f_0 should be used, due to the non-local transport of the less collisonal electrons, this approximation does illustrate how the heat flux can be changed dramatically by preventing f_1 from attaining unphysically large values. In flux limiting f_1 before performing the integration, $\int v^5 \cdot f_1 dv$, the diffusion value for f_1 at all velocities at which it is applicable ($f_1 < f_0$) was used, and the upper bound $f_1 = f_0$, was used only where it is required. The commonly used "free-streaming" limit is obtained by using the upper bound value for f_1 for the entire velocity range, independent of whether the diffusion result is applicable or not. As will be shown, this picture has to be applied only for high velocities (above $\sim 2.2 \cdot v_{th}$ for large λ_0/L), and therefore one obtains a more restrictive upper bound to the heat flux than the "free-streaming" limit.

In order to carry out this limiting procedure self-consistently, $f_1(v)$ was found simultaneously with the neutralizing electric field. Using a flux limited f_1, without self-consistently determining the electric field, results in non-zero currents, and for $\lambda_0/L \gtrsim 0.05$, negative net Q's.

The results of the above treatment are compared to the S-H theory in Fig.14. In Fig.14a, $\lambda_0/L = 0.002$, f_1 exceeds its maximum value only at $v \simeq 3 \cdot v_{th}$, and since Q is insensitive to Q(v) in this range, the limiting procedure does not significantly change Q from the S-H heat flux. In contrast, for $\lambda_0/L = 0.1$ (Fig.14b) the assumptions of S-H theory are violated, as illustrated by f_1 which exceeds f_0 near $v \simeq 2\ v_{th}$. Limiting f_1 sharply reduces the heat flux, corresponding

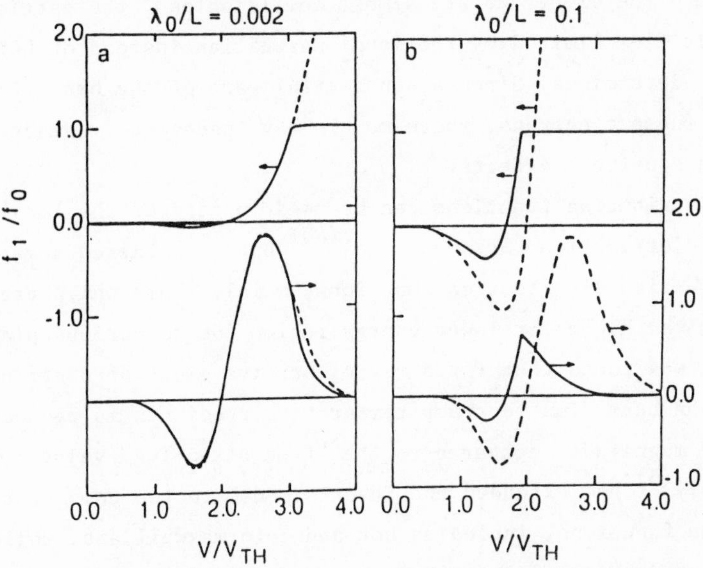

Fig.14: Spitzer-Harm (---) and self-consistently flux limited (——) particle (f_1/f_0) and heat ($Q(v)$) fluxes[38].

to an effective flux limiter of the order of 0.1. Limiting the positive portion of f_1 also results in a substantial reduction in the return current needed to preserve charge neutrality, and hence a reduction in the required electrical field.

The model described above has been extended by Kishimoto and Mima[39]. Higher orders in the perturbation expansion with respect to λ_0/L are considered. This perturbation approach is used for velocities up to the point, v_c, where the λ_0/L expansion breaks down. Above this velocity v_c, which is about the velocity where f_1 exceeds f_0 in the previous model, a similar flux limitation of f_1 to the local maxwellian, f_0, was used. The reduction in the heat flux was similar to that obtained in the previous model. Higher order terms, which contain the second and third spatial derivatives of the temperature, lead to a multivalued behavior of the heat flux with respect to the temperature gradient. Therefore these higher order terms can exhibit some non local properties of the heat flux.

The main weakness of both models described is that they assume that f_0 is the local maxwellian distribution. Therefore at the high velocity region, where the electrons are almost collisionless, the particle flux, f_1, is flux limited by the local maxwellian instead of being non-locally determined. Since a substantial part of the heat flux is carried by these electrons, these models may inadequately describe part of the non-local effects.

Other distribution functions can be used instead of the local maxwellian distribution. Strauss et al[40] have considered a variety of non maxwellian distribution functions, mainly those which are depleted in the higher or lower energy region due to various plasma effects. It was found that for a representative group of distribution functions the heat flux in steep temperature gradients is reduced by an order of magnitude, compared to the "free streaming" value.

Shkarofsky[5] has extended the Shvarts model to two component distribution functions, including hot and cold maxwellians. While the author used this model to study the effects of fast electrons on thermal transport inhibition, through the self consistent electrical field, the same model can be used also to study non-local distribution function effects.

3.3 Simplified Non-Local Models

In the presence of steep temperature gradients, when non-local transport is important, adequate modeling of the thermal electron transport requires a numerical solution of the full Fokker-Planck equation. A short review of the work done recently on these numerical solutions will be given in 3.4. In order to get some qualitative feeling of the non-local behaviour of the transport process, it is worthwhile to look at some simplified, somewhat hueristic, analytical models.

The inhibition of the heat flux due to non-local transport effects can be understood qualitatively by examining the transport of the heat carrying electrons. Since these electrons have velocities of about $2.5 \cdot v_{th}$ their mean free path will be about 40 times larger than the thermal electron mean free path, λ_0. Therefore, in the presence of large temperature gradients, when λ_0/L is of the order of 0.1, the mean free path of the heat carrying electrons will be larger than the gradient length. These electrons will travel a distance of some gradient lengths before slowing down and therefore their spatial gradient,

f/λ, will be less steep than the temperature spatial gradient, f/L. This will result in a reduced heat flux, $Q \sim \lambda \cdot \frac{df}{dx}$, in the gradient direction.

If the whole thermal transport process is cast into the naive picture of the transport of single velocity electrons, at about $2.5 \cdot v_{th}$, the steady state transport equation, neglecting electric field effects, will be:

$$\mu \cdot v \cdot \frac{\partial f}{\partial x} = \frac{f_{M.B.} - f}{\tau} \qquad (3.8)$$

where the collision operator has been taken to be of the Krook type. Using the diffusion approximation, $f = f_0 + \mu f_1$, and taking the first two moments of (3.8) one gets[41]:

$$\frac{v}{3} \cdot \frac{df_1}{dx} = \frac{f_{M.B.} - f_0}{\tau_{ee}}$$

$$v \cdot \frac{df_0}{dx} = - \frac{f_1}{\tau_{ei}} \qquad (3.9)$$

where τ_{ee} ($= \lambda_{ee}/v$), the electron electron collision time, was used as the maxwellization time scale, and τ_{ei} ($= \lambda_{ei}/v$), the electron ion collision time, was used as the isotropization time scale. Substituting for f_1 in the first equation of (3.9) the diffusion equation results:

$$\tilde{\lambda} \cdot \frac{d}{dx} \left(\tilde{\lambda} \cdot \frac{df_0}{dx} \right) = f_0 - f_{M.B.} \qquad (3.10)$$

where $\tilde{\lambda} = \sqrt{\lambda_{ee} \cdot \lambda_{ei} / 3}$ is the effective range of the electrons in the diffusion approximation, corresponding to a random walk description of the transport process. The solution of (3.10) is:

$$f_0(x) = \int_{-\infty}^{\infty} f_{M.B.}(x') \cdot \frac{1}{2\tilde{\lambda}} \exp\left(- \frac{|x-x'|}{\tilde{\lambda}} \right) \cdot dx' \qquad (3.11)$$

and the flux is:

$$f_1(x) = - \lambda_{ei} \cdot \frac{df_0}{dx} = \int_{-\infty}^{\infty} \left(-\lambda_{ei} \frac{df_{M.B.}(x')}{dx'} \right) \cdot \frac{1}{2\tilde{\lambda}} \exp\left(- \frac{|x-x'|}{\tilde{\lambda}} \right) \cdot dx' \qquad (3.12)$$

The result is a convolution[42] of the local "Spitzer Härm" flux, $f_1(x) \equiv -\lambda_{ei} \cdot \frac{df_{M.B.}}{dx}$, with a delocalization kernel which smears the flux exponentially over a distance of the order of $\tilde{\lambda} = \sqrt{(Z+1)/3} \cdot \lambda_{ei}$ (we used $\lambda_{ee} \simeq (Z+1) \cdot \lambda_{ei}$). Taking the velocity of the heat carrying electrons as $2.5 \cdot v_{th}$, one obtains that $\tilde{\lambda}$ is about $23 \cdot \sqrt{Z+1} \cdot \lambda_0$, where $\lambda_{ei} = (2.5)^4 \cdot \lambda_0$ was used, λ_0 is the thermal electron-ion collision mean free path.

Even though the above model is very crude and depends upon simplified assumptions it does illustrate the general nature of the non-local heat flux. The heat flux at a given point is non-locally determined by the gradients in a region with a typical size of the order of $\tilde{\lambda}$. For a gentle temperature gradient, $\tilde{\lambda} \ll L$, (3.12) simplifies and gives the local (S-H) heat flux, while for steeper gradients a strong non-local contribution arises.

The same model can be extended to other, more realistic, approximations of the Fokker-Planck equation. When a velocity dependent Krook collision operator, $(\partial f(v)/\partial t)_c = (f_{M.B.}(v) - f(v))/\tau(v)$, is used to approximate the Fokker-Planck operator, a procedure identical to that described above can be followed for each velocity. The integrated heat flux, $Q = \int v^5 \cdot f_1(v) dv$, can then be calculated, resulting in a similar delocalization integral:

$$Q(x) = \int_{-\infty}^{\infty} Q_{S-H}(x') \cdot K(x',x)\, dx' \tag{3.13}$$

where $Q_{S-H}(x')$ is the local Spitzer-Härm heat flux and the kernel $K(x',x)$ contains only an integration over the velocity variable, and depends upon the distance between x' and x, measured in units of the thermal mean free path at x', $\lambda_0(x')$. Performing the velocity integration numerically $K(x',x)$ was found to be well approximated by the simple exponential expression:

$$K(x',x) \simeq \frac{1}{2\,\tilde{\lambda}(x')} \cdot \exp\left(-\frac{|x-x'|}{\tilde{\lambda}(x')}\right) \tag{3.14}$$

where $\tilde{\lambda}(x') = 29\sqrt{Z+1} \cdot \lambda_0(x')$. The result obtained before in the naive single velocity calculation, $\tilde{\lambda} = 23\sqrt{Z+1} \cdot \lambda_0$ is in surprisingly good agreement with this model.

A more realistic approach is to use the high velocity limit of the Fokker-Planck collision operator, which is adequate to describe the heat carrying electrons. In that limit the coefficients $D_{\parallel}$, $D_{\perp}$ and C in (3.2) become simple analytical functions, independent of the details of the distribution function. Neglecting the electric field and assuming a high Z plasma, the steady state Fokker-Planck diffusion equation becomes[42-44]:

$$\lambda_{ee}(v) \cdot \frac{\partial}{\partial x} \left(\lambda_{ei}(v) \cdot \frac{\partial f_0}{\partial x} \right) = - v \cdot \frac{\partial}{\partial v} \left(\frac{T}{mv} \cdot \frac{\partial f_0}{\partial v} + f_0 \right) \quad (3.15)$$

By approximating f_0 in the parallel diffusion term by $f_{M.B.}$ the right hand side of (3.15) becomes $v \cdot \frac{\partial}{\partial v}(f_{M.B.} - f_0)$, and the equation can be solved[42] by following the same method as described above. The Green's function now involves an additional integration, over the velocities, since the collision operator now couples both velocity and space. The heat flux can be expressed just as in equation (3.13), with a more complicated expression for K(x',x), involving a double integral over velocities. This kernel can be approximated by the exponential form (3.14) with $\tilde{\lambda}(x') = 33\sqrt{Z+1} \cdot \lambda_0(x')$.

Luciani et al[42] have found that this delocalization model is in very good agreement with the heat flux deduced from a numerical solution of the Fokker-Planck equation, as shown in Fig.15.

When electric field effects were included in the solution, the same kernel was found to fit the heat flux. The only change, caused by the electric field, was that the local heat flux, Q_{S-H}, was multiplied by 0.4, which is the usual Spitzer correction factor, ε. This can be hueristically explained by noticing that the self consistent electric potential $e\phi$ is of the order of kT, and therefore the reduction factor for the almost collisionless, heat carrying electrons, $\exp(-e\phi/kT)$, may reasonably be of the order of 0.4.

An attempt to include the effects of the electric field in the underlying equation was recently made by Williams et al[45]. Using the same Green's function, as in the last case above, they recieved similar delocalization formulas for both the current and the heat flux, including seperate contributions of the temperature gradient and the electric field. In order to calculate the heat flux self-consistently, the

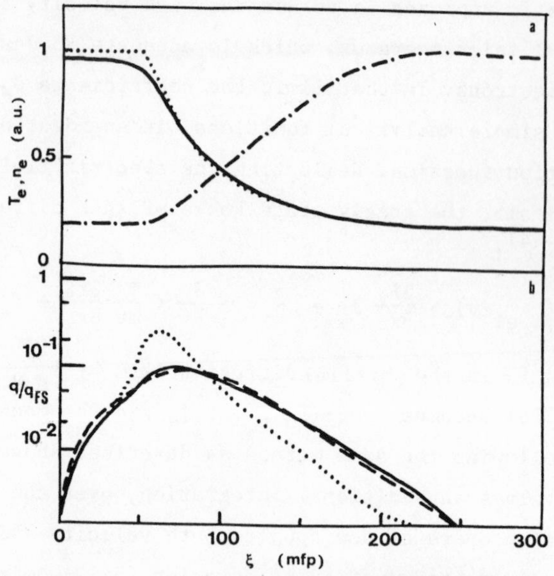

Fig.15: (a) Initial (••••) and final (——) temperature profiles, and density profile (-·-·-). (b) Final heat flux profiles, from the Fokker-Planck simulation (——), from S-H law (••••), and from the delocalization model (----)(42).

zero current condition should be applied, resulting in an integral equation for the electric field which must be solved, before the delocalized heat flux can be calculated.

In these non-local models no flux limitation of f_1 is needed. Since the gradient of f_0 is of the order of $f_0/\tilde{\lambda}$ the flux f_1 $(=-\lambda_{ei} \cdot \frac{\partial f_0}{\partial x})$ is of the order of $f_0/\sqrt{Z+1}$. Thus, at least for high Z plasmas, f_1/f_0 always remains small. As is demonstrated in Fig.15, the heat flux in the steep temeperature gradient region is an order of magnitude smaller than the free streaming value, corresponding to an effective flux limiter of about 0.1. In the low temperature region, the heat flux exceeds the S-H value due to non-local transport from the hot region.

Such semi-analytical models, which are in accordance with Fokker-Planck Solutions, result in delocalization formulas involving only macroscopic variables. Thus they are much more economic to employ in hydrodynamic simulation codes, and enable the study of coupled hydro-transport effects[42].

3.4 Numerical Solution of the Fokker-Planck Equation

The first solution of the Fokker-Planck equation, as applied to the propagation of a heat front in a plasma, was given by Bell[46]. The equation was solved for a uniform density case using the Legendre polinomial expansion method to describe the angular dependence of f. The result of this calculation is shown in Fig.16, where the heat flux, Q, normalized to the free streaming value, $Q_f = n_e \cdot kT \cdot (kT/m_e)^{\frac{1}{2}}$, is plotted as a function of L/λ_0 along the temperature profile. It is shown that in the presence of steep temperature gradients, the heat flow at any point is not simply a function of the local condition but is determined by the temperature profile over a region which is a few mean free paths thick. For a given λ_0/L, the thermal heat flux cannot be expressed by a unique expression of the local variables, but depends on the whole temperature profile. As was qualitatively shown in 3.3, in the hot part of the plasma and the heat front, the flux is less than

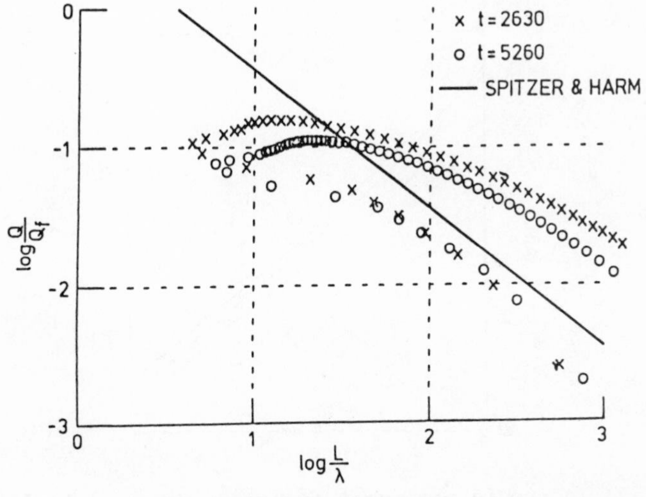

Fig.16: Plot of the heat flow against inverse temperature gradient from the numerical solution of the Fokker-Planck equation for a propagating heat front, compared with the Spitzer-Härm conductivity, for Z = 4. Here $\lambda = \frac{9}{4}\lambda_0$ [46].

the Spitzer-Härm flux and its maximum value is about ten percent of the free-streaming value. At the foot of the heat front, the heat flow is actually higher than the classical flux because of the presence of long-mean-free path electrons which originated in the hot part of the plasma and have not lost their energy through collisions.

Using a similar method, Matte and Virmont(47) obtained similar results for a stedy state cathode problem, where electrons are transported between hot and cold boundaries, in which the size of the cathode was changed to vary λ_0/L. Fig. 17 shows typical shapes of the isotropic part of the distribution function, f_0, along the cathode. Near the cold boundary the shape is close to a bi-maxwellian distribution with the main component at the background cold temperature and a hot tail at the hot boundary temperature. The magnitude of the tail decreases with the distance from the hot boundary, leading to a small preheat foot ahead of the main heat front. A similar non-local behaviour has been observed by Mason using a Monte-Carlo approach(54).

Solving the Fokker-Planck equation for a cathode problem may help in studying the basic physics of electron transport, but the conditions are somewhat far from those of laser fusion in which laser energy is

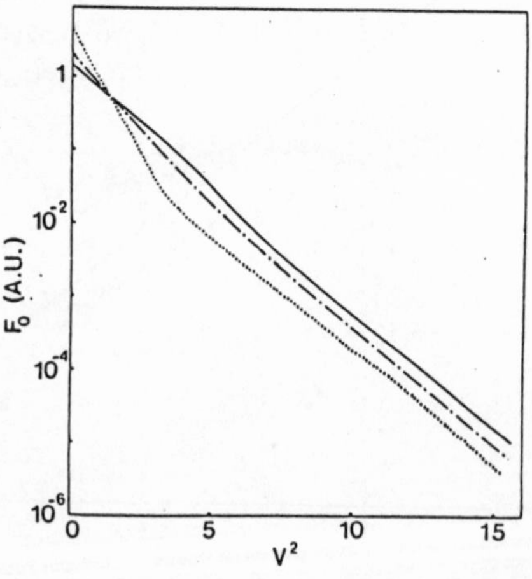

Fig.17: Isotropic component f_0 versus square velocity v^2, near the hot wall (——), at the center (-·-·-), and near the cold wall (····). The slab parameters are $L/\bar{\lambda} = 5$, $T_h/T_c = 4$. v^2 is in units of $k(T_h+T_c)/m_e$(47).

continuously deposited predominantly in the leading edge of the heat front, causing the steepening of the temperature profile there. Albritton[48] has used a simplified high Z diffusion approximation of the Fokker-Planck equation to study the heat flux in laser plasmas under various conditions of laser illumination. In simulating the inverse bremsstrahlung absorption process the Longdon[49] effect has been taken into account. Since the laser energy is absorbed by the lower velocity electrons, a self-consistent treatment of the absorption process and the electron-electron collisions has been used in order to calculate the actual absorption rate. This rate is lower than that predicted using inverse-bremsstrahlung opacities derived with the assumption of a Maxwell-Boltzmann distribution. Calculations were carried out for 1-μm laser light at intensities of 10^{14} and 10^{16} W/cm^2, an exponential density profile with a scalelength of 100μm, and Z = 10 and 40. At a laser intensity of 10^{14} W/cm^2, the peak flux was 0.14 of the free streaming flux. The distribution function near the critical surface is depleted near zero velocity due to the laser absorption and a hot tail appears near the bottom of the heat front.

The Fokker-Planck equation was recently solved for conditions which included inverse-bremsstrahlung and ion motion[50]. An initial density profile which consisted of two plateaus of density 4×10^{19} and 2×10^{22} cm^{-3} joined by an exponential density profile with a gradient length of 46μm was illuminated by a constant 1.06μm laser pulse of intensity 3×10^{14} W/cm^2. After 600 ps the results of the Fokker-Planck solution were compared with those of a hydrodynamic code which used the simplest flux limited conductivity. The absorption fraction was controlled in the hydrocode to match that of the Fokker-Planck results. The heat front was best modeled with f = 0.08 (see Fig.18), in which case the coronal temperatures were too high. For f = 0.15, the coronal temperatures agreed with the detailed solution but not the heat front excursion.

This last result shows that solving the transport equation coupled with the hydrodynamic motion is important in understanding the thermal transport behaviour. Under laser plasma interaction conditions, coupled hydro-transport effects change the conditions in the heat front,

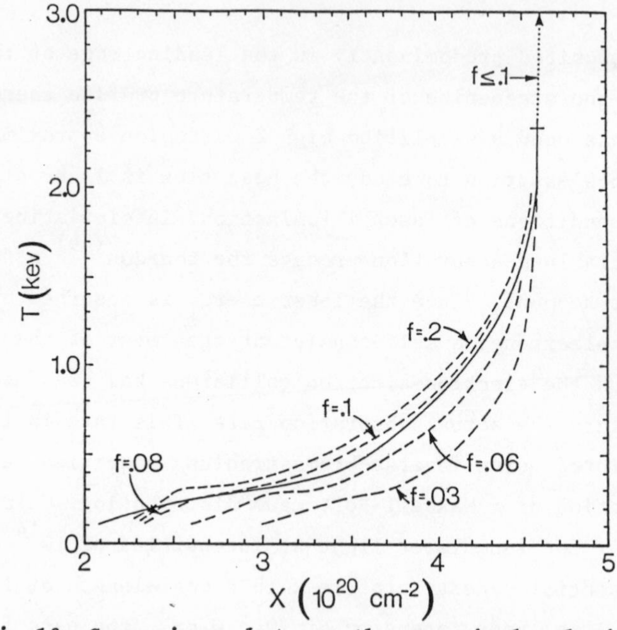

Fig.18: Comparisons between the numerical solution of the Fokker-Planck equation (——) and results of the flux-limited model (----) for a moving ion problem at 1.05μm laser illumination. Temperature versus Langrangian coordinate $X = \int_0^x n_e(x')\,dx'$, *at 600 ps*[50].

leading to a reduced effective flux limiter, compared with that inferred from steady state cathode like problems.

Recently Bell[51] has modeled the electron energy transport in a steadily ablating laser produced plasma. He found that at high laser intensities when the plasma flow diverges as it flows away from the solid target, the temperature and density profiles are steepened and a further reduction of the heat flux is observed, compared to that obtained when no motion effects are taken into account.

Hydrodynamic effects on non-local electron transport have also been considered using the delocalization formula discussed in 3.3[42], and hybrid type models[52,53] which, in accordance with the models in 3.3, solve the transport equation by following only the electrons relevant to the heat transport process (velocities greater than $\sim 2 \cdot v_{th}$).

An effective flux limiter in the range of 0.06–0.1 has been

inferred from these solutions which take into account the hydrodynamic effects[50-53].

3.5 Magnetic Field Effects

In two-dimensional situations, the existence of large magnetic fields ($\sim$1MG), perpendicular to the plane containing the temperature and density gradients, has been known for some time[25] and various authors have modeled the reduction of thermal conductivity caused by such fields[55].

Like in the one dimensional case, electron transport in magnetic fields can be treated well by fluid equations in the case of small gradients and must be treated by full numerical simulation or by simplistic heuristic models in the case of large gradients or collisionless plasmas.

The extension of the Spitzer-Härm solution for small gradients to include the effects of magnetic fields is described by Braginskii[56]. The electrons in a magnetic field have curved trajectories, and therefore tend to stay localized, even in a collisionless plasma. This effect will add a new typical length to the system, the Larmor radius of a typical electron, which will tend to reduce the effective mean free path of the electron in the direction perpendicular to the field.

When applying Braginskii's model outside its domain of validity (electron mean free path $\simeq$ 0.01 of the temperature scale length) it was hoped that the localization of electron orbits provided by the magnetic fields will guarantee the validity of the model. However, there are still situations where the assumption of such a local model is questionable.

Recently[57], the classical treatment of thermal heat transport in the presence of magnetic fields has been modified to include effects associated with steep temperature gradients, using the Shvarts model[38] in three dimensions. The effects of a magnetic field are described in terms of the parameter β_0, which is the ratio of the collision mean free path, λ_0, to the Larmor radius ($\beta_0 = \omega_c \cdot \tau_{ei} = \lambda_0/R_L$). Fig.19 shows f in the temperature gradient direction, x, and in the perpendicular direction, y, for different values of the

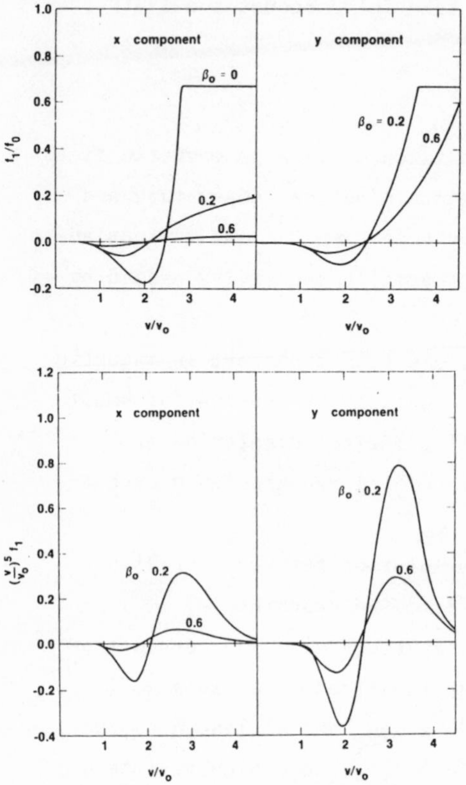

Fig.19: Distribution functions for $\lambda_0/L = 0.1$. Upper plots f_1/f_0, lower plots normalized heat flow $(v/v_0)^5 f_1$, for x and y directions. Note that $v_0 = (kT/m)^{\frac{1}{2}}$(57).

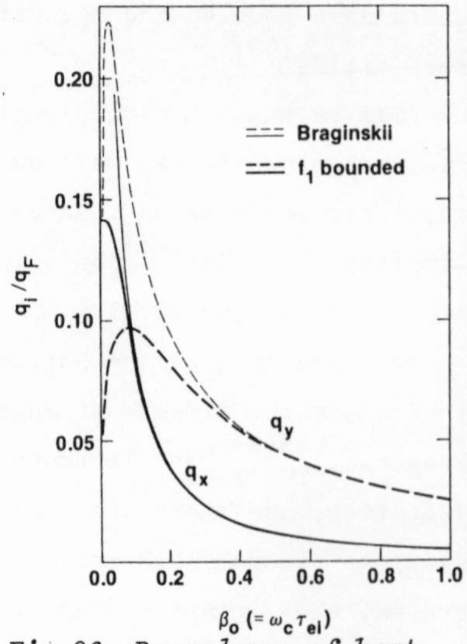

Fig.20: Dependence of heat flux on $\beta_0 = \lambda_0/R_L$, for $\lambda_0/L = 0.1$, obtained when f_1 is bounded by the local maxwellian(57).

magnetic field, which is in the z direction. Here λ_0/L was taken as 0.1 and a zero net current is assumed. Also shown is the heat flow velocity distribution. The figure shows that even modest magnetic fields eliminate the need to flux limit f_1 in the x direction, because of their localizing effect. Fig.20 shows the resulting heat flux, and compares it with the results of Braginskii's model. For $\beta_0 \gtrsim 0.2$ a strong reduction of both components of the heat flux accurs due to the inhibition of the more energetic heat carrying electrons. The effective flux limiters corresponding to these magnetic fields are of the order of 0.05 or less. The transverse heat flux exceeds the longitudinal component, and has a peak at very low β_0. This is expected since the former should reach its maximum when β of the heat carrying electrons, $\beta_0 \cdot (v/v_0)^3$, is of the order of π. This is based on the simple physical picture in which these electrons travel half a Larmor orbit between collisions. Indeed, for $\beta_0 = 0.1$ and using $v/v_0 = 3.2$ (the maximum in Fig.19, corresponding to the heat carrying electrons) one finds $\beta \simeq 3$.

To our knowledge, no direct correlation between inferred values of the flux limiter and measured or calculated magnetic fields, has been reported. Thus, the flux inhibition caused by magnetic fields can only be speculated upon.

Multi-dimensional full Fokker-Planck simulations which include collisions, have not yet been reported. Nevertheless, two dimensional, time dependent collisionless transport, with self consistent fields has been simulated[26]. The extremely fast electrons generated by long wavelength lasers are shown to generate megagauss magnetic fields even at relatively low intensities.

The simulations show a substantial heat flux inhibition due to magnetic fields, while the energy is being passed to fast ions or convected laterally to the edges of the target, an effect which has been seen both numerically[26] and in experiments[27]. The electron lateral displacement extends in time together with the magnetized region forming a "magnetized plasma wave". Such calculations should be extended to the more collisional regions which are of relevance to laser fusion plasmas.

ACKNOWLEDGEMENTS

I wish to acknowledge the contribution of N. Fried in preparing the manuscript. Without his devoted assistance in discussing, writing and preparing this manuscript the work would not have been finished. I would like to thank my colleagues at LLE, Rochester, especially J. Delettrez and S. Skupsky, for keeping me informed of their recent work. Thanks are due to Mrs. M. Speyer for carefully typing the manuscript.

REFERENCES

1. C.E. Max, C.F. McKee and W.C. Mead, Phys. Fluids, 23, 1620 (1980).
2. L. Spitzer and R. Härm, Phys. Rev. 89, 977 (1953); L. Spitzer, Jr., Physics of Fully Ionized Gases, Wiley Interscience, New York (1962)
3. W.L. Kruer, Comments on Plasma Phys. 5, 69 (1979).
4. R.J. Bickerton, Nucl. Fusion 13, 457 (1973); W.M. Manheimer, Phys. Fluids 20, 265 (1977); D.R. Gray, J.D. Kilkenny, M.S. White, P. Blyth and D. Hull, Phys. Rev. Lett. 39, 1270 (1977).
5. I.P. Shkarofsky, Phys. Rev. Lett. 42, 1342 (1979); Phys. Fluids 26, 3131 (1983).
6. W.C. Mead, Lawrence Livermore Report, UCRL-85246 (1980).
7. J. Delettrez, in "Workshop on Physics of Laser Fusion", Vancouver, Canada (1985).
8. B. Yaakobi, T. Boehly, P. Bourke, Y. Conturie, R.S. Craxton, J. Delettrez, J.M. Forsyth, R.D. Frankel, L.M. Goldman, R.L. McCrory, M.C. Richardson, W. Seka, D. Shvarts and J.M. Soures, Opt. Comm. 39, 175 (1981).
9. B. Yaakobi, J. Delettrez, L.M. Goldman, R.L. McCrory, R. Marjoribanks, M.C. Richardson, D. Shvarts, S. Skupsky, J.M. Soures, C. Verdon, D.M. Villeneuve, T. Boehly, R. Hutchison and S. Letzring, Phys. Fluid 27, 516 (1984).
10. R. Benattar, C. Popovice, R. Sigel and J. Virmont, Phys. Rev. Lett. 12, 766 (1979).
11. W.B. Fechner, C.L. Shepard, G.E. Busch, R.J. Schroeder and J.A. Tarvin, Phys. Fluids 27, 1552 (1984).
12. F. Amiranoff, R. Fabbro, E. Fabre, C. Garban, J. Virmont and M. Weinfeld, Phys. Rev. Lett. 43, 522 (1979).

13. R.C. Malone, R.L. McCrory and R.L. Morse, Phys. Rev. Lett. 34, 721 (1975).

14. F.C. Young, R.R. Whitlock, R. Decoste, B.H. Ripin, D.J. Nagel, J.A. Stamper, J.M. McMahon and S.E. Bodner, Appl. Phys. Lett. 30, 45 (1977).

15. P.M. Campbell, R.R. Johnson, F.J. Mayer, L.V. Powers and D.C. Slater, Phys. Rev. Lett. 39, 274 (1977).

16. M.D. Rosen, D.W. Phillion, V.C. Rupert, W.C. Mead, W.L. Kruer, J.J. Thomson, H.N. Kornblum, V.W. Slivinsky, G.J. Caporaso, M.J. Boyle and K.G. Tirsell, Phys. Fluids, 22, 2020 (1979).

17. R.A. Hass, W.C. Mead, W.L. Kruer, D.W. Phillion, H.N. Kornblum, J.D. Lindl, D. MacQuigg, V.C. Rupert and K.G. Tirsell, Phys. Fluids 20, 322 (1977).

18. R.S. Craxton and R.L. McCrory, Laboratory for Laser Energetics Report, LLE-108 (1980).

19. C. Garban-Labaune, E. Fabre, C.E. Max, R. Fabbro, F. Amiranoff, J. Virmont, M. Weinfeld and M. Richard, Phys. Rev. Lett. 48, 1018 (1982).

20. W. Seka, R.S. Craxton, J. Delettrez, L. M. Goldman, R. Keck, R.L. McCrory, D. Shvarts, J.M. Soures and R. Boni, Opt. Commun. 40, 437 (1982).

21. E.B. Goldman, J. Delettrez and E.I. Thorsos, Nucl. Fusion 19, 555 (1979).

22. B. Yaakobi and T.C. Bristow, Phys. Rev. Lett. 38, 350 (1977).

23. W.C. Mead, E.M. Campbell, K.G. Estabrook, R.E. Turner, W.L. Kruer, P.H.Y. Lee, B. Pruett, V.C. Rupert, K.G. Tirsell, G.L. Stradling, F. Ze, C.A. Max and M.D. Rosen, Phys. Rev. Lett. 47, 1289 (1981).

24. M.H. Key, in "Laser Plasma Interactions 2", edited by R.A. Cairns, SUSSP publications, Dept. Physics, Univ. of Edinburgh (1983).

25. J.A. Stamper, K. Papadopoulos, R.N. Sudan, S.O. Dean, E.A. McLean and J.M. Dawson, Phys. Rev. Lett. 26, 1012 (1971).

26. D.W. Forslund and J.U. Brackbill, Phys. Rev. Lett. 48, 1614 (1982).

27. M.A. Yates, D.B. van Hulsteyn, H. Rutkowski, G. Kyrala and J.U. Brackbill, Phys. Rev. Lett. 49, 1702 (1982).

28. W.C. Mead, E.M. Campbell, W.L. Kruer, R.E. Turner, C.W. Hatcher, D.S. Bailey, P.H.Y. Lee, J. Foster, K.G. Tirsell, B. Pruett, N.C. Holmes, J.T. Trainor, G.L. Stadling, B.F. Lasinski, C.E. Max and F. Ze, Phys. Fluid 27, 1301 (1984).

29. T.J. Goldsack, J.D. Kilkenny, B.J. MacGowan, P.F. Cunningham, C.L.S. Lewis, M.H. Key and P.T. Rumsby, Phys. Fluid 25, 1634 (1982).

30. M. Nakai, H. Azechi, F. Mizutani, M. Tanaka, S. Uchida, N. Miyanaga T. Yabe, K. Nishihara, K. Mina, T. Yamanaka and C. Yamanaka, ILE Research Report 8513P (1985).

31. J.A. Tarvin, W.B. Fechner, J.T. Larsen, P.D. Rockett and D.C. Slate Phys. Rev. Lett. 51, 1355 (1983).

32. M.D. Rosen, Comments on Plasma Phys. and Cont. Fusion 8, 165 (1984).

33. A. Hauer, W.C. Mead, O. Willi, J.D. Kilkenny, D.K. Bradley, S.D. Tabatabaei and C. Hooker, Phys. Rev. Lett. 53, 2563 (1984).

34. B. Yaakobi, O. Barnouin, J. Delettrez, L.M. Goldman, R. Marjoribanks, R.L. McCrory, M.C. Richardson and J.M. Soures, J. Appl. Phys. 57, 4354 (1985).

35. S. Skupsky, (Private Communication).

36. I.P. Shkarofsky, T.W. Jonston and M.P. Bachynski, "The Particle Kinetics of Plasmas", Addison-Wesley, Reading, MA. (1966).

37. D.R. Gray and J.D. Kilkenny, Plasma Phys. 22, 81 (1980).

38. D. Shvarts, J. Delettrez, R.L. McCrory, C.P. Verdon, Phys. Rev. Lett. 47, 247 (1981).

39. Y. Kishimoto and K. Mima, J. Phys. Japan 52, 3389 (1983).

40. M. Strauss, G. Hazak and D. Shvarts, J. Phys. D. 17, 327 (1984).

41. S. Skupsky, in "Transport Workshop", Laboratory for Laser Energetics, Rochester (1983).

42. J.F. Luciani, P. Mora and J. Virmont, Phys. Rev. Lett. 51, 1664 (1983); J.F. Luciani, P. Mora and R. Pellat, Phys. Fluid 28, 835 (1985).

43. N. Fried, D. Ofer, D. Shvarts. (Private Communication).

44. E. Lindman and K. Swartz, in the report of "CECAM Workshop on Transport, Interactions and Instabilities in Laser Plasma", Orsay, France (1983).

45. E.A. Williams, J.R. Albritton, K. Swartz and I.B. Bernstein, presented at the 15th Annual Anomalous Absorption Conference, Banff, Canada (1985).

46. A.R. Bell, R.G. Evans and D.J. Nicholas, Phys. Rev. Lett. 46, 243 (1981).

47. J.P. Matte and J. Virmont, Phys. Rev. Lett. 49, 1936 (1982).

48. J.R. Albritton, Phys. Rev. Lett. 50, 2078 (1983).

49. A.B. Langdon, Phys. Rev. Lett. 44, 575 (1980).

50. J.P. Matte, T.W. Johnson, J. Delettrez and R.L. McCrory, Phys. Rev. Lett. 53, 1461 (1984).

51. A.R. Bell, Phys. Fluids 28, 2007 (1985).

52. R.T. Wright, J. Phys. D. 14, 805 (1981).

53. D. Shvarts, J. Delettrez and R.L. McCrory, in the report of "SECAM Workshop on Transport, Interactions and Instabilities in Laser Plasmas", Orsay, France (1981).

54. R.J. Mason, Phys. Rev. Lett. 47, 652 (1981).

55. J.B. Chase, L.M. LeBlanc and J.R. Wilson, Phys. Fluids 16, 1142 (1973); D.G. Colombant, K.G. Whitney, D.A. Tidman, N.K. Winsor and J. Davis, Phys. Fluids 18, 1687 (1975); R.S. Craxton and M.G. Haines, Phys. Rev. Lett. 35, 1336 (1975).

56. S.I. Braginskii, in "Reviews of Plasma Physics", ed. M.A. Leontovich, Consultants Bureau, New York, vol. 1 p. 205 (1965).

57. M. Strauss, G. Hazak, D. Shvarts and R.S. Craxton, Phys. Rev. A. 30, 2627 (1984).

COMPRESSION AND HYDRODYNAMICS

M.H. Key

Rutherford Appleton Laboratory Chilton Oxfordshire England

1. INTRODUCTION

These notes give a brief update of material presented in more detail in the proceedings of the Summer Schools in 1979 and 1982 (referred to here by their dates). Substantial progress has been made in the three years since the last school and my aim here is to outline the main development and to refer to some more detailed information.

2. ABLATION AT LASER IRRADIATED SOLID SURFACES

Compression by laser irradiation begins with ablation pressure generated at a laser irradiated surface, which depends on energy transport by electrons and photons, (see 1979 and 1982).

3. THERMAL ELECTRONS

The major part of the energy transport is by thermal electrons.

Mach number and heat flow: A simple (ideal gas) model is helpful. Laser intensity I is absorbed at a plane across which outward flow of plasma advects energy with an intensity W. The energy flow inwards is by heat conduction of intensity S. In equilibrium it follows that I = S = W, by equating inward and outward energy flow on each side of the plane. Hence the heat flow can be expressed as

$$S = 3\,\rho\, c^3\, (0.8M + 0.2M^3) \qquad (1)$$

where ρ is the plasma density, c the velocity of sound and M the Mach number of the flow, the first term in the bracket being enthalpy and the second kinetic energy. Hence a Mach 1 flow in the region of heat conduction requires $S/S_M = 0.05$, where S_M is the free streaming limit (Eq (6) 1982).

It can be shown that, if the maximum heat flow is unrestricted, the ablation process leads to plasma flow velocity in the conduction region with $M > 1$, using eg. analytic models for spherical[(1)] and planar[(2)] geometry. The flow geometry is determined by the separation D of the absorption and ablation fronts, which is a function of irradiance and wavelength[(1,2)].

The planar approximation is appropriate when $D \ll \phi$ or r, where r is the target radius or ϕ the focal spot diameter. For unrestricted heat flow, M = 1 and $S/S_M = 0.05$[(2)]. (Restriction of the heat flow to lower values leads to a flow at Mach 1 on the laser side of the absorption plane but with a change to $M < 1$ and increased density on the target side[(2)].

In spherical geometry ($D > r$ or ϕ), for unrestricted heat flow, the spherical divergence between the ablation front and absorption region accelerates the flow to $M > 1$, requiring $S/S_M > 0.05$. Restriction of the heat flow to lower values limits the Mach number in the conduction region to that given by Eq (1) and leads to a density change across the absorption plane, as in the planar case.

A practically important question is how any limit on heat flow affects the ablation pressure for a given irradiance, and this can be shown to be a rather weak effect for parameters of practical interest,[1,2] (1982),

The Spitzer Harm theory of heat conduction in a plasma is invalid for $S/S_M \geqslant 0.02$, ie $M \geqslant 0.4$ (1982) and thus always invalid for the ablation process. Solution of the Fokker Planck equation in steep temperature gradients (1982 and D Schwartz ibid) shows that heat flow saturates at $S/S_M < 0.05$ to 0.1 and that the heat flow at any point is a 'non local' quantity. Electrons of energy around 3kT are the main heat carrying group and have mean free paths $(\alpha\ (\text{energy})^2)$ an order of magnitude greater than electrons of energy kT. [A recent very useful mathematical model[3] describes heat flow in terms of a convolution where each point in the temperature gradient is a source of heat flow over a finite region of about $30\ell_e$, where ℓ_e is the range of an electron of energy kT]. Another recent analysis combines for the first time a Fokker Planck treatment of heat conduction with a numerical hydrodynamic model of ablation in spherical geometry[4]. It gives calculated results (Fig 1) which illustrate the previuos discussion and can be compared with experiment. The spatial plateau in the Mach number at $M \sim 1.2$ in Fig 1 correlates with the maximum calculated heat flow $S/S_M \sim 0.06$ through Eq (1). The difference between Spitzer Harm and FP treatment of heat flow is however seen to be relatively small and mainly in the temperature profile, which has a local maximum and steeper gradient near the point of absorption.

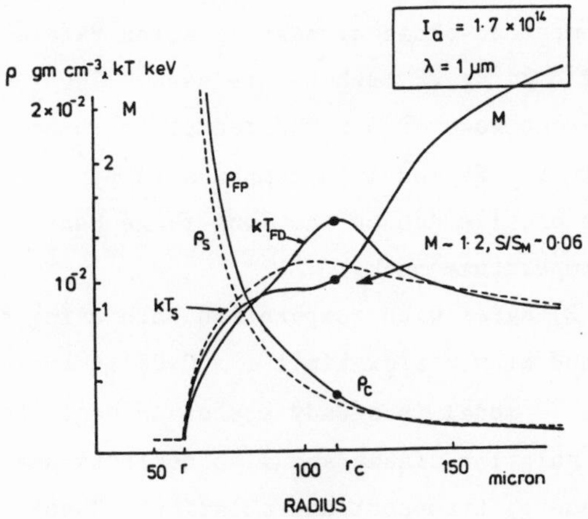

Fig 1 Computed steady state spatial variation of density ρ, temperature kT and Mach number M for a spherical target at an absorbed irradiance of $1.7 \times 10^{14}\ W\ cm^{-2}$ and wavelength 1μm. Solid lines using a Fokker Planck heat flow model. Dotted lines using Spitzer Harm heat flow (with permission from Ref 4).

Comparison with experiment requires accurate experimental data free from the complication of edge effects (discussed in the 1982 notes). Spherical geometry therefore gives potentially superior data for accurate comparsion with theory and recent work has given new results of importance both for this reason and for more sophisticated diagnostic measurements [5, 6, 7, 8]. From the most recent work [8] Fig 2 shows experimentally measured mass ablation rates $\dot{m}$, from time resolved X-ray spectroscopy of layer burn through, compared with 'LASNEX' hydrocode calculations using the 'flux limiter' f_e (1982 notes). The calculations illustrated in Fig 1 have been used here for an additional comparison with experiment shown in Fig 2.

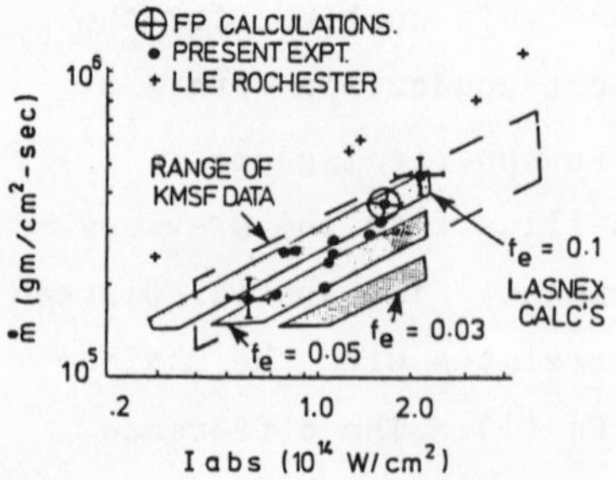

Fig 2 Experimentally measured mass ablation rate as a function of absorbed irradiance together with computed results using the LASNEX code for various values of the flux limiter f_e (with permission from Ref 8. A computed point using the full FP model illustrated in Fig 1 is also shown.

It can be concluded from the theorerical modelling [4] that there is negligible difference in the computed values of mass ablation rate $\dot{m}$ and pressure P whether Spitzer Harm or FP methods are used. The temperature profiles are different however but the reduction in heat flow at the temperature peak in the FP model is compensated by enhancement at the foot of the profile due to the long range heat carrying electrons from the temperature peak.

The experimental data (Fig 2) agree with computation both using the FP model (where $S/S_M < 0.06$) and with a flux limit $f_e \sim 0.08$ in the LASNEX code modelling [8]. The FP model is steady state and neglects the transient build up of the ablation plasma and also neglects any contribution of radiation to energy transport and ablation. These factors possibly explain why there is a difference in computed $\dot{m}$ and P between simple Spitzer Harm and flux limited modelling in LASNEX but not in the FP model. Care should be exercised in detailed comparison with experiment because of discrepancies between some experiments, (6, 7, 8).

4. RADIATION

There is considerable interest in the role of radiation in ablation because of its use in indirect compression (introduction ibid). High Z targets convert up to 50% of the laser intensity into soft X-rays with a spectrum similar to that of a black body as illustrated in Fig 3[9]. (A black body radiates intensity $I_B \sim \sigma T_B^4$ with a spectral peak at $h\nu \sim 3kT_B'$ and for $kT_B' \sim 100$eV, $I_B \sim 10^{13}$ W cm^{-2}]. The soft X-rays have a rather small penetration depth, illustrated in Fig 3, and transport energy diffusively. They ablate material at low temperature $T' < T_B'$ and therefore create a rather large mass ablation rate and pressure relative to laser irradiation. This can be estimated from simple arguments based on the three hydrodynamic conservation equations, the equation of state and one further relationship defining the coupling, as in the analysis of laser ablation[11] (1979). Assuming ablation occurs with a temperature at the sonic point $kT_s > kT_B$ gives the relationship,

$$\dot{m} > \left(\frac{(1+\gamma)}{c}\right)^2 \left(\frac{m_p}{k}\right)^{\frac{1}{4}} \sigma I_B^{3/4} \qquad (2)$$

or more simply

$$\dot{m}/(1\text{grm cm}^{-2}\text{ sec}^{-1}) > 2.1 \times 10^5 \, (I_B/10^{13}\text{W cm}^{-2})^{3/4} \qquad (3)$$

which may be compared with the significantly lower values for laser ablation, eg[8]

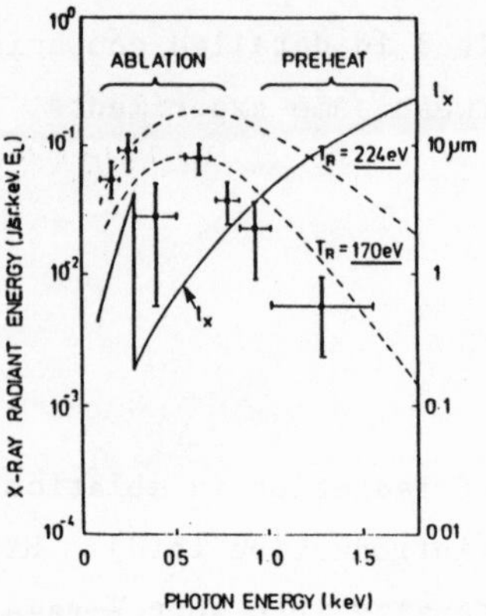

Fig 3 Emitted spectrum for a gold target together with Black body spectra for 2 temperatures (with permission from Ref 9). The range ℓ_x of the X-ray photons in cold CH polymer is also shown.

A more accurate model[10] assumes self regulation using the 'Rosseland' mean free path of Black body emission photons to define the absorption depth in a manner analogous to self regulation by inverse Bremsstrahlung absorption in laser ablation[11].

Accurate computation of radiation transport in a plasma is a difficult problem which is approximated in sophisticated codes like LASNEX [see R G Evans' lecture notes ibid] and is being progressively more accurately calculated in specialised atomic physics and radiative transfer codes[12]. Unfortunately no single calculation has combined FP (or convolution approximation) treatment of heat flow with detailed hydrodynamic, atomic physics and radiation transfer modelling. Without such calculations detailed comparison with experiments (except for very low Z targets) will be subject to uncertainties because radiation modifies energy transport especially in the low temperature precursor region of the heat front. Experimental information on energy transport by radiation is sparse, probably because much work has not been published due to classification rules in the USA particularly, though some results have been reported[13].

5.
WAVELENGTH SCALING

Aside from the question of ablation by soft X-rays (which is in a sense an extreme case of using short wavelength), the scaling of ablation parameters with laser wavelength remains topical, Fig 4 adds more recent data to a compilation of mass ablation rates at the laser fusion relevant absorbed irradiance of 10^{14} W cm^{-2}. It emphasises the discrepancies between experiments but suggests λ^{-1} scaling of $\dot{m}$ and thus $\lambda^{-\frac{1}{2}}$ scaling of ablation pressure[11]. The advantages of short wavelength are more pronounced than this implies, since higher intensity can be used at shorter wavelengh while maintaining constant hot electron temperature at constant $I\lambda^2$. (1982 notes).

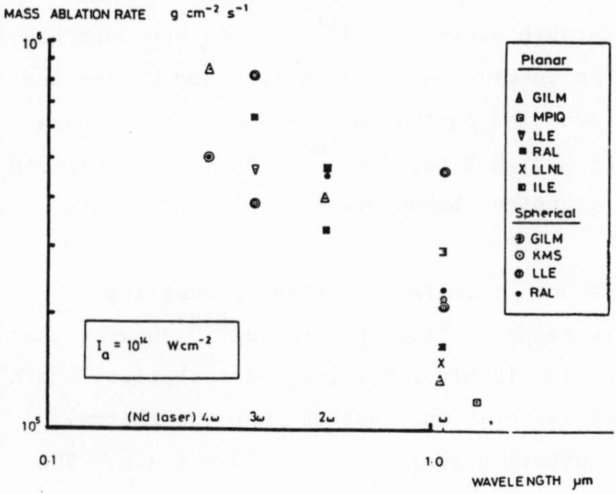

Fig 4 Mass ablation rates measured in both planar and spherical geometry at an absorbed irradiance of 10^{14}W cm^5 for various wavelengths. The data were obtained at laboratories indicated in the key.

6.

PREHEATING AND 'EXPLODING PUSHER' COMPRESSION

Preheating by 'hot' electrons, radiation and shocks is discussed in the 1979 and 1982 notes. Implosions in which preheating from a short pulse at λ = 1μm causes explosion of a thin spherical shell, with negligible acceleration of the centre of mass, can produce the 10KeV ion temperature needed for thermonuclear fusion and hence rather large yields of thermonuclear neutrons. Density and ρr are low however, so efficient fusion burn is impossible. Recent work giving 'record' 10^{12} yields of neutrons using longer pulses and short wavelength (0.53 and 0.35μm) irradiation of thin wall large aspect ratio targets is of interest here[14, 15]. The centre of mass of the imploding shell is accelerated to high velocity by ablation pressure, but the imploding shell decompresses to low density due to preheating, so that the density in the compressed core is typically less than 1 gm cm^{-3} but the high implosion velocity raises the temperature to circa 10KeV.

7.
THERMAL SMOOTHING

Experimental study of thermal smoothing (1982 notes) has continued with new results being obtained using pulsed radiography[16] and streak camera shock breakthrough observations[17]. The new work has added results for $\lambda = 0.35\mu m$ to previous data at 1.06 and 0.53μm and has also provided some information on the smoothing effect of energy transport by soft X-rays from high Z targets[16]. Data on wavelength scaling to short λ and on transient behaviour is still incomplete however.

Theoretical work has advanced with the addition of numerical modelling of thermoelectric magnetic field generation[18] showing that its effect on thermal smoothing is minor for long perturbation length and (contrary to elementary expectation) that it improves thermal smoothing slightly for perturbation lengths on the 10μm scale. The increased thermal smoothing arises from a negative feedback process in which perturbations are attenuated by the inverse of the magnetic field generating thermal instability (discussed below) which occurs for parallel gradients, $\nabla n // \nabla T$, at densitites below critical, but operates in reverse in the thermal conduction region where ∇n and ∇T are anti parallel.

8.
THERMAL INSTABILITY

Experimental observations using optical probing techniques to study the fine structure of plasma ablating from laser irradiated targets have revealed breakup of the plasma flow into fine scale (~ 20μm) jets surrounded by magnetic fields[19] Fig 5. The jets follow the plasma flow direction and are more pronounced on high Z targets.

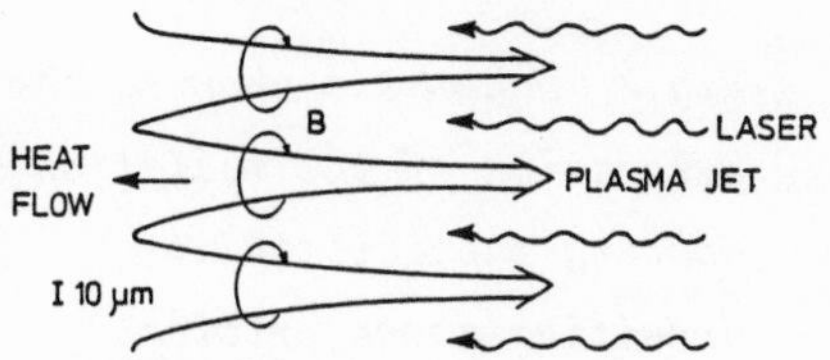

Fig 5 Schematic of plasma jets created by heat flow instability in laser irradiated targets.

The theoretical explanation of these observations is still not fully resolved. The magnetic field generating thermal instability[20] occurring where ∇n is parallel to ∇T at $n < n_c$ has been shown to have a growth rate $\gamma \sim 10^{12}$ sec^{-1} for a wavelength of 15μm where $kT \sim 1KeV$ and $Z \sim 10$. The mechanism is illustrated in Fig 6, which shows how a perturbation in temperature induces a thermoelectric magnetic field which deflects heat flow (the Righi Leduc effect) to enhance the temperature perturbation).

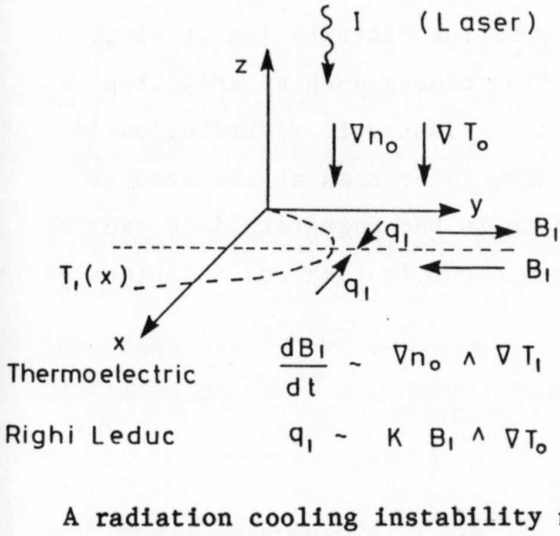

Fig 6 Schematic showing the mechanism of the magnetic field generating thermal instability at densities below critical density. Subscript 1 denotes the perturbed variable and subscript 0 the principal variables.

A radiation cooling instability may occur in high Z plasmas of high density ($n > n_c$), where radiation is a major cause of cooling[21], Fig 6. The radiated power scales as $n^2 T^{\frac{1}{2}}$. Thus, if pressure equilibrium is maintained, a reduction T_1 in temperature causes an increase N_1 in density such that $T_1 / T_0 = -N_1 / N_0$.

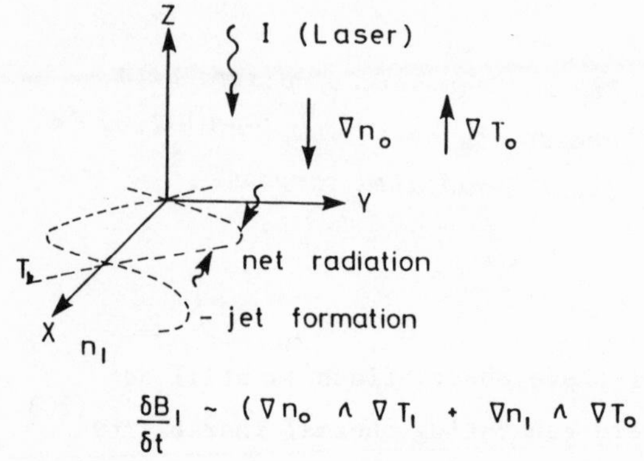

Fig 7 Schematic showing the generation of the radiation cooling instability at densities above critical. Subscript 1 denotes the perturbed variables and subscript 0 the principal variables.

The change in radiation is greatest where the density increases, so that a cool region cools more rapidly, leading to unstable collapse of the temperature and increase of the density. Magnetic fields then grow through the thermoelectric effect. The growth rate can be as large as 10^{10} sec^{-1} for a 50μm perturbation at $kT \sim 0.5keV$, $Z \sim 50$, $\eta_e \sim 10^{23}$.

The Weibel instabilty can occur when the distribution of electron velocities is not isotropic. Heat flow causes such an anisotropy and, if any magnetic field is present, this anisotropic distribution is rotated Fig 8, so that net current flow is created at the zero of magnetic field. The current flow adds to the magnetic field causing unstable growth. The instability can occur in both collisionless and collisional plasmas[(22)].

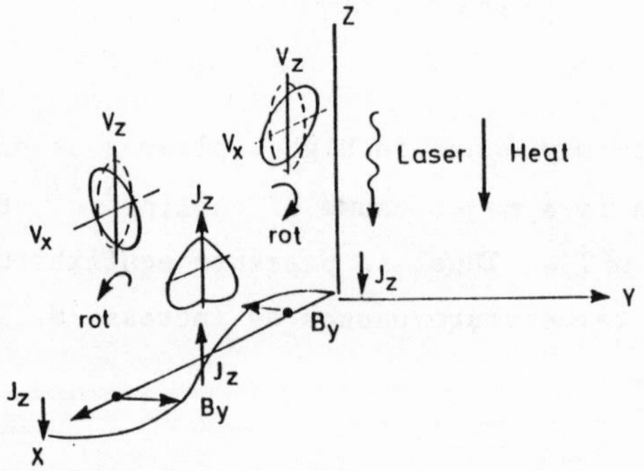

Fig 8 Schematic showing the mechanism of the Weibel instability.

It is not yet established which if any of these effects cause the observed jets, nor whether they occur at $\eta_e > $ or $< \eta_c$, whether they lead to local variations of ablation pressure or how and at what level saturation of the instability occurs.

9. RAYLEIGH-TAYLOR INSTABILITY

The limit on aspect ratio composed by the RT instabilty is discussed in the 1979 and 1982 notes, which include reference to initial experimental studies. These have since been several experimental results which have been obtained using pulsed radiography methods to study the growth of perturbations. These include observations of two dimensional patterns in transmission radiography of an accelerated plane foil(23) and side on observations of bubble and spike growth in both plane(24) and spherical(25) targets.

There is also been further theoretical work(26) though this has not significantly changed the main features of understanding arrived at by 1982. The new experimental data are broadly in agreement with the theory. There is a significant lack of experimental data on the development of the instability at the pusher/fuel interface in the deceleration phase of implosions.

10. ABLATIVE COMPRESSION

The continuing study of ablatively driven implosions (see a recent review(27)) has lead to higher density in direct drive compression. 10 gm^{-3} has been observed by X-ray radiography of polymer shells driven by 0.53μm radiation(28) and the dynamics of the implosion have been studied by streak radiography(29). Better symmetry has been obtained with 12 beam irradiation(30). Significant progress has been made towards experiments at higher energy and shorter wavelengths notably with the Omega 24 beam facility operating at 0.35μm 3kJ, the

Gekko XII laser at 0.53μm, 10kJ and Nova at 20kJ 0.35μm.

Indirect drive experiments carried out at 0.53 and 0.35μm with the LLNL 2-beam prototype of NOVA (see Fig 2 Introduction ibid) have yielded higher densities, measured by radiochemical analysis of neutron activation of the imploding shell, with up to 40 gm cm^{-3} being reported[31]. Short wavelength is very important since the long scale length of plasmas in a cavity induces strong Raman and 2 plasmon decay instability at $\lambda > 1\mu m$, causing severe preheat and reduced compression[31].

Experiments have yielded ρr information from the energy loss of escaping fusion particles (for ρr up to 10^{-2} gm cm^{-2}[32]) and from neutron activation of the pusher for ρr up to 10^{-1} gm cm^{-2}[33].

The large laser installations Nova and Gekko XII are on a scale sufficient to begin to study targets of ρr values approaching the ignition level (0.3 gm cm^{-2}).

The prospects for ICF seem reasonable in that no insuperable new difficulties have emerged and the move to short wavelength has made matters considerably more favourable.

11.
ACKNOWLEDGMENTS

I am grateful to my colleagues in UK Universities and at the Rutherford Appleton Laboratory, involved in the study of laser produced plasmas at the SERC Central Laser Facility, for advice and assistance in preparing this paper. Particular thanks are due to A R Bell and R G Evans.

REFERENCES

1. C. Max, C. McKee and W. Mead, Phys. Fluids, 23, 1620 (1980).
2. R. Fabbro, C. Max, E. Fabre, Phys. Fluids, 28, 1463 (1985).
3. J. Luciani, P. Mora, J Virmont, Phys. Rev. Lett., 51, 1664 (1983).
4. A. Bell, Phys. Fluids (in press).
5. T. Goldsack, J. Kilkenny, B. McGowan, P. Cunningham, C. Lewis, M. Key, P. Rumsby, Phys. Fluids, 25, 1634 (1982).
6. J. Tarvin, W Fechner, J Lasen, P Rockett, D Slater, Phys. Rev. Lett., 51, 1355 (1983).
7. B. Yaakobi, J. Delettrez, L. Goldman, R. McRory, R. Majoribanks, M. Richardson, D. Schwartz, S. Skupsky, J. Soures, C. Verdon, D. Villenueve, T. Boehly, R. Hutchinson, S. Letzring, Phys. Fluids, 27, 516 (1984).
8. A. Hauer, W. Mead, O. Willi, J. Kilkenny, D. Bradley, S. Tabatabaei, C. Hooker, Phys. Rev. Lett., 53, 2563 (1984).
9. T. Mochizuki, T. Yabe, Osaka University Report No. 10, 15 (1984).
10. T. Yabe, S. Kiyokawa, T. Mochizuki, S. Sakabe, C. Yamanaka, Japan J. Appl. Phys., 22, L88 (1983).
11. M. Key, W. Toner, T. Goldsack, J. Kilkenny, S. Veats, P. Cunningham, C. Lewis, Physics Fluids, 26, 2011 (1983).
12. D. Duston, R Clark, J Davis, Phys. Rev. A, 31, 3220 (1985).
13. J. Mizui, M. Yamaguchi, S. Taguki, K Hishihara, Phys. Rev. Lett., 47, 1000 (1981).
A. Ng, D. Partennik, L. Dasilva, D. Pusini, Phys. Fluids, 28, 2915 (1985).
14. M. Richardson (to be published in Laser Information and related Plasma Phenomena, H. Hora, G. Miley Eds. Plenum Press
15. C. Yamanaka, S. Nakai (submitted to Nature).
16. J. Buchet, M. Decroisette, P. Holstein, M. Louis Jaquet, B. Meyer, A. Saleres, Phys. Rev. Lett., 52, 823 (1984).
17. A. Rankin, J. Kilkenny, A. Cole, M. Key, Rutherford Appleton Laboratory Report RL 83-043 5.7 (1983).
18. A. Bell, Rutherford Appleton Laboratory Report RAL 85-047, A5.22, (1985).

19. O. Willi, P. Rumsby, S. Sartang, IEEEJ. Quant. Electr. QE17, 1909 (1981).

20. P. Tidman, R. Shanney, Phys. Fluids, 17, 1207 (1974).

21. R. Evans, J. Phys. B, 14, L173 (1981).

22. M. Ogasawara, A. Hirao, J. Phys. Soc. Japan, 50, 990 (1981).
E. Epperlain, Plasma Phys. & Contr. Fusion, 27, 1027 (1985).

23. J. Gran, M. Emery, S. Kacenjar, C. Opal, E. McLean, S. Obenschain, B. Ripin, A. Schmitt, Phys. Rev. Lett., 53, 1352 (1984).

24. R. Whitlock, M. Emergy, J. Stamper, E. McLean, S. Obenschain, M. Peckerar, Phys. Rev. Lett., 52, 819 (1983).

25. J. Wark, J. Kilkenny, A. Cole, M. Key, P. Rumsby, Appl. Phys. Lett. (in press).

26. W. Manheimer, D. Colombant, Phys. Fluids, 27, 985 (1984) (and references cited).
M. Emery, J. Gardner, J. Boris, Appl. Phys. Lett., 41, no. 9, 808 (1982).

27. G. McCall, Plasma Phys., 25, 237 (1983).

28. M. Key, Plasma Phys. and Contr. Fusion, 26, 1383 (1984).

29. M. Key, P. Rumsby, P. Cunningham, C. Lewis, M. Lamb, Opt. Comm., 44, 343 (1985).

30. C. Lewis et al, Rutherford Appleton Laboratory Report RAL 85-047, A2.1 (1985).

31. J. Holtzrichter, E. Campbell, J. Lindl, E. Storm, Science, 220, 1045 (1985).

32. D. Walch, D. Harris, A. Bennish, G. Miley, Appl. Phys. Lett., 45, 1284 (1984).
B. McGowan, J. Kilkenny, M. Key, P. Rumsby, W. Toner, A. Fews, D. Henshaw, Opt. Comm, 48, 256 (1983).

33. E. Campbell et al, J. Appl. Phys., 51, 6062 (1980).

PLASMA PROCESSES IN NON-IDEAL PLASMAS

Richard M. More

Lawrence Livermore National Laboratory, Livermore, CA 94550

1.
INTRODUCTION

The traditional subject of fusion research is hot ionized hydrogen, magnetically confined at low densities $n_i \leq 10^{14}/cm^3$. The basic properties of these <u>ideal</u> <u>plasmas</u> were discovered by Langmuir, Landau, Spitzer, Rosenbluth and others (Spitzer,[1] Krall and Trivelpiece[2]).

However, plasmas produced by laser irradiation of high-Z solid targets (e.g., aluminum foils) are <u>non-ideal</u> because of their high density, partial ionization and great optical depth (Fig. 1). In these lectures we examine energy-conversion and energy-transport in non-ideal plasmas from a fundamental atomic viewpoint.

Most of the material is adapted from unpublished Livermore Laboratory reports[3,4] and the paper should be considered a tutorial discussion rather than a comprehensive review.

Although we emphasize the straightforward task of finding and verifying useful formulas for plasma processes, the fact is that one opens up a wide range of novel and interesting physical phenomena as one begins to consider nonequilibrium events in non-ideal plasmas. As improved laser experiments provide better understanding of the novel physical conditions in these plasmas, we will uncover a rich new field of scientific investigation.

1.1 Ideal and Nonideal Plasmas

The main features of non-ideal plasmas are: partial ionization, strong Coulomb interactions, and the possibility of free-electron degeneracy.

The ideal plasma is completely ionized ($Q = Z$) whereas non-ideal plasmas are usually partially ionized ($Q < Z$), where Q is the average ion charge and Z is the nuclear charge or atomic number. The energy required for ionization is appreciable; for aluminum the ionization potentials are .44, 2.1 and 2.3 keV respectively for lithium-like, helium-like and hydrogen-like ions. In general, the ionization potential $I \simeq (Q/n)^2$ 13.6 eV, where n is the principle quantum number of the outermost

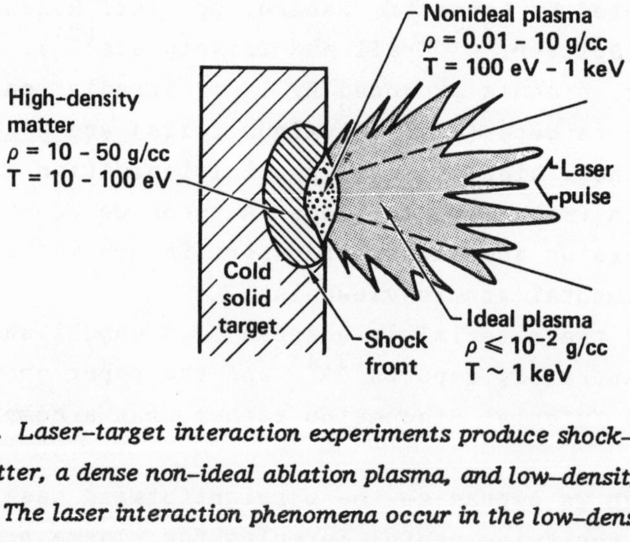

Figure 1. Laser-target interaction experiments produce shock- compressed solid matter, a dense non-ideal ablation plasma, and low-density blowoff plasma. The laser interaction phenomena occur in the low-density ideal plasma; however heat conduction, ionization, pressure generation and hydrodynamic phenomena occur at nonideal conditions.

bound electron. Obviously it is difficult to fully ionize high-Z plasmas, so they are usually nonideal or at least not completely ionized.

In the ideal plasma the electrical interactions between particles are weak in the sense that the screening cloud around one charge is attached by a binding energy $\simeq Z^2 e^2 / D$ (where D = Debye length) much less than the thermal energy ~ kT. This implies that a large number of particles participate in the screening:

$$N \simeq \frac{4\pi}{3} D^3 \; n_i \gg 1$$

The situation changes at higher density (or lower temperature); as the screening length decreases, the screening energy rises and ultimately exceeds kT. When this occurs, one has a strongly-coupled plasma which is non-ideal in the sense that electrical interactions exceed the ion ideal-gas energy (Brush et al.,[5] Hansen,[6] Baus and Hansen[7]).

In this case the first few neighbor ions place themselves so as to screen any given charge. The ion screening length is the average interparticle separation rather than the Debye length D. In practical formulas, one often describes dense plasma screening by replacing the Debye length by the ion sphere radius $R_o = (3/4\pi\, n_i)^{1/3}$. This prescription must be examined carefully to see where it works.

Ideal plasmas emit radiation by bremsstrahlung and radiative recombination processes; there is also some line radiation from hydrogen-like or helium-like ions. If there are high-Z impurities in a magnetically confined plasma, the bremsstrahlung radiation becomes an important cooling mechanism. Because radiation escapes freely from low-density plasmas it is important as a diagnostic: plasma density and/or temperature can be inferred from careful analysis of emitted or scattered radiation (Griem[8]).

In partially-ionized plasmas, radiation is strongly absorbed by bound electrons and does not easily escape. The photon population builds up toward a black-body distribution and can represent a significant specific heat. Radiation can also carry heat by diffusive heat-conduction. For these reasons we are interested in radiation absorption cross-sections, especially photoelectric and line absorption.

Electron heat conduction is limited by multiple small-angle Coulomb scattering. Because of the strong energy-dependence of the Coulomb cross-section (E^{-2}), the plasma conduction coefficients depend strongly on temperature; the electrical conductivity $\sigma \propto T^{3/2}$, and the electron thermal conductivity $\kappa \propto T^{5/2}$. The power-laws result from averages over a Maxwell distribution of free electron velocities and if the actual distribution is non-ideal (i.e., non-Maxwellian) the conduction coefficients will be changed.

The ideal plasma pressure is $p_e = Z\, n_i\, kT_e$, $p_i = n_i\, kT_i$, ideal-gas formulas for electrons and ions respectively. For the low-density case these formulas are valid as long as the distribution functions are Maxwellian, and in the absence of anisotropic effects associated with large magnetic fields. The dense plasma case is more interesting and we will examine it in detail (section 2).

If the free electrons are degenerate, the Maxwell distribution

$$f(\varepsilon) \propto \exp(-\varepsilon/kT)$$

is replaced by the Fermi-Dirac distribution,

$$f(\varepsilon) = 1/(1 + \exp(\varepsilon - \mu)/kT)$$

This causes averages over the electron distribution to be significantly altered. For example,

Quantity			Nondegenerate		Degenerate
Pressure	p	$\propto$	$n_e T$	$\rightarrow$	$n_e T_f \propto n_e^{5/3}$
Conductivity	σ	$\propto$	$T^{3/2}$	$\rightarrow$	$T_f^{3/2}$
Thermal conductivity	κ	$\propto$	$T^{5/2}$	$\rightarrow$	$T\, T_f^{3/2}$

In this table we indicate the density-temperature dependence in terms of the Fermi temperature,

$$kT_F = \frac{\hbar^2}{2m}(3\pi^2 n_e)^{2/3} \simeq 7.86 \text{ eV}\left(\frac{n_e}{10^{23}/\text{cm}^3}\right)^{2/3}$$

where n_e is the density of free electrons. The criterion for degeneracy is $T \leq T_F$.

The most complicated physics of ideal plasmas involves plasma waves and their role in plasma instabilities. Wave dispersion relations are complicated because of the strong influence of magnetic fields on single-particle motion. In an ideal plasma the wave amplitudes can grow much larger than thermal-equilibrium amplitudes (energy per mode $\simeq$ kT in equilibrium).

In dense plasmas the collision rate is very large and this appears to maintain wave populations at the thermal level. The collision rate is also large enough to greatly reduce the effects of magnetic fields. For these reasons, plasma waves are probably not very important in dense plasmas.

Next we must ask: are laser-produced plasmas nonideal?

Expanding laser-heated blowoff plasma is very strongly ionized, reaches electron temperatures ~ .5 – 2 keV, and has low densities ~ $10^{20}/\text{cm}^3$; this plasma is comparatively ideal and supports interesting wave phenomena which are important in absorption. Likewise the DT fuel in fusion targets is almost always an ideal plasma.

Where then is the non-ideal plasma? Plasma in the laser absorption region is non-ideal if the material is not fully-ionized, i.e., if $Z > 10$ at most laser intensities (Figure 1). The absorbed energy is conducted into the target to an ablation region where plasma conditions are usually non-ideal. For solid targets the high-density region compressed by shock-waves from the ablation front is always a non-ideal plasma. In fusion targets, any imploding pusher or fuel capsule will be strongly non-ideal.

1.2 Plasma Processes

Instead of trying to formulate a general definition of plasma processes, it is simplest to list some quantities of interest:

Equation of state $p(\rho, T)$, $E(\rho, T)$
Radiation absorption and emission coefficients, K_ν, j_ν
Stopping power for fast charged particles, dE/dx.
Electron-ion heat exchange time τ_{ei}
Electrical and thermal conductivity σ, κ
X-ray scattering cross-section $d\sigma/d\Omega$
X-ray refractive index $n(\nu)$

This list should give the idea: we want to describe the macroscopic or hydrodynamic behavior of the plasma with coefficients which summarize microscopic or atomic plasma processes. In many cases there is an ideal plasma formula to begin with, and we want to extend it to the non-ideal case.

From the preceding discussion, we can put forward a standard recipe for non-ideal plasma formulas:

STANDARD RECIPE

---

1. Partial ionization : replace Z by $Q(\rho,T)$
2. Strong coupling : replace D by R_o
3. Degenerate free electrons : replace temperature power-laws T^n by $T^m\ T_f^{n-m}$ for some m.

If this recipe gave the correct answer, our subject would be rather simple. Of course, the interesting cases are more difficult; the large effects of bound electrons are not even hinted at by the standard recipe. In many cases the standard recipe fails badly; those cases generate the interesting new physics of non-ideal plasmas.

As we examine each plasma process, we will ask several questions: Is the process important? When, where and why? How is the process affected by non-ideal plasma conditions? What exact or rigorous relations are known? Are there intuitive or qualitative connections to other plasma processes? What formulas should we use in practical applications?

2.

EQUATION OF STATE

The thermodynamic quantities pressure and energy directly govern the hydrodynamic motion of plasmas. For example, if we know the energy deposited by a laser pulse and know the energy function $E(\rho, T)$ then we can calculate the temperature T. From that we can evaluate the pressure $p(\rho, T)$; of course the pressure determines the expansion speed of the plasma blowoff, determines the strength of the recoil shock launched into the solid and thereby sets the implosion velocity in spherical target geometry. The full resources of theory and experiment have been concentrated upon the thermodynamic functions and so they are the best-understood of all plasma material properties.

The recent demonstration of x-ray laser gain in a laser-irradiated selenium foil gives a clear example of the importance of plasma thermodynamic properties.[9] In this experiment, it is necessary to control plasma density gradients in order to avoid refractive loss of the soft x-rays. LASNEX calculations of the hydrodynamic flow, verified by interferometric measurements of plasma densities, proved to be sufficiently accurate for a priori design of the selenium experiments.

In thermal equilibrium the pressure p, energy E and entropy S can be calculated from the Helmholtz free energy $F(\rho, T)$,

$$p = \rho^2 \; \partial F/\partial \rho$$

$$S = - \; \partial F/\partial T$$

$$E = F + TS \qquad (2\text{-}1$$

In Eq. (2-1), ρ is the mass density (g/cm^3) and F, E, and TS are energies per unit mass of material (e.g., Mbar-cm^3/gram = 10^5 Joules/gram). Because the mixed second derivatives of F are equal, there is an automatic connection between pressure and energy, the condition of thermodynamic consistency:

$$\rho^2 \frac{\partial E}{\partial \rho} = p - T \frac{\partial p}{\partial T} \qquad (2\text{-}2$$

This is an exact and general property of the thermal equilibrium equation of state. Eq. (2-2) can be used as a check if the pressure and energy are obtained from separate tables or formulas.

In numerical hydrodynamic calculations, Eq. (2-2) is often used to transform the energy equation; in this case it is a practical question whether the equation of state is consistent.

The simplest equation of state (EOS) is the non-interacting ideal-gas approximation: pressure and energy are taken to be sums of independent contributions from electrons, ions and photons (Zel'dovich-Raizer[10]):

$$p = p_e + p_i + p_r$$

$$E = E_e + E_i + E_r \tag{2-3}$$

For most laser-plasma target conditions, the electron pressure and energy are dominant and are approximately given by the <u>partially-ionized gas</u> EOS,

$$p_e = n_e kT_e = Q\rho kT_e/AM_p$$

$$E_e = (\tfrac{3}{2} QkT_e + E_i(Z, Q))/AM_p \tag{2-4}$$

Here Q = average ion charge, Z = nuclear charge, A = atomic weight, M_p = atomic mass unit, ρ = mass density (g/cm^3) = $n_i AM_p$, n_i = ion number density, n_e = Qn_i = electron number density, T_e = electron temperature, and $E_i(Z, Q)$ = sum of first Q ionization potentials.

E_e is the electron energy per gram, i.e., the kinetic energy of Q free electrons per atom plus the energy E_i required to ionize the atom Q times. Usually, the ion and radiation parts of the EOS are less important.

Immediately we see that the "standard recipe" of the introduction fails for the equation of state. One cannot simply take the ordinary ideal-gas formula and replace the nuclear charge Z by the ion charge Q; that recipe would omit the ionization energy $E_i(Z, Q)$.

For aluminum plasma at a representative density of .01 g/cm^3 (~ 10^{21} electrons/cm^3), one reaches full ionization at a temperature ~ 200 eV. The kinetic energy of the free electrons is then ~ (3/2)(13)(200 eV) ≃ 4 keV per atom. At the same time, the ionization energy is 6.6 keV/atom, i.e., 2/3 of the total plasma energy goes to produce the ionization. This should be carefully noted before one attempts to describe laser plasmas by ideal-gas hydrodynamic flows.

2.1 Foil Expansion Example

This point is illustrated by comparing plasma flow with real and ideal equations of state.

We consider the planar expansion of an aluminum foil which begins as a homogeneous plasma of density ~ 0.1 g/cm^3 and temperature 50 eV. At these initial conditions the average ionization state is $Q \simeq +6$.

First we examine the ideal-gas expansion, neglecting ionization changes. The flow is then a self-similar adiabatic rarefaction, which can be represented in Eulerian form,

$$v(x,t) = \begin{cases} 0 & x < -c_0 t \\ \frac{3}{4}\left(c_0 + \frac{x}{t}\right) & -c_0 t < x < 3c_0 t \end{cases} \tag{2-5}$$

$$\rho(x,t) = \begin{cases} \rho_0\left[1 - \frac{v(x,t)}{3c_0}\right]^3 & x < 3c_0 t \\ 0 & x > 3c_0 t \end{cases} \tag{2-6}$$

For adiabatic flow, pressure $p \propto \rho^{5/3}$ and temperature $T \propto p/\rho \propto \rho^{2/3}$; $c_0 = \sqrt{\frac{5}{3}\frac{p}{\rho}}$ is the sound speed. The spatial coordinate x is measured from the initial front surface of the foil. In reality the foil expands in both directions: one rarefaction enters from each side and the analytic solution breaks down after they meet in the center of the foil.

Figures (2, 3) show the LASNEX numerical solution of this problem, based on ideal-gas pressure and energy and neglecting nonhydrodynamic processes such as heat conduction or radiative emission. LASNEX reproduces the analytic solution to high accuracy (~ 0.1 %). The figures also show LASNEX calculations using a tabular equation of state for aluminum.

Figure 2 compares densities $\rho(x,t)$ for real and ideal flows at t = 70 nsec. The density profiles are very similar, although the nonideal foil expands more rapidly.

Figure 3 compares temperatures $T(x,t)$ for the same case. Here there is a large difference; the realistic equation of state gives a higher temperature (as much as a factor 4 higher in the low-density region).

This substantial difference results from energy released during recombination in the expanding foil. Each electron which recombines releases ~ 5 kT to the remaining free electrons. In this case the energy associated with ionization is very important to the plasma hydrodynamics.

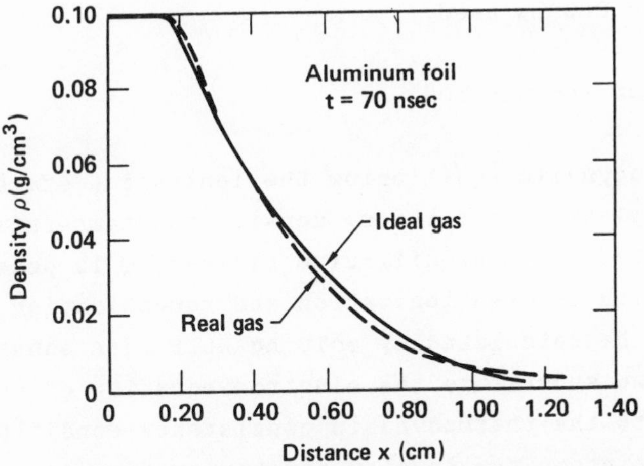

Figure 2. Density profile for expanding aluminum foil which begins at density 0.1 g/cm^3 and temperature 50 eV. The initial free surface is at x = 0.5 cm in the figure. Although the density profiles are very similar, the nonideal foil has expanded somewhat more rapidly.

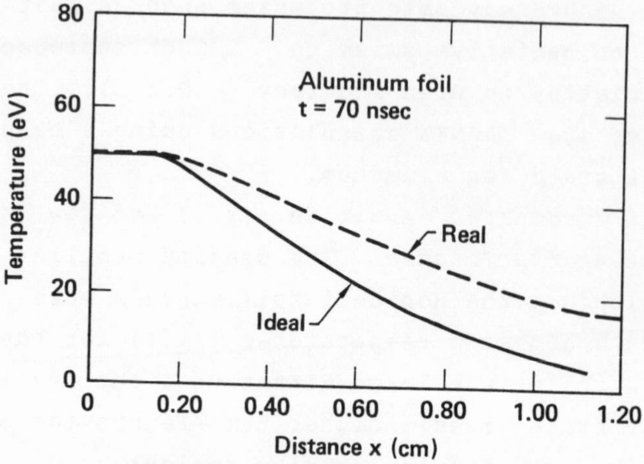

Figure 3. Temperature profile for the same foil expansion. The non-ideal foil is considerably hotter due to significant energy released by recombination occurring in the expansion.

Another illustration of the importance of EOS on foil dynamics is given by Schmalz and Meyer-ter-Vehn[11]; ideal-gas results are substantially changed when a realistic EOS is used.

2.2 Ionization and the EOS

In thermodynamic equilibrium the ionization state depends only on plasma composition, density and temperature: $Q = Q(Z, \rho, T)$. In nonequilibrium plasmas, Q is determined by competition between ionization and recombination processes, and must be calculated by solving NLTE rate equations.

If we substitute the electron equation of state, Eq. (2-4) into the thermodynamic consistency condition we find a constraint on the density and temperature dependence of the equilibrium (LTE) ionization state Q:

$$\frac{T(\partial Q/\partial T)}{\rho(\partial Q/\partial \rho)} = \frac{3}{2} + \frac{I(Q)}{kT} \qquad (2\text{-}7$$

where $I = dE_i/dQ$ = ionization potential.

If the ionization state $Q(\rho, T)$ is determined from

$$Q = a \frac{T^{3/2}}{\rho} \exp(- I(Q)/kT) \tag{2-8}$$

then Eq. (2-7) is satisfied and the electron equation of state will be thermodynamically consistent. Equation (2-8) can be understood as a simplifed version of the Saha equation, and is a rough but not unreasonable approximation.[10]

In order to calculate the ionization more accurately, one should add several physical corrections: thermal excitation of the ions and the specific heat associated with excited states; pressure ionization of states with large quantum numbers; possible degeneracy of the free electrons; and continuum lowering resulting from the electrostatic potentials of neighbor ions. These corrections to Eq. (2-8) are increasingly important at higher density.[3]

Eqs. (2-4) require the ionization model of Eq. (2-8) for thermodynamic consistency. What happens if one employs a different or more accurate ionization theory? Unless one simultaneously improves Eqs. (2-4) for pressure and energy, they are no longer consistent.

2.3 Ion and Radiation EOS

For the hot plasma ($T > 5$ eV) the ions are usually described by the ideal-gas model,

$$p_i = \rho k T_i / A M_p$$

$$\varepsilon_i = \frac{3}{2} k T_i / A M_p \tag{2-9}$$

The ion temperature T_i is often lower than the electron temperature because ions are not heated directly by the laser but only indirectly via collisions with free electrons. For plasmas with $Q > 10$, the ion contribution to the total pressure or energy is not very important.

Even in this case there are reasons to look carefully at the ion equation of state. For certain applications we may need to know the ion temperature relatively accurately. For example, the ion temperature determines the Doppler line width, so it has a direct influence on the gain of x-ray lasers.

Another reason to examine the ion contributions is that the ion pressure is dominant at low temperatures (i.e., $kT < 1$ eV) where the plasma is not yet strongly ionized.

If the black-body radiation has an equilibrium (Planck) distribution then its contributions to the EOS are given by

$$p_r = \frac{\pi^2}{45} \frac{(kT_r)^4}{(\hbar c)^3} = 45.7 \text{ Mbar } \left(\frac{kT_r}{1 \text{ keV}}\right)^4$$

$$E_r = 3\, p_r \tag{2-10}$$

where the radiation energy here is given as energy/volume rather than energy/gram (1 Mbar = 10^5 Joule/cm^3). From these formulas we find that radiation pressure becomes important only at high temperatures > 1 keV. It is difficult to produce equilibrium radiation at such high temperatures, so in practice black-body radiation pressure is not very important in laser interaction experiments.

2.4 Two Temperature Equation of State

Electrons and ions interact through electrical forces and in dense plasmas the interaction energies are large enough to be important. On the other hand, the interactions do not always equalize the two species temperatures.

Electrical (screening) energy is exchanged between electrons and ions in very short times, of the order of the ion plasma oscillation time. The heat transfer occurs in collisions and this is slow because of the great difference in electron and ion masses. One may say that the species exchange energy as electrical work on a short time-scale but exchange heat only on a slower collisional time-scale.

This leads to an interesting nonequilibrium state: the electrons and ions have separate, well-defined temperatures Te, Ti but one cannot separate the energy into additive electron, ion parts because the electrical interaction energy involves both species simultaneously. How is one to describe this situation in a precise way?

Our two-temperature theory[3,4,12,13,14] is based on the idea that electron velocities greatly exceed ion velocities when $T_i << M_i T_e/m_e$. This means that one can neglect the ion motion while working out the electron statistical mechanics, i.e., one treats a sysem of electrons at temperature T_e interacting with fixed ions at constant positions $\{R_i\}$.

We then form a probability distribution for classically-defined ion positions $\{R_i\}$ and momenta $\{P_j\}$ as

$$P(\{R_i\}, \{P_j\}) = \frac{1}{Z} \exp - \left\{ \sum_j \frac{P_j^2}{2M} + F_e(\{R_i\}, T_e) \right\} / kT_i \qquad (2\text{-}11)$$

The electron free energy F_e $(\{R_i\}, T_e)$ appears as a potential energy for ion motion. The Helmholtz free energy F_e is taken as the potential because the electrons remain essentially isothermal for ion motions on the atomic size-scale important for the equation of state. In principle this free energy is calculated for fixed ion positions $\{R_i\}$, neglecting ion thermal velocities (Born-Oppenheimer approximation).

F_e is assumed to be the result of a quantum-statistical calculation; for each ion configuration we consider the gas of electrons moving in the presence of fixed classical point charges, described by the Hamiltonian:

$$H = \sum_i \frac{p_i^2}{2m_e} + \frac{1}{2}\sum_{ij} \frac{e^2}{|r_i - r_j|} - \sum_{ij} \frac{Qe^2}{|r_i - R_j|} + \frac{1}{2}\sum_{ij} \frac{Q^2 e^2}{|R_i - R_j|} \tag{2-12}$$

Here $\hat{r}_i$, $\hat{p}_j$ = electron coordinate, momentum (operators), R_i = ion coordinate. The ion-ion Coulomb energy is included to prevent the expression from diverging for a large system. The electron free energy F_e is then determined by

$$\exp\left(-\frac{F_e}{kT_e}\right) = Z_e = \text{Trace}\left[e^{-H/kT_e}\right] = \sum_s \exp\left(-\frac{E_s}{kT_e}\right) \tag{2-13}$$

where E_s = energy eigenvalue of H ; Z_e = electron partition function calculated for fixed ion positions. This prescription is exactly the usual statistical mechanics of an electron gas (temperature = T_e) in the external potential provided by the ions. We have the obvious formulas:

$$S_e(\{R_i\}, T_e) = -\partial F_e/\partial T_e \tag{2-14}$$

$$E_e(\{R_i\}, T_e) = F_e + T_e S_e \tag{2-15}$$

These give the electron entropy and energy as functions of T_e and the complete set of ion coordinates $\{R_i\}$.

Obviously F_e cannot be exactly calculated. For many purposes it will be a reasonable approximation to assume

$$F_e \simeq F_0(\rho, T_e) + \frac{1}{2}\sum_{i,j} \frac{Q^2 e^2}{|R_i - R_j|} e^{-|R_i - R_j|/D_e(\rho, T_e)} \tag{2-16}$$

where $D_e = (kT_e/4\pi\, n_e\, e^2)^{1/2}$ is the Debye length for electron screening only, and F_0 is the ideal-gas electron free energy. With this approximation to F_e, we can examine the probability P.

The ion probability distribution of Eq. (2-11) is normalized by

$$Z = \frac{1}{N!} \cdot \int \frac{d^{3N} R_i \, d^{3N} P_j}{(2\pi \hbar)^{3N}} \exp - \left\{ \sum_j \frac{P_j^2}{2M} + F_e \right\} / kT_i \qquad (2\text{-}17)$$

The partition function Z factors into $Z_i^o \bullet Z_{conf}$ because F_e is independent of $\{P_j\}$. Z_i^o is the ideal-gas ion partition function, and Z_{conf} is the configurational term,

$$Z_{conf} = \int \frac{d^{3N}R_i}{V^N} e^{-F_e/kT_i} = \int \frac{d^{3N}R_i}{V^N} [(Z_e(\{R_i\}, T_e))]^{T_e/T_i} \qquad (2\text{-}18)$$

The last formula emphasizes the novel physical content of the two-temperature theory: the ion phase-space is weighted in a special way because any movement of the ions forces the electrons to follow in order to maintain approximate neutrality.

It is useful to define a total free energy as

$$F = - kT_i \log Z (T_e, T_i, \rho) \qquad (2\text{-}19)$$

The ion entropy is now defined from P by the Shannon-Weaver-Kinchin method: entropy $S_i = - k < \log P >$, giving

$$S_i = - (\partial F/\partial T_i)_{\rho, T_e} \qquad (2\text{-}20)$$

The electron entropy, averaged over ion configurations, can be shown to be

$$S_e = \langle S_e \rangle = - (\partial F/\partial T_e)_{\rho, T_i} \qquad (2\text{-}21)$$

The total energy, defined as the average of the Hamiltonian, is readily shown to be

$$E = F + T_e S_e + T_i S_i \qquad (2\text{-}22)$$

The energy E and entropies S_e, S_i depend upon both temperatures (T_e, T_i) as well as the density.

From these equations, we obtain an interesting

generalized thermodynamics. First, there are the basic differential laws,

$$dF = (p/\rho^2)\, d\rho - S_e dT_e - S_i dT_i \tag{2-23}$$

$$dE = (p/\rho^2)\, d\rho + T_e dS_e + T_i dS_i \tag{2-24}$$

We can immediately derive several Maxwell relations, e.g.,

$$\left(\frac{\partial p}{\partial T_e}\right)_{\rho, T_i} = -\left(\rho^2 \frac{\partial S_e}{\partial \rho}\right)_{T_e, T_i} \tag{2-25}$$

and a generalized condition of thermodynamic consistency,

$$\rho^2 \left(\frac{\partial E}{\partial \rho}\right)_{T_e, T_i} = + p - T_e \left(\frac{\partial p}{\partial T_e}\right)_{\rho, T_i} - T_i \left(\frac{\partial p}{\partial T_i}\right)_{\rho, T_e} \tag{2-26}$$

The specific heat becomes a matrix; for changes at constant density, we have

$$dQ_e = T_e dS_e = C_{ee}\, dT_e + C_{ei} dT_i$$

$$dQ_i = T_i dS_i = C_{ie}\, dT_e + C_{ii} dT_i \tag{2-27}$$

with a general symmetry relation

$$\frac{C_{ie}}{T_i} = \frac{C_{ie}}{T_e} \tag{2-28}$$

which follows from the Maxwell relation

$$\left(\frac{\partial S_e}{\partial T_i}\right)_{\rho, T_e} = \left(\frac{\partial S_i}{\partial T_e}\right)_{\rho, T_i} \tag{2-29}$$

This theory does not try to separate pressure and energy into electron and ion parts; the pressure and energy actually cannot be separated because the two species interact and continually exchange electrical energy (in the form of screening).

We can understand the matrix specific heat of Eq. (2-27) if we consider a sudden heat pulse applied to a plasma by absorption of laser energy. The laser energy is absorbed by free electrons, and raises their temperature. This temperature rise induces an immediate change in the electron screening length D_e which has the effect of making the ion interactions effectively stronger. The ions are spatially distributed with pair-distributions that correspond to the original (weaker) ion repulsion and must readjust their spatial correlations to reach equilibrium with the stronger forces. In general this readjustment releases a certain amount of thermal energy and promptly raises the ion temperature. These changes require only a few ion plasma oscillations; the collisional heat transfer occurs on a longer time-scale.

In this section we have given a relatively detailed presentation of the two-temperature thermodynamic model because it illustrates the new nonequilibrium phenomena encountered in non-ideal plasmas. There are many similar questions, not yet answered, concerning the thermodynamic description of nonequilibrium ionization phenomena in nonideal plasmas.

3.

RADIATION PROCESSES

Most of our experimental knowledge of laser-plasma dynamics is derived from the emitted x-rays: they indicate the temperature and density of the plasma and give information about the time-dependence of events in the plasma, especially when analyzed with high-speed streak cameras (Ahlstrom[15]). Imploded fuel density can be measured by analysis of Stark broadening of lines emitted by impurity ions in the fuel, an important spectroscopic diagnostic technique (Griem,[8] deMichelis and Mattioli[16]). The continuum emission reflects the energy-spectrum of free electrons (Lamoureux, Moller and Jaegle[17]) as well as spatial profiles of density and temperature. Energetic x-rays signal the presence or absence of suprathermal electrons at high energies.[18]

Radiation also affects target hydrodynamics. Hot regions radiate strongly and the emitted photons transport energy to cooler parts of the plasma or permit energy to escape altogether. Radiation travels more rapidly than matter and this raises its leverage on the plasma dynamics.[10]

Using semiclassical methods, H. A. Kramers[19] showed that the x-ray photoelectric absorption cross-section has the form

$$\sigma \propto Z_{eff}^{4} \lambda^{3} \qquad (3-$$

where Zeff is an effective nuclear charge and λ is the x-ray wavelength. This theory describes the main interaction of x-rays with matter in terms of atomic electrical forces and is the practical application of Bohr's atomic model to radiative processes. We will examine Kramers' theory and extend it to atoms in partially-ionized plasmas.

There are actually nine radiative cross-sections whic we can classify according to the incident particle or projectile (electron or photon) and the state of the electron which participates (free or bound) (Fig. 4).

The nine cross-sections are related by detailed balance, which connects absorption and emission rates, and by continuation in energy, i.e., each process involving bound electrons is described by extrapolation of the formu for the corresponding process involving free electrons.

While Kramers' theory describes a classical electron colliding with a point nucleus, these connection formulas are more general. The connections can be expected to remain valid as we incorporate additional physics -- partial ionization, quantum effects, relativity effects, and effects of the dense plasma environment. Thus the connection formulas represent a powerful tool for construction of the complete theory of nonideal plasmas.

Bremsstrahlung Emission

Bremsstrahlung radiation is emitted when electrons are scattered by atoms, ions or fully-stripped nuclei (Fig. 5a). In Kramers' theory, a free electron of energy $\varepsilon_o = 1/2\ m\ v_o^2$ is assumed to approach a point charge (nucleus) + Ze with impact parameter b. The electron accelerates as it approaches, gaining a kinetic energy Ze^2 / r ; at small radii this kinetic energy gain can exceed the initial energy ε_o. During the approach, the angular momentum $\hbar\ell = m\ v_o b$ is constant; the distance of closest approach is

$$r_{min} \simeq \begin{cases} b & Ze^2/r_{min} \ll \varepsilon_o \\ \frac{1}{2}\ a_o(\ell + \frac{1}{2})^2/Z & Ze^2/r_{min} \gg \varepsilon_o \end{cases} \quad (3\text{-}2)$$

Kramers' calculation is most accurate in the case where $Z\ e^2 / r_{min} \gg \varepsilon_o$; for the other case one should use the Born approximation. In Kramers' case the electron follows a hyperbolic path with strong curvature; the closest portion of the orbit at radius $\simeq r_{min}$ is traversed with a velocity $v_{max} \simeq (2Z\ e^2 / m\ r_{min})^{1/2}$.

The electron radiates as a result of its acceleration, and the strongest emission occurs during the peak acceleration. The emitted power is determined by the Fourier transform of the acceleration; the result is a combination of Hankel functions.[20] This expression can be summed over impact parameters of the incident electron to produce the total differential cross-section for emitting a photon

$$\frac{d\sigma^{(A)}}{dh\nu} = \frac{8\pi}{3\sqrt{3}}\ \frac{Z^2\ \alpha^3\ a_o^2}{h\nu}\ \left(\frac{e^2/a_o}{\varepsilon_o}\right)$$

$$= \frac{32\pi^2}{3\sqrt{3}}\ \frac{Z^2\ e^6}{m^2\ c^3\ v_o^2\ h^2\ \nu} \quad (3\text{-}3)$$

Radiative cross-sections:

	Electron emits	Electron absorbs	Photon is absorbed
FF	A Bremsstrahlung	B Inverse Bremsstrahlung	C I.B. or joule heating
BF	D Radiative recombination	E Photoelectric effect	F Photo ionization
BB	G Line emission	H Line absorption	I Line absorption

Figure 4. The nine radiative cross-sections can be classified as FF (free-free), BF (bound-free) and BB (bound-bound) transitions. The cross-hatched processes (E, G, H) will be treated as collision processes in which the incident electron has negative energy. By this connection technique the bound-electron rates are deduced from the FF cross-section.

The bremsstrahlung emission is a continuous spectrum. In Eq. (3-3), $\alpha = e^2/\hbar c = 1/137.04$ is the fine-structure constant and $a_o = \hbar^2/me^2 = .529\ 10^{-8}$ cm is the Bohr radius.

To get a feeling for the numbers, we should compare the cross-section for emission to the cross-section for the electron to be scattered by the nucleus. The ratio of the total cross-sections is

$$\frac{\sigma_{rad}^{(A)}}{\sigma_{scat}} \propto \alpha \frac{\varepsilon_o}{mc^2} \ll 1$$

showing that radiation is emitted in a small minority of collisions.

The rate of emission of photons from a small volume of plasma is obtained by integrating the cross-section (3-3)

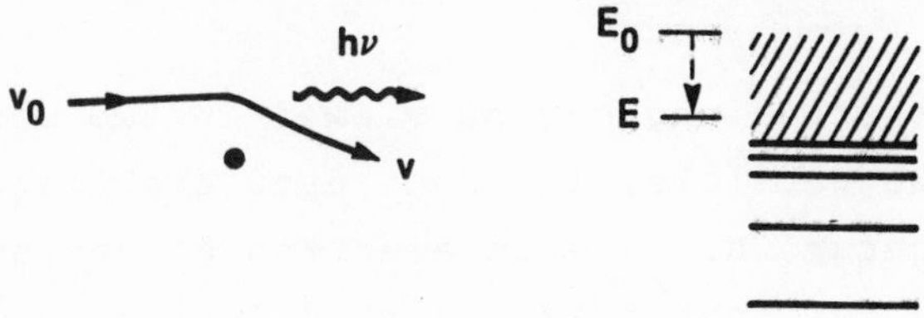

Figure 5a. Schematic illustration of bremsstrahlung emission of photon of energy $h\nu = E_0 - E$ during an electron's collision with a point charge Ze.

over the flux of electrons hitting each ion:

$$\frac{dN}{d^3r dt\, d h\nu} = n_i \int \frac{2d^3 p_0}{h^3} |v_0| f(\varepsilon_0) \frac{d\sigma^{(A)}}{d h\nu} (n_\nu + 1)[1 - f(\varepsilon)] \quad (3\text{-}4)$$

Here n_i is the number-density of target nuclei, $f(\varepsilon_0)$ is the distribution function for the incident electron, and $p_0 = mv_0$ is its momentum. The rate includes a factor $(n_\nu + 1)$ which incorporates stimulated emission, where n_ν is the number of photons per state. After the photon is emitted, the electron's energy is reduced to $\varepsilon = \varepsilon_0 - h\nu$. The Fermi-Dirac factor $[1 - f(\varepsilon)]$ prevents transitions into fully-occupied final states; for nondegenerate plasma conditions, $f(\varepsilon) \ll 1$ and this factor can be ignored.

Neglecting the degeneracy correction, we can easily perform the indicated integral and convert it to an energy emission rate by multiplying by $h\nu$:

$$\frac{dE}{d^3r dt\, d h\nu} = \frac{16}{3}\sqrt{\frac{2\pi}{3}}\, \frac{Z^2 e^6 n_e n_i}{\hbar m^2 c^3 \sqrt{kT_e/m}}\, \frac{e^{-h\nu/kT_e}}{(1 - e^{-h\nu/kT_r})} \quad (3\text{-}5)$$

In this formula, T_e is the electron temperature and T_r is the radiation temperature assumed in $n_\nu = (\exp(h\nu/kT_r - 1)^{-1}$. Equation (3-5) is the usual bremsstrahlung emission formula.

3.2 Free-Free Absorption

Inverse bremsstrahlung can be viewed in two ways because there are two particles incident upon the target nucleus, a photon of energy $h\nu$ and an electron of energy $\varepsilon = 1/2\ mv^2$. If the photon is absorbed during the encounter, the electron gains energy $h\nu$ to reach a final energy $\varepsilon_0 = \varepsilon + h\nu$ (The notation makes this the time-reversed process to the emission).

From the viewpoint of the incident electron, the number of absorption events per cm^3-sec-keV is

$$\frac{dN}{d^3r dt\ dh\nu} = n_i \int \frac{2\ d^3p}{h^3} |v|\ f(\varepsilon)\ \sigma^{(B)} \{n_\nu \frac{dN}{dh\nu}\}[1-f(\varepsilon_0)] \quad (3-6)$$

This rate is formed from the flux (number arriving/cm^2 sec) of incident electrons of energy $\varepsilon = p^2/2m = 1/2\ mv^2$, the electron's absorption cross-section $\sigma^{(B)}$, and a density of ambient photons at energy $h\nu$,

$$n_\nu \frac{dN_\nu}{dh\nu} = n_\nu \frac{(h\nu)^2}{\pi^2(\hbar c)^3} \quad (3-7)$$

Equation (3-6) includes the degeneracy factor $[1 - f(\varepsilon_0)]$ which reduces the rate if the final state is occupied. The cross-section $\sigma^{(B)}$ has units cm^5 which are corrected to the usual cm^2 by the photon density of Eq. (3-7).

According to the principle of detailed balance, in thermal equilibrium the absorption rate equals the emission rate. With equilibrium distribution functions $n_\nu^0 = 1/(\exp(h\nu/kT) - 1)$ and $f^0(\varepsilon) = 1/(1 + \exp(\varepsilon - \mu)/kT)$, the two rates are equal

$$\varepsilon\ \sigma^{(B)} \frac{dN}{dh\nu} = \varepsilon_0 \frac{d\sigma^{(A)}}{dh\nu} \quad (3-8)$$

This formula gives

$$\sigma^{(B)}\ (\varepsilon,\ h\nu) = \frac{8\pi^3}{3\sqrt{3}} Z^2\ a_0^5 \left(\frac{e^2/a_0}{h\nu}\right)^3 \frac{e^2/a_0}{\varepsilon} \quad (3-9)$$

If we are interested in getting a useful approximate cross-section, Eq. (3-9) is the answer; however if we are interested in building a structure for the theory, Eq. (3-8) is more important. It gives a connection between emission and absorption cross-sections which remains valid when we generalize the theory to include non-ideal plasma effects.

The same absorption process can be regarded as a limitation on the photon mean free path. In this case we write the rate of absorption as

$$\frac{dN}{d^3r\,dt\,dh\nu} = n_I \left\{c n_\nu \frac{dN}{dh\nu}\right\} \sigma_\nu^{(C)} \tag{3-10}$$

The factor in brackets is the incident photon flux, i.e., number of photons arriving per cm^2-sec-keV.

For any plasma the absorption rate of Eq. (3-10) must exactly equal the rate of Eq. (3-6); the two formulas describe the same process in different notation. This means that $\sigma_\nu^{(C)}$ is determined by an integral of $\sigma_\nu^{(B)}$ over the electron distribution.

The photon cross-section $\sigma_\nu^{(C)}$ is usually reported as an opacity coefficient defined as the photon absorption cross-section per gram,

$$K_\nu^{FF} = \sigma_\nu^{(C)}/AM_p \tag{3-11}$$

With this notation, the Kramers' absorption opacity is

$$K_\nu^{FF} = \frac{16\pi^2}{3\sqrt{3}} \frac{Z^2 e^6 n_e}{hc(2\pi m)^{3/2} AM_p \nu^3 \sqrt{kT}}$$

$$= 2.78 \frac{cm^2}{g} \left(\frac{Z}{A}\right)^2 \frac{Q\rho}{(h\nu)^3 \sqrt{kT}} \tag{3-12}$$

Equation (3-12) applies to nondegenerate plasmas. In the numerical version of the formula, the density ρ is in g/cm^3 and $h\nu$, kT are in keV.

3.3 Non-Ideal Plasma Effects

Kramers' formulas are derived for nonrelativistic classical motion of an electron colliding with an isolated point charge. There are a number of obvious physical corrections:

a.) Even for collisions with a point charge ion, the quantum theory gives a correction to the classical emission cross-section called the <u>Gaunt factor</u> (Karzas and Latter[21]).

b.) For collisions with a high-Z point charge there are significant relativistic corrections. These corrections can be large even for $\varepsilon_o \ll mc^2$ if the kinetic energy $\simeq Ze^2 / r_{min}$ is relativistic.

c.) For collisions with partially-stripped ions, the emission cross-section is likely to vary between a Kramers' cross-section determined by the ion charge Q for soft photons to that determined by the nuclear charge Z which applies for hard photons produced at small radii. The bound-electron screening correction is discussed by Lamoureux and Pratt[22] and by Kogan and Kukushkin.[23]

Numerical calculations of bremsstrahlung from partially stripped ions including these effects are reported by Pratt et al.[24]

d.) One dense-plasma effect is plasma degeneracy represented by the final-state factor [1-f] in Eq. (5). The effect of degeneracy on the net emission rate turns out to have simple analytic form in the Kramers approximation.[25]

e.) Another dense-plasma effect is the plasma refractive index which affects the dispersion relation of the outgoing photon. This effect is interesting because it alters the detailed-balance equations, but the results are not quantitatively large in typical laboratory plasmas.[26]

f.) Another dense-plasma effect is screening of the ion potential by the exterior plasma. This has been calculated analytically by use of the Born approximation[27] for the scattering; however the Born approximation is very inaccurate in the case of greatest practical interest (the Kramers case mentioned above).[28] In reality, most of the radiated photons originate at small radii where plasma screening is not large. This is verified by numerical studies of bremsstrahlung emission in electron collisions with screened ions.[29,30]

g.) A more sophisticated viewpoint on the calculation of radiation from dense plasmas describes the effect of the environment with ion pair-correlation functions.[31] The result is interference between scattering by adjacent ions. This calculation is also limited to the Born approximation and predicts its main effect for low-energy electrons for which the Born approximation is usually invalid.

h.) There is considerable interest in the possibility of significantly non-Maxwellian electron distributions in actual laser-plasma experiments. The predicted changes in bremsstrahlung spectrum, following from Eq. (3-4), are quite substantial and may help identify non-Maxwellian distributions in experimental continuum emission spectra (Lamoureux, Moller and Jaegle[17]).

Of these corrections to Kramers' formulas we will concentrate on the large effect of screening by bound electrons, i.e., the choice of effective charges Q which replace the nuclear charge Z when the plasma is not fully-ionized. Actually this question is most important for recombination and line emission because those processes are strongly dominant in partially-ionized plasmas. We examine Kramers treatment of recombination and line emission in order to develope simple approximations which apply to partially-ionized plasmas.

3.4 Radiative Recombination

Radiative recombination occurs when a free electron radiate more energy than its initial kinetic energy ε_0 so that it is trapped at negative energies in a bound state (Fig. 5b). To calculate radiative recombination, Kramers extrapolates the bremsstrahlung emission cross-section of Eq. (3-3) to events in which $h\nu > \varepsilon_0$. All transitions to energies near the quantized final energy E_n are grouped together as

$$\frac{d\sigma^{(D)}}{dh\nu} = \frac{d\sigma^{(A)}}{dh\nu} \frac{\partial E_n}{\partial n} \delta(\varepsilon_0 - h\nu - E_n)$$

$$= \sigma^{RR}(\varepsilon_0)\ \delta(\varepsilon_0 - h\nu - E_n) \qquad (3\text{-}13$$

For hydrogenic ions, $E_n = -Z^2e^2/2a_on^2$ and this extrapolation procedure gives the Kramers cross-section for radiative recombination into the shell of principal quantum number n,

$$\sigma^{RR}(\varepsilon_0) = \frac{8\pi}{3\sqrt{3}} \frac{Z^2\ \alpha^3\ a_o^2}{h\nu} \left(\frac{e^2/a_o}{\varepsilon_0}\right) \frac{\partial E_n}{\partial n} \propto \frac{Z^4}{n^3\ \varepsilon_0 h\nu} \qquad (3\text{-}14$$

The rate of radiative recombination is then given by an integral over the flux of electrons incident upon the target ion,

$$\frac{dN}{d^3r dt\ dh\nu} = n_i \int \frac{2d^3p_o}{h^3} |v_o|\ f(\varepsilon_0) \frac{d\sigma^{(D)}}{dh\nu} (n_\nu+1)[1-\frac{P_n}{D_n}] \quad (3\text{-}15$$

Again we include factors for stimulated emission and for possible occupation of the final state(s). P_n is the number of electrons assumed to occupy the shell of energy E_n and $D_n = 2n^2$ is the number of states available, so that $[1 - P_n/D_n]$ is the fractional number of vacant states in the shell.

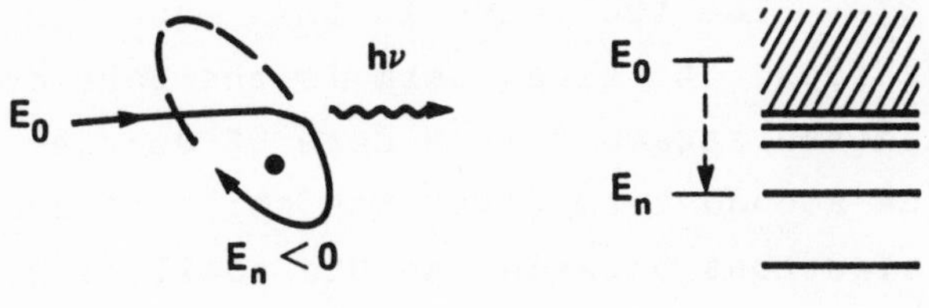

Figure 5b. Schematic representation of radiative recombination.

5 Bound Electron Screening

We will describe radiation from partially-ionized plasmas by a screening model based upon hydrogenic[32-35] formulas, i.e., an approximation in which energy-levels are taken to depend only on the principal quantum number n.

The model is based upon the use of effective charges Q_n which summarize the screening effect of bound electrons inside to shell n. The effective charge Q_n is determined in terms of electron populations P_n by a linear relation,

$$Q_n = Z - \sum_{m \leq n} \sigma_{nm} P_m \tag{3-16}$$

The screening coefficients σ_{nm} enter the theory as a fixed set of parameters.[35] The interpretation of σ_{nm} is that it gives the fraction of the charge of shell m which resides inside shell n. Eq. (3-16) includes only the effect of electrons inside the average radius of shell n, so that Q_n is the effective charge which determines the electric field at this radius, ($r_n = a_o n^2/Q_n$). Q_n also determines the acceleration experienced by an electron at radius $\simeq r_n$.

In terms of these effective charges, the electron eigenvalues are approximately given by

$$E_n = - \frac{Q_n^2 e^2}{2a_o n^2} + \sum_{m \geq n} \sigma_{mn} \left(\frac{Q_m e^2}{a_o m^2}\right) P_m \tag{3-17}$$

E_n depends upon the populations P_m and screened charges Q_m. The two terms in this equation have simple interpretations: the first term is the Bohr-model energy for an electron attached to a core of charge Q_n, whereas the second term gives the shift in energy produced by other electrons outside the nth shell (i.e., inner and outer screening respectively).

The hydrogenic screening model follows intuitively as a direct extension of the Bohr theory, and this interpretation is developed in detail elsewhere.[14,35,36] For present purposes we note the useful approximate formula

$$\frac{\partial E_n}{\partial n} \simeq \frac{Q_n^2 e^2}{a_o n^3} \quad (3\text{-}18)$$

which follows from Eqs. (3-17) if we neglect (small) contributions $\partial\sigma_{nm}/\partial n$.

Given this rough description of the energy-levels of many electron ions we can extend Kramers' recombination cross-section to the non-hydrogenic case. The level-spacing of Eq. (3-18) is used in Eq. (3-13); the main charge factor Z is replaced by the effective charge Q_n on the grounds that the acceleration occurs at small radii. This yields

$$\sigma^{RR} = \frac{8\pi}{3\sqrt{3}} \frac{Q_n^2 \alpha^3 a_o^2}{h\nu} \left(\frac{e^2/a_o}{\varepsilon_o}\right) \left(\frac{\partial E_n}{\partial n}\right)$$

$$\propto Q_n^4/n^3 \varepsilon_o h\nu \quad (3\text{-}19)$$

This formula gives a simple and useful description of the radiative recombination cross-section. It has been compared to much more elaborate quantum calculations and appears to give acceptable agreement. One example of these comparisons is reported by Huebner et al.[37]; these authors find good agreement if one employs an inner-screening effective charge in Eq. (3-19). Other choices of effective charge, based on the atomic potential $V(r)$

rather than electric field dV/dr, prove to be unsatisfactory.

It is also worth comment that Eq. (3-19) applies relatively close to the threshold, i.e., for $\varepsilon_0 < |E_n|$. At higher energies the appropriate effective charge is essentially the nuclear charge; this latter case is much less important in practise because the thermal distribution emphasizes the region $\varepsilon_0 \leq kT \leq |E_n|$.

Photoelectric Effect

Next we consider the <u>photoelectric effect</u>, in which a bound electron of principal quantum number n absorbs a photon of energy $h\nu$ and makes a transition to a free state of energy $\varepsilon_0 = E_n + h\nu$. This process is readily described by a photon absorption cross-section $\sigma_\nu^{(F)}$; the rate is

$$\frac{dN}{d^3r\,dt\,dh\nu} = n_i\, P_n \left\{ c n_\nu \frac{dN}{dh\nu} \right\} \sigma_\nu^{(F)} \, [1 - f(\varepsilon_0)] \qquad (3\text{-}20)$$

In Eq. (3-20) we again take the photon viewpoint: the photoelectric rate is written as the photon flux times a photon cross-section $\sigma_\nu^{(F)}$. The absorption rate (3-20) is related to the radiative recombination emission rate (3-15) by detailed balance, which implies

$$2n^2 c \frac{dN}{dh\nu} \sigma_\nu^{(F)} = \frac{1}{\pi^2} \frac{2m\,\varepsilon_0}{\hbar^2} \frac{1}{\hbar} \sigma^{RR} (\varepsilon_0;\, h\nu) \qquad (3\text{-}21)$$

This connection then gives the photoelectric opacity per target electron,

$$K_\nu^{BF} = \frac{\sigma_\nu^{(F)}}{AM_p} = \frac{8\pi}{3\sqrt{3}} \frac{\alpha\, a_0^2}{AM_p} \left(\frac{e^2/a_0}{h\nu}\right)^3 \frac{Q_n^4}{n^5} \qquad (3\text{-}22)$$

$$= \frac{12.0\, Q_n^4}{A(h\nu)^3\, n^5} \; \frac{cm^2}{g}$$

This is the Kramers formula generalized to include effective charges for the bound electron. Eq. (3-22) is the detailed version of Eq. (3-1) and is in reasonable agreement with experimental data on x-ray absorption for x-ray energies $h\nu \geq 1$ keV. Again we draw attention to th connection formula, Eq. (3-21), which remains valid even when we improve the physical model for the photoelectric process.

There is another way to think about the photoelectric process which turns out to be useful. In this case we consider a bound electron in a strongly elliptic orbit (e.g., one with small angular momentum). When the electron approaches the nucleus it is strongly accelerated and is able to absorb an ambient photon. In this viewpoint, the bound electron's interaction with the nucleus is essentially a collision process, and the absorption is physically identical to the inverse bremsstrahlung absorption process.

This viewpoint leads to a prediction of the photoelectric cross-section from the inverse bremsstrahlung cross-section, another useful connection. Evidently it does not matter whether the electron approaches from a grea distance with a slightly positive or slightly negative energy because the absorption probability is entirely determined by the strong acceleration which occurs near the nucleus.

The trick is then to describe the interaction of bound electrons with the nucleus as if these interactions were collisions.

3.7 Equivalent Flux of Bound Electrons

We assume there are P_n electrons in the initial state n statistically distributed over the subshell states, so that

$$P_{n\ell} = 2(2\ell + 1)\,[P_n/2n^2] \qquad (3\text{-}23$$

The electrons in states of small angular momenta (elliptical orbits) approach the nucleus with a frequency $\nu_n = (1/h)\, \partial E_n/\partial n$ giving a collision rate:

$$\frac{dN_\ell}{dt} = P_{n,\ell}\, \nu_n \tag{3-24}$$

Now we define an effective incident electron flux,

$$\phi_n = \left(\frac{P_n}{2n^2}\right) \frac{2m|E_n|}{\pi^2 \hbar^2} \frac{1}{\hbar} \frac{\partial E_n}{\partial n} \tag{3-25}$$

If the idea of a uniform incident flux makes sense, then a geometrical calculation of the collision rate should agree with Eq. (3-24). With an impact parameter determined by $mv_n b = \hbar(\ell + 1/2)$, the geometric calculation is[36]

$$\frac{dN_\ell}{dt} = \phi_n\, 2\pi b\, db$$

One readily verifies that this agrees with Eq. (3-24).

To summarize: the bound electron(s) in state n are equivalent to a uniform incident flux ϕ_n in terms of their rate of close approaches to the nucleus. This viewpoint makes most sense for the electrons of small angular momentum which are dominant in the radiative processes.

.8 Photoeffect as an Electron Collision

Using the concept of bound electron flux, we can tentatively write an equation for the rate of photoelectric absorption events in terms of a cross-section $\sigma^{(E)}(E_n, h\nu)$ for a bound electron to absorb an ambient photon during its close approach to the nucleus (or ion core):

$$\frac{dN}{d^3r\,dt\,dh\nu} = n_i\, \phi_n\, \sigma^{(E)}\, n_\nu \frac{dN}{dh\nu} [1 - f(\varepsilon)] \tag{3-26}$$

In this picture the photoelectric process is physically equivalent to inverse bremsstrahlung and so the cross-sections should have the same functional form,

$$\sigma^{(E)} = \sigma^{(B)} \qquad (3\text{-}27)$$

The charge Z appearing explicitly in $\sigma^{(B)}$ is replaced by the effective charge Q_n of the bound state on the grounds that the acceleration occurs at small radii.

This reasoning yields

$$\sigma^{(E)}(E_n, h\nu) = \frac{8\pi^3}{3\sqrt{3}} Q_n^2 a_o^5 \left(\frac{e^2/a_o}{h\nu}\right)^3 \frac{e^2/a_o}{|E_n|} \qquad (3\text{-}28)$$

It is readily shown that the photoelectric rate determined by this cross-section via Eq. (3-25) agrees completely with the rate predicted by $\sigma^{(F)}$ via Eq. (3-20). Thus the two connection formulas (detailed balance and energy continuation) give identical results.

3.9 Line Emission

Next we consider line radiation, and find that the methods developed so far lead us to something new and useful.

Line emission occurs when a bound electron in an upper state n_o spontaneously emits a photon of energy $h\nu = E_{n_o} - E_n$ and falls into a lower level E_n. The rate for this process is normally written in terms of the Einstein rate coefficient $A_{n_o,n}$ as:

$$\frac{dN}{d^3r dt dh\nu} = n_i \sum_n \sum_{n_o} P_{n_o} A_{n_o \rightarrow n} (n_\nu + 1) [1 - \frac{P_n}{D_n}] \qquad (3\text{-}29)$$

The emission rate includes the number of excited electrons in the initial state n_o, a stimulated emission factor, and a degeneracy factor which corrects for possible occupation of the final state. The Einstein rate coefficient is determined by

$$A_{n_o \rightarrow n} = \frac{8\pi^2 e^2 \nu^2}{mc^3} f^{emiss}_{n_o \rightarrow n} I(h\nu) \qquad (3\text{-}30)$$

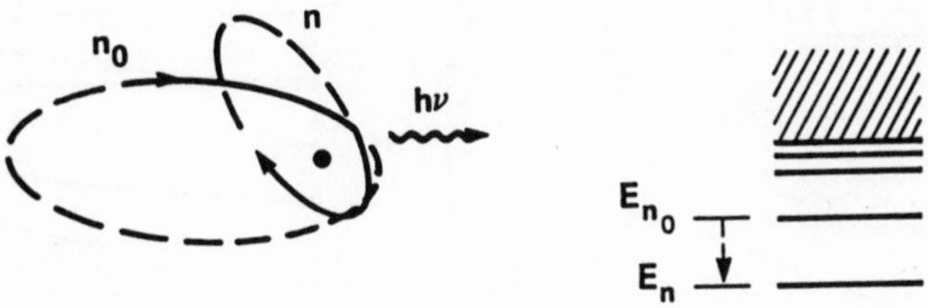

Figure 5c. Schematic representation of line emission. Comparison with Figs. 5a and 5b suggests that the line transition can be regarded as a generalized collision process.

where the line-profile $I(h\nu)$ and emission oscillator-strength $f^{emiss}_{n_o \to n}$ obey

$$\int_{-\infty}^{\infty} I(h\nu)\, dh\nu = 1 \tag{3-31}$$

$$f^{emiss}_{n_o \to n} = \frac{n^2}{n_o^2} f^{abs}_{n,n_o} = \frac{n^2}{n_o^2} \left[\frac{32}{3\pi\sqrt{3}} \frac{n\, n_o^3}{(n_o^2 - n^2)^3}\right] \tag{3-32}$$

The expression for f^{abs}_{n,n_o} is the Kramers formula, and applies to hydrogenic atoms only. As with all the other radiative processes there should be corrections for quantum effects omitted by the Kramers theory, for relativistic effects and above all for the effects of screening by bound electrons. The results we quote below indicate that quantum and relativistic corrections can change f by a factor of two while the screening effects are much larger (factors of ten in some cases).

To obtain a better formula for line emission which includes the bound-electron screening effects, we again use the method of continuation in energy. In this case, the line emission is considered to be an extrapolation of radiative recombination, i.e., a collisional emission in which the initial state is the upper boundstate n_o.

From this viewpoint the rate of emission is written in

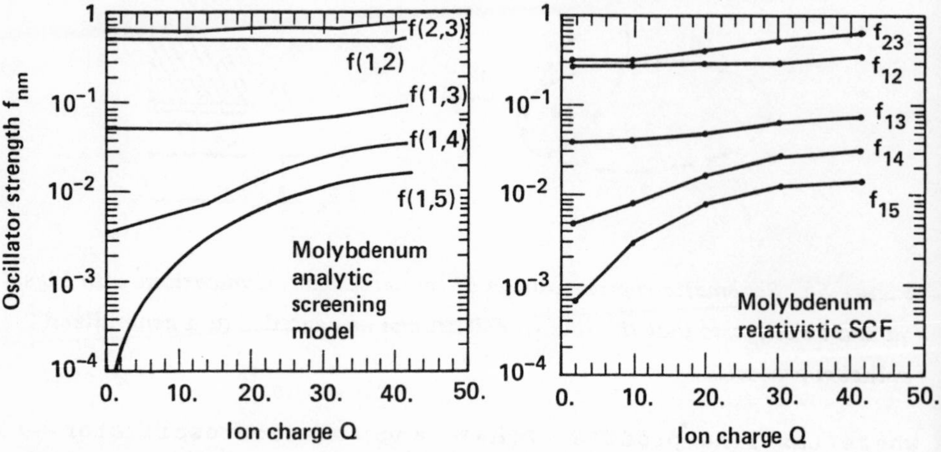

Figure 6a, 6b: Absorption oscillator-strengths for molybdenum ions, $f_{n,n''}$ averaged over initial states and summed over final states. The analytic calculation uses the screening coefficients of reference (35). The relativistic self-consistent field calculations were performed by D. A. Liberman.

terms of the flux of incident electrons ϕ_{n_0} in the upper st a line-emission cross-section $\frac{d\sigma^{(G)}}{dh\nu}$ and the usual final sta factors,

$$\frac{dN}{d^3r\ dt\ dh\nu} = n_i\ \phi_{n_0}\ \frac{d\sigma^{(G)}}{dh\nu}\ (n_\nu + 1)\ [1 - \frac{P_n}{D_n}] \qquad (3\text{-}33$$

In this case the proposed continuation formula is simply

$$\frac{d\sigma^{(G)}}{dh\nu} = \frac{d\sigma^{(D)}}{dh\nu} \qquad (3\text{-}34$$

When Eq. (3-33) is forced back into the usual form of Eq. (3-29) we find a new expression for non-hydrogenic oscillator strengths,[3,36]

$$f^{abs}_{n,n_o} = \frac{4}{3\pi\sqrt{3}}\ \frac{Q_n^4\ Q_{n_0}^2}{n^5\ n_0^3} \left(\frac{e^2/a_0}{h\nu}\right)^3 \qquad (3\text{-}35$$

Equation (3-35) gives a remarkably simple and general

prediction of shell-averaged oscillator-strengths for arbitrarily charged ions. The effective charges Q_n and energy $h\nu = E_{n_o} - E_n$ are determined by the screening model of Eqs. (3-16,17), using the screening coefficients of reference 35.

Equation (3-35) reduces to the hydrogenic form of Eq. (3-32) when $Q_{n_o} = Q_n$. Of course Eq. (3-32) is not exact for hydrogenic ions; there are relativistic and quantum corrections of the order of 50 % (Bethe and Salpeter,[38] Rose[39]).

However Eq. (3-35) manages to nicely reproduce the larger effect of screening by bound electrons (Fig. 6). In the example shown, certain oscillator-strengths change by factors ~ 10 during ionization and Eq. (3-35) follows these changes to 50% accuracy.

Equation (3-35) agrees with a semi-empirical scaling law for 1s-np transitions recently discovered by Benka and Watson.[40] Benka and Watson examine the ratio of rates for transitions ($n_1\ell_1 \to n\ell$) to transitions ($n_o\ell_o \to n\ell$). They express this ratio as

$$\frac{\text{rate } (n_1\ell_1 \to n\ell)}{\text{rate } (n_o\ell_o \to n\ell)} = \frac{P_{n_1\ell_1}}{P_{n_o\ell_o}} \left[\frac{Z^*(n_1\ell_1)}{Z^*(n_o\ell_o)}\right]^a \cdot \left\{\begin{matrix}\text{hydrogenic}\\ \text{ratio}\end{matrix}\right\} \quad (3\text{-}36)$$

and are somewhat surprised to find that the expected value a = 4 gives a poor match to experimental and/or relativistic Hartree-Fock ratios, whereas a = 2 succeeds to ~ 20% for all Z. In Eq. (3-35) the ℓ-dependence is averaged out and the predicted ratio becomes

$$\frac{\text{rate } (n_1 \to n)}{\text{rate } (n_o \to n)} = \frac{P_{n_1}}{P_{n_o}} \frac{Q_{n_1}^2}{Q_{n_o}^2} \frac{h\nu_{n_1 \to n}}{h\nu_{n_o \to n}} \cdot \left\{\begin{matrix}\text{hydrogenic}\\ \text{ratio}\end{matrix}\right\} \quad (3\text{-}37)$$

Because the energy difference $h\nu$ is dominated by the charge Q_n of the lower state, Eq. (3-35) correctly predicts $a \simeq 2$.

The physical ingredients of Eq. (3-35) are recognized by various previous workers: the dominance of screening at the inner turning point and the factors $\partial E_n/\partial n$ which are equivalent to wave-function normalization factors. It is also true that more accurate oscillator-strengths are readily obtained from quantum calculations. However Eq. (3-35) represents a significant advance for fusion applications in which a high priority is placed upon simplicity and generality.

The oscillator strength f_{nm} is also used in calculations of excitation and ionization by electron impact. With Eq. (3-35) the rates for inner-shell excitation of heavy ions can be greatly improved over hydrogenic values.

We conclude this section with the practical formula for radiative line opacity which follows immediately from Eqs. (3-30) to (35):

$$K_\nu^{BB} = \sum \frac{\pi e^2 h}{mc\, AM_p} f_{n \to n_0} P_n \{1 - \frac{P_{n_0}}{D_{n_0}}\} I(h\nu)$$

$$= 6.61 \times 10^4 \frac{cm^2}{g} \frac{f_{nn_0} P_n \left\{1 - P_{n_0}/2n_0^2\right\}}{A} I_{nn_0}(h\nu) \qquad (3\text{-}38$$

In this equation the line-shape $I(h\nu)$ is normalized as in Eq. (3-31) and the absorption oscillator-strength is to be taken from Eq. (3-35).

4.

ENERGY-EXCHANGE COEFFICIENTS

Electrons and ions lose energy as they move through a plasma and this process is described variously as slowing, stopping, energy-loss or energy-deposition. The rate of energy loss is measured by $dE/dt = v\, dE/dx$, where the stopping power dE/dx is proportional to the target density. The energy-loss is caused by electrical interactions and is therefore closely related to other plasma processes. In this section we will show that the plasma opacity K_ν can be used to calculate the stopping power dE/dx, which in turn gives rise to an approximate formula for the collisional heat transfer between electrons and ions.

Theory shows that fast-ion energy-loss should increase in moderately hot targets, i.e., the ion range (in cm^2/g) in plasma is shorter than the range in the cold solid. This prediction refers to partially-ionized plasmas and is another failure of the "standard recipe" of section I.

There have been several recent efforts to measure temperature dependent stopping in laboratory plasmas, and the experimental results appear to support the idea of range-shortening without sufficient accuracy to significantly constrain the theory. Because most of the experimental work is not yet published we do not attempt to review it here (see references 41, 42).

Why is it necessary to understand ion energy-loss? One critical application is in heavy-ion fusion research. Because of the expense of a high-current heavy-ion accelerator, one must be confident of understanding beam-target coupling before embarking on a large construction project. The performance of an ion fusion experiment would depend vitally upon theoretical deposition calculations; if the actual stopping power (dE/dx) fell below theoretical expectations, the target temperature would be reduced with consequent degradation in all aspects of target performance.

Another important application arises in laser-driven fusion experiments with long-wavelength lasers, where a substantial fraction of the laser energy goes to generate energetic suprathermal electrons with energies of 10 - 200 keV. These electrons preheat the fuel and otherwise redistribute the absorbed laser energy, usually in harmful fashion. Suprathermal electrons are detected by energetic bremsstrahlung x-ray emission; the efficiency of emitting x-rays is essentially the ratio [n_i hν $d\sigma^{(A)}/dh\nu$] / [dE/dx]. For a given (measured) x-ray signal, a change in the theoretical formula for energy-loss thus implies a change in the inferred hot-electron population.(43)

Fusion reactions produce 3.5 MeV alpha particles and these heat the plasma through the energy-loss mechanism. Alpha energy deposition occurs mainly in fully-ionized non-degenerate low-Z fuel where the usual plasma formula is likely to be adequate. However some alpha energy is deposited in the high-Z pusher of a small fusion target (Skupsky(44)) and this again raises the question of non-ideal plasma effects. In this case a change in the theoretical expression for dE/dx alters the predicted ignition of marginal targets.

4.1 Ion Stopping in Plasma Targets

There is a large body of theory and experiment concerning ion ranges in cold solid matter.(45,46) At the other extreme the fully-ionized plasma may be regarded as understood, at least for low-intensity beams of fully-stripped projectiles. We want a more comprehensive theory which goes between the limits of cold and hot matter

Nardi, Peleg and Zinamon(47) examined the consequences of ionization for ion stopping, which occurs through electrical energy transfer to electrons of the target material. In solid targets, the energy transfers must exceed the 10-100 eV average binding of electrons. In a heated target, however, some electrons are ionized into free states where they can accept much smaller energy transfers < 1 eV. This change increases the stopping power of a partially-ionized target. At higher temperature, the energy-loss should again decrease when the average thermal velocity of free electrons becomes comparable to the ion speed.

Basic Stopping Formula

Our discussion of charged-particle stopping begins with the expression for ion energy-loss to target electrons,

$$- \frac{dE}{dx} = \frac{4\pi (Z_1^*)^2 e^4}{m v_o^2} (n_F L_F + n_B L_B) \tag{4-1}$$

In this equation, Z_1^* = instantaneous charge of the projectile, m = electron mass, v = ion velocity, n_F, n_B = free, bound electron number densities, and L_F, L_B = stopping number for free, bound electrons. Equation (4-1) is derived in standard textbooks.[48] To evaluate the formula for hot plasma targets we need to know the projectile effective charge Z_1^* and the target stopping numbers L_F, L_B.

Projectile Charge-State

The projectile charge-state Z_1^* appears squared in Eq. (4-1), so it has a strong effect on the answer. The simplest assumption is that Z_1^* is determined by the nuclear charge Z_1 of the projectile and the projectile velocity v according to an empirical formula proposed by Betz,[46]

$$Z_o(v) = Z_1(1 - \exp(- v/Z_1^{2/3} v_o)) \tag{4-2}$$

where $v_o = e^2/\hbar$ is the Bohr velocity. The formula suggests a Thomas-Fermi scaling law, and indeed one obtains a very similar (but not identical) result by requiring that

$$1/2\ m\ v^2 = I_{TF}(Z_1, Z_1^*) = Z_1^{4/3} f(Z_1^*/Z_1)$$

This equation says that the ion strips until its ionization potential equals the impact energy (seen in the ion frame) of collision with an electron at rest in the target. The implication is that the ion charge state is determined by the threshhold for ionization which evidently prevails against recombination processes.

This logic predicts that the charge state Z_1^* should be independent of the target material or even of the target temperature (at least for high-velocity projectiles). Experiments on solids are in reasonable agreement with this expectation, but experiments on gas targets give somewhat different charge states probably indicative of multi-step ionization.(46)

In any case it is clear that a low-velocity ion will strip to a charge state greater than or equal to the charge it would have as an impurity at rest in the given plasma, which we may express

$$Z_1^* \geq Q(Z_1, \rho, T) \tag{4-4}$$

The most interesting uncertainty concerning Z_1^* is whether high-velocity heavy ions which arrive in a low charge state ($Z_1^* \cong 1$-5) will be stripped to the equilibrium charge $Z_0(v)$ before traversing an appreciable target thickness.(3,49) The reason for concern on this point is the small cross-section for inner-shell impact ionization of heavy ions. If the charge remains smaller than $Z_0(v)$, the range is lengthened and the energy deposition occurs over a larger target volume.

It has also been suggested that charge states may actually be determined by an interplay of ionization-recombination processes including charge transfer from target atoms.(50) The electrons involved in charge transfer are outer electrons, easily ionized, and so the charge transfer rate (and hence Z_1^*) should be sensitive to the target temperature.

4.4.1 Free Electron Stopping

Charged-particle stopping in a uniform electron gas

has been thoroughly studied by Skupsky,[51] Arista and Brandt,[52] More,[53] and Deutsch et al.[54] These calculations employ the random phase approximation (RPA) developed by Lindhard for stopping by a degenerate electron gas, in this case extended to arbitary electron density n_e and temperature T. At high temperatures where electrons are nondegenerate, this theory becomes the usual plasma dielectric function method.

These calculations based upon linear-response theory assume a low-charge projectile (e.g., a proton). Both binary collisions and collective loss mechanisms are included without introducing any arbitrary cutoff. We merely quote the results of these calculations here; for high velocities ($v \gg \bar{v}_e$).

$$L_F \cong \log \left(\frac{2mv^2}{\hbar\omega_p}\right) \tag{4-5}$$

(ω_p is the electron plasma frequency.) Equation (4-5) applie equally to degenerate or nondegenerate electrons. However the validity criterion ($v \gg \bar{v}_e$) depends on temperature, because $\bar{v}_e$ is the appropriate average thermal velocity of target electrons. The corrections to Eq. (4-5) form a series of inverse powers of $(\bar{v}_e/v)$ with coefficients given by Deutsch <u>et al</u>.[54]

When the ion slows to velocities comparable with $\bar{v}_e$ the stopping is greatly reduced. In the nondegenerate case, the low-velocity stopping-power is

$$L_F \simeq \left(\frac{v}{\bar{v}_e}\right)^3 \frac{1}{3}\sqrt{\frac{2}{\pi}} \log \frac{2kT}{\hbar\omega_p} \tag{4-6}$$

In the degenerate case one has

$$L_F \simeq \frac{1}{2}\left(\frac{v_o}{v_F}\right)^3 \frac{1}{(1-\frac{x}{3})^2}\left[\log\left(\frac{1}{x}+\frac{2}{3}\right) - \frac{1-x/3}{1+2x/3}\right] \tag{4-7}$$

where

$$x = 1/\pi a_o k_F$$

with $\quad k_F = mv_F/\hbar$

$$= (3\pi^2 n_e)^{1/3}$$

The final result of this free-electron stopping theory is a stopping-power which decreases monotonically with increasing temperature. One only obtains the correct (opposite) result when one considers the effects of ionization.

4.4.2 Bound-Electron Stopping

For high-velocity ions ($v > Z_2 e^2/\hbar$) the Bethe theory (based on the Born approximation) gives a stopping number closely analogous to Eq. (4-5),

$$L_B = \log (2mv^2/\bar{I}) \tag{4-8}$$

where $\bar{I}$ is the average excitation/ionization energy, informally written as

$$\log \bar{I} = (\sum_j f_{ij} \log(E_i - E_j))/\sum_j f_{ij} \tag{4-9}$$

This notation may be somewhat misleading because the summation on the right-hand side must cover all possible initial and final states, including bound and free states. f_{ij} is the oscillator strength for the transition in question. A somewhat more precise expression for $\bar{I}$, within the average-atom model, is

$$\log \bar{I} = \frac{\sum_{n,m} P_n(D_m - P_m) f_{nm} \log(E_m - E_n) + \sum_n \int d\varepsilon\, P_n f_{n\varepsilon} \log(\varepsilon - E_n) g(\varepsilon)(1 - f(\varepsilon))}{\sum_{n,m} P_n(D_n - P_m) f_{nm} + \sum_n \int d\varepsilon\, P_n f_{n\varepsilon} g(\varepsilon)(1 - f(\varepsilon))} \tag{4-10}$$

In this equation, n and m are symbolic one-electron quantum numbers; it would be a very poor approximation to neglect subshell splitting in the evaluation of Eq. (4-10), at least for states having the same principal quantum number. For the continuum contribution, $g(\varepsilon)$ is the density of states, $f_{n\varepsilon}$ is the photoelectric oscillator strength and $f(\varepsilon)$ is the Fermi distribution.

A great deal of atomic data is required to directly evaluate Eq. (4-10). For ions of Al, Kr and Au, McGuire[55] has carried out calculations using the Born approximation for excitation and ionization cross-sections evaluated with generalized oscillator strengths. His results are shown in Figure 7.

At present there is no suitable experimental data for stopping by hot (ionized targets) which would make it possible to directly determine $\bar{I}(Z,Q)$; however there is a large amount of data for stopping by cold solid targets giving $\bar{I}(Z,0)$ for neutral atoms ($Q = 0$). For $Z \geq 10$ the experimental results are reproduced to about 10 % accuracy by the well-known formula[46,56]

$$I(Z,0) \cong a\ Z \tag{4-11}$$

with $a = 10$ eV. The value of $\bar{I}(Z,0)$ is only weakly dependent on the state of chemical bonding of the valence electrons.

A result much like Eq. (4-11) is predicted by the theory of Bloch[57] according to which the atomic photoinduced transitions (line and photoelectric absorption) scale with a local plasma frequency of the atomic electron gas:

$$\log \bar{I} = \frac{\int n(r) \log [\hbar\omega_p(r)]\, d^3r}{\int n(r)\, d^3r} \tag{4-12}$$

where $n(r)$ = electron density within the atom, and $\omega_p(r) = (4\pi e^2\, n(r)/m)^{1/2}$ is the local electron plasma frequency. If Eq. (4-12) is evaluated with the electron density resulting from the Thomas-Fermi theory, the TF Z-scaling property gives a result of the same functional form as Eq. (4-11). The Thomas-Fermi theory predicts a numerical coefficient $a \cong 6.5$ eV, somewhat smaller than the experimental result for neutral atoms.

In collaboration with J. Harte the author has evaluate Eq. (4-12) for Thomas-Fermi ions and found the result to be accurately reproduced by[14,43]

$$\bar{I}(Z, Q) = aZ \frac{\exp\left[1.29\ x^{(.72 - .18\ x)}\right]}{\sqrt{1 - x}} \qquad (4\text{-}13)$$

where $x = Q/Z$ is the fractional ionization. This one formula describes all possible ionization states of all elements to the accuracy of Thomas-Fermi theory. With the empirical coefficient $a = 10$ eV, it does quite well in comparison with the quantum calculations (Fig. 7).

It is observed by Peek[58] that Eq. (4-12) and hence Eq. (4-13) fail in the hydrogenic case ($Q = Z-1$) where we expect to find $\bar{I} \sim Z^2$. The objection is correct, but of course in that limit the bound electron stopping is negligable.

Based on the comparison in Fig. 7 it appears that Eq. (4-13) gives a good enough account of high-velocity ion stopping for most practical applications. In particular it should be used in preference to other formulations[59] whic (incorrectly) replace $\bar{I}(Z,Q)$ by the ionization potential $I(Z,Q) \propto Z^{4/3}\ f(x)$.

Equation (4-13) also can be used to determine the fast electron stopping in laser fusion targets. Use of this formula has removed a long-standing discrepency between LASNEX calculations of suprathermal bremsstrahlung and experimental data for x-ray production by electron beams in cold targets.[43]

Stopping and Virtual Photons

A qualitative consistency must exist between opacity and the charged particle stopping-power; in fact dE/dx proves to be a average of the photon absorption cross-section like the Planck or Rosseland means.[36]

This connection is seen by representing the electromagnetic field of a fast charged particle as a superposition of virtual photons having the frequency distribution

$$N(h\nu) = \frac{2\alpha}{\pi h\nu} Z_1^2 \left(\frac{c}{v}\right)^2 \log\left(\frac{av}{2\pi b_o \nu}\right) \qquad (4\text{-}14)$$

In this well-known expression v is the ion speed (assumed $<<$ c), c is the speed of light, $\alpha = e^2/\hbar c$ is the fine-structure constant, a is a constant near unity and b_o is the appropriate minimum impact parameter.[48]

The energy-loss results from absorption of these virtual photons by bound or free electrons of the target plasma. Strictly speaking, virtual photons do not obey the dispersion relation $\omega = c \mid \vec{k} \mid$ of the free photon, but this does not matter to the extent that one calculates bound-electron absorption in the dipole approximation. Then the bound-electron opacity K_ν^{BB} of Eq. (3-38) generates an energy-loss of the form

$$-\left(\frac{dE}{dx}\right)_{BB} = \int h\nu\, N(h\nu)\, \rho K_\nu^{BB}\, dh\nu$$

$$= \frac{4\pi\, Z_1^2 e^4}{mv^2} n_I \sum_{n,m} f_{nm}^{abs} P_n\left(1 - \frac{P_m}{D_m}\right) \log\left(\frac{av}{2\pi b_o \nu_{nm}}\right) \qquad (4\text{-}15)$$

This expression is equivalent to the Bethe formula for high-velocity bound-electron stopping, Eqs. (4-8, 4-9), when the appropriate minimum impact parameter $b_o = \hbar/mv$ is selected.

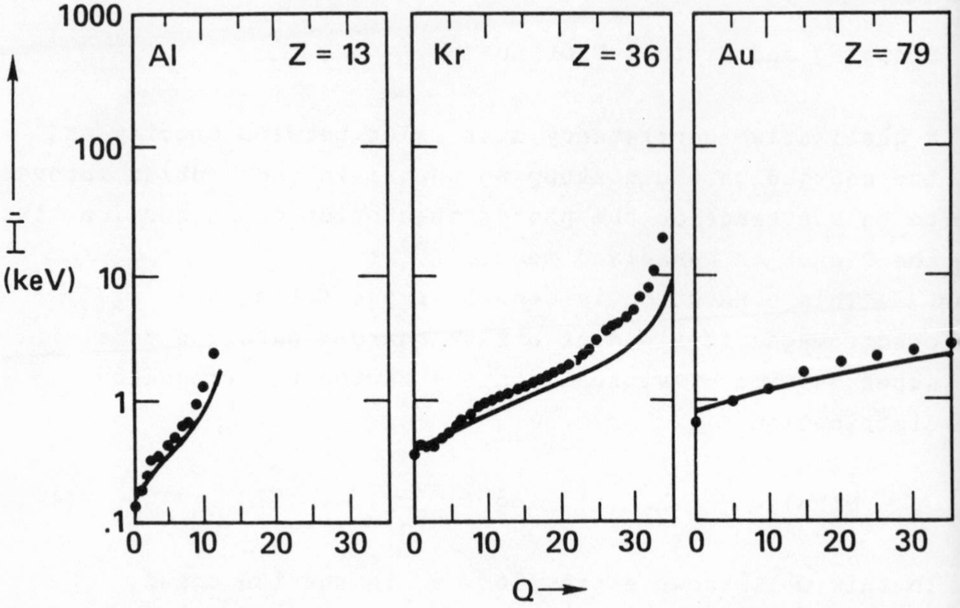

Figure 7. Average Ionization/Excitation potential for ions of aluminum, krypton and gold. The points are generalized oscillator strength calculations of the Bethe sum (E. J. McGuire). The solid curve is Eq. (4–13). The agreement appears good enough for practical applications.

4.6 Problems of the Bethe Method

The Bethe theory expressed in Eq. (4-8, 4-9) for bound-electron stopping is valid at large projectile energies where the projectile moves more rapidly than any of the target electrons. At lower energies, the bound electrons become ineffective and their contributions to L_B decrease.

A bound electron of principal quantum number n has an average orbital velocity $v_n = Q_n e^2/n\hbar$ where Q_n is the appropriate screened charge for shell n. The contribution of this shell to L_B decreases when $v < v_n$.

This important effect is the <u>shell correction</u>.[45] A tractable formula for the shell correction exists only for the K-shell contribution, which is first to be modified as the ion velocity falls below $Ze^2/\hbar$. At lower velocities the Bethe formulation becomes inconvenient.

In addition, the significance and/or definition of $\bar{I}$ becomes rather mysterious for very dense plasmas where one cannot sharply distinguish bound and free states of closely packed ions. It is not clear that any practical modification of Eq. (4-10) would apply to degenerate plasmas.

Inhomogeneous Electron-Gas Theory

There is another approach to stopping calculations which is somewhat more powerful than the Bethe theory.[60] This method is based on the inhomogeneous electron-gas model: the idea that the electrons inside an atom can be treated as if they were a free electron gas of varying density, so that the stopping from electrons at density $n(r)$ is $L_F(n(r), T; v)$. In this case the electron density $n(r)$ is taken from complete Thomas-Fermi or quantum cell-model calculations at the given plasma density-temperature conditions. Then one calculates an atomic stopping number as

$$n_I L_{at} = n_F L_F + n_B L_B = \int n(r)\, L_F(n(r), T; \nu)\, d^3r \qquad (4\text{-}16)$$

The free-electron stopping number L_F is assembled from high- and low-velocity forms given above in Eqs. (4-5 to 4-7).

This method includes shell corrections in the sense that within the K-shell region, the electron density $n(r)$ is very large and the local Fermi velocity $\sim n^{1/3}$ is correspondingly large. For low ion velocities we see from Eq. (4-7) that the stopping contribution is inhibited.

Several density-temperature effects including thermal or pressure ionization and continuum lowering are automatically built into the Thomas-Fermi theory. There are no adjustable parameters.

The predictions of this method are rather good for cold matter, as has been shown in detail by Chu and Powers(and Anderson and Zeigler.[(56)] Logically, the method should become more accurate for elevated temperatures where the electron gas becomes more uniform throughout the atomic volume.

A representative calculation by this method is given in Fig. 8.

4.8 Electron-Ion Heat Exchange

We conclude this section with another example of a useful connection between plasma process coefficients: a calculation of the electron-ion energy-exchange rate by use of the low-velocity ion stopping formula of Eq. (4-6).

We consider a plasma in which ions are hotter than electrons so that $kT_i >> kT_e$, but assume that $kT_i/M_i << kT_e/m$ so that the electrons still move faster than ions. In this case Eq. (4-6) is the appropriate stopping number.

The ion energy-loss to the electron gas is now physically the same process as the electron-ion heat exchange, and we have[(36)]

$$\frac{1}{\tau_{ei}} = \frac{1}{T_i}\frac{dT_i}{dt} = \frac{1}{\frac{3}{2}\,kT_i}\left\langle v\,\frac{dE}{dx}\right\rangle \tag{4-17}$$

In this equation the average is taken over a Maxwell distribution of ion velocities v. The energy per ion is taken to be the ideal-gas energy $3/2\ kT_i$. The resulting relaxation time,

$$\tau_{ei} = \frac{3}{8\sqrt{2\pi}}\;\frac{m_e M_i V_e^3}{Z^2 e^4 n_e \log\left(\frac{2kT_e}{\hbar\omega_p}\right)} \tag{4-18}$$

agrees with the form originally obtained by Landau.[(62)]

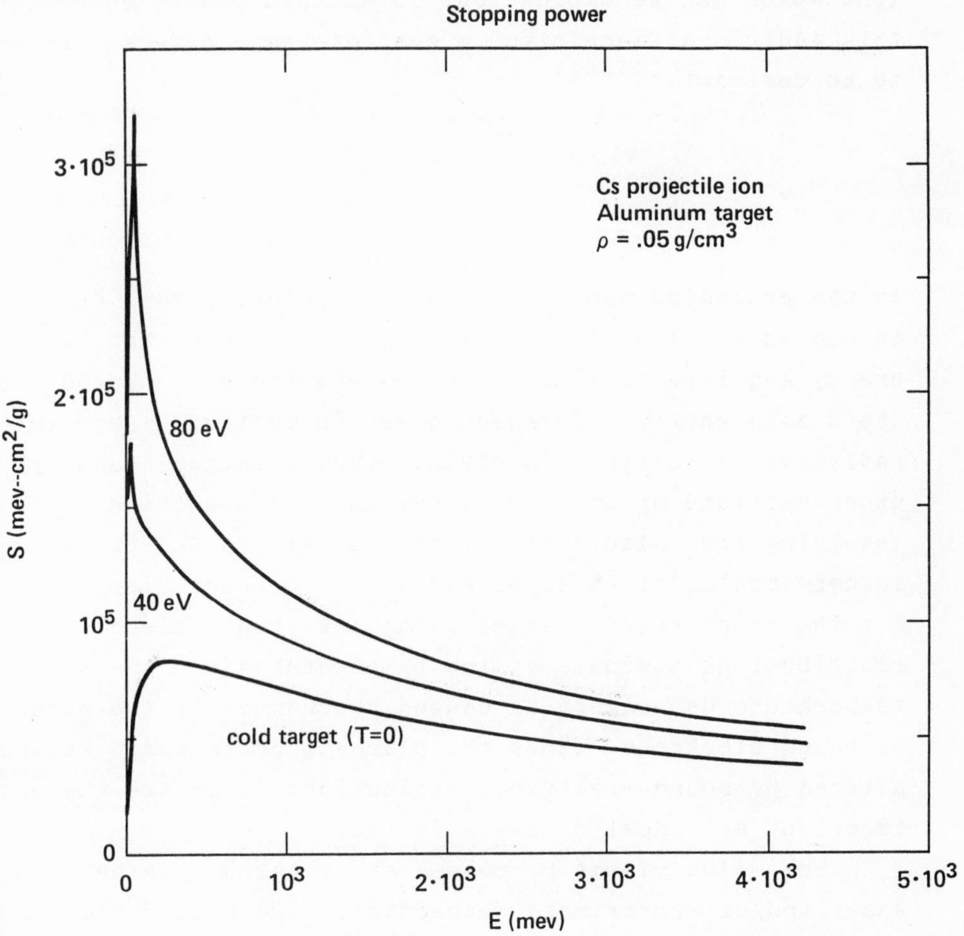

Figure 8. Stopping of cesium ions in aluminum showing temperature-dependent energy-loss. E is the projectile energy and S is $(1/\rho)dE/dx$; the curves are labelled by the target temperature. The calculations were performed by the inhomogeneous electron-gas method.

The collisional heat-transfer described by Eq. (4-18) is only part of the story. One can show that there is an additional mechanism for heat transfer in non-ideal plasmas associated with electrical energy transfer occuring in ionization and recombination; at certain plasma conditions this additional mechanism, a sort of Raman effect, appears to be dominant.[36,63]

5.

CONCLUSIONS

In the preceding pages we examined typical plasma processe in non-ideal plasmas. In section 2, we found that the energy required to liberate bound electrons can be 50 % of the plasma energy. Formulas given in section 3 show that radiative transitions involving bound electrons have cross-sections up to 100 times the cross-sections involving free electrons; of course this is due to the larger accelerations experienced by the bound electrons. For the stopping of charged particles, bound electron contributions dominate at low temperatures and the main temperature dependence is caused by changes in the number of bound electrons. Thus the plasma processes are strongl altered by bound-electron contributions which are the most important non-ideal plasma effects.

The major plasma processes are related by a network o exact and/or approximate connections. We have found connections between pressure and energy in Eqs. (2-2, 2-26 formulas which connect absorption and emission cross-sections in Eqs. (3-8, 3-21) and also quantitative connections between bound-electron and free- electron cross-sections in Eqs. (3-13, 3-27, 3-34). In addition we found that the bound electron contribution to stopping pow dE/dx can be calculated from the radiative opacity K_ν, and that the low-energy stopping is essentially equivalent to the electron-ion heat exchange rate.

Figure 9 suggests other connections between plasma processes. The plasma x-ray index of refraction $n(\nu)$ can be calculated from the absorption coefficient K_ν by

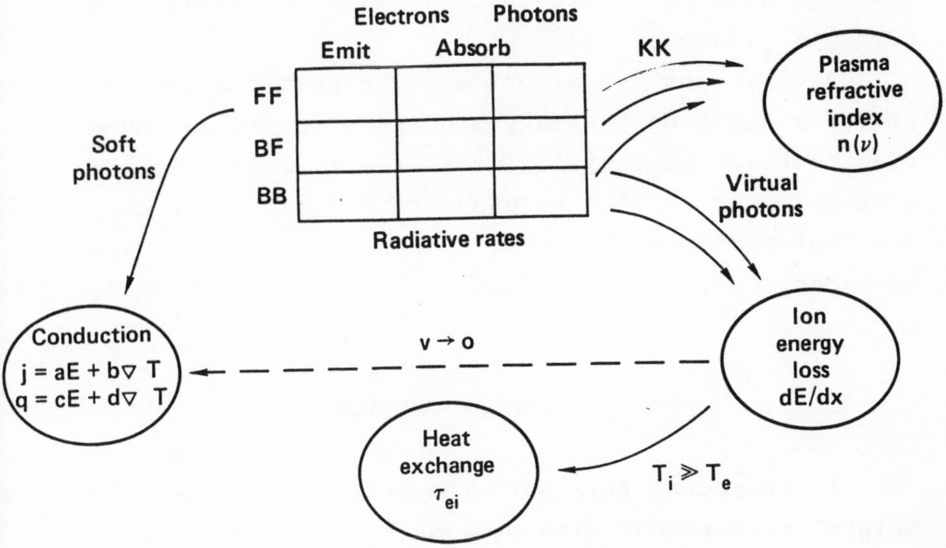

Figure 9. A graphical summary of connections between plasma process coefficients.

the Kramers-Kronig relation. The low-velocity ion stopping is described by a friction coefficient which gives the drag force on an ion moving slowly through an electron gas. The same friction coefficient describes the resistance an ion fluid exerts on a current-carrying electron gas and is therefore essentially equivalent to the plasma electrical resistivity.

Inverse bremsstrahlung absorption of very soft photons ($h\nu \to 0$) is physically equivalent to Joule heating or AC resistivity, so that the soft photon limit of the bremsstrahlung cross-section is related to the elastic scattering cross-section.

Most of these connections are extensions of basic laws known from condensed matter or other fields of physics. For non-ideal plasmas the connections point the way toward a firmer theoretical structure. For example, any new mechanism introduced in radiative absorption calculations has implications for the other plasma processes (e.g., dE/dx, τ_{ei}, etc.).

A great deal of effort will be required before one ha properly explored the many processes involving bound electrons and nonequilibrium states in dense non-ideal plasmas. Our ability to understand x-ray laser or inertial-fusion experiments will improve as this work advances.

ACKNOWLEDGEMENT

In preparing this work the author has benefitted from helpful discussions with many scientists especially including G. B. Zimmerman, D. S. Bailey and J. A. Harte of the LASNEX code group.

Work performed under the auspices of the U.S. Department of Energy by the Lawrence Livermore National Laboratory under contract number W-7405-ENG-48.

REFERENCES

1. L. Spitzer, Jr., Physics of Fully Ionized Gases, 2nd Rev. Ed., Interscience Publishers, New York (1962).
2. N. A. Krall and A. W. Trivelpiece, Principles of Plasma Physics, McGraw-Hill Book Company, New York (1973).
3. R. M. More, Atomic Physics in Inertial Confinement Fusion, preprint UCRL-84991, Lawrence Livermore National Laboratory (1981).
4. Laser Program Annual Report, Lawrence Livermore National Laboratory report UCRL-50021 (1978 to present).
5. S. Brush, H. Sahlin and E. Teller, J. Chem. Phys. 45, 2102 (1966).

6. J.-P. Hansen, Phys. Rev. A8, 3096 (1973); J.-P. Hansen, Laser-Plasma Interactions, SUSSP, Edinburgh Physics Department, 1980.
7. M. Baus and J.-P. Hansen, Physics Reports 59, 1 (1980).
8. H. R. Griem, Plasma Spectroscopy, McGraw-Hill, New York (1964); H. R. Griem, Handbook of Plasma Physics, Eds. M. N. Rosenbluth and R. Z. Sagdeev, Vol. 1, p. 73, North-Holland Publishing Co., 1983.
9. M. D. Rosen, P. L. Hagelstein, et al., Phys. Rev. Lett. 54, 106 (1985); D. L. Matthews, P. L. Hagelstein, et al., Phys. Rev. Lett. 54, 110 (1985).
10. Ya. B. Zel'dovich and Yu. P. Raizer, Physics of Shock Waves and High-Temperature Hydrodynamic Phenomena, Vol. 1, Ed. W. D. Hayes and R. F. Probstein, Academic Press, New York (1966).
11. R. F. Schmalz and J. Meyer-ter-Vehn, Phys. Fluids 28, 932 (1985).
12. R. More, Lawrence Livermore National Laboratory report UCRL-84379 (1980).
13. D. A. Boercker and R. M. More, Phys. Rev. A, to be published.
14. R. M. More, Atomic and Molecular Physics of Controlled Thermonuclear Fusion, p. 399, Ed. by C. Joachain and D. Post, Plenum Publishing Corp. (1983).
15. H. Ahlstrom, "Diagnostics of Experiments on Laser Fusion Targets at LLNL," in Physics of Laser Fusion, Vol. II, UCRL-53106 (1982).
16. C. DeMichelis and M. Mattioli, Rep. Prog. Phys. 47, 1233 (1984).
17. M. Lamoureux, C. Möller and P. Jaegle, Phys. Rev. A30, 429 (1984).
18. P. H. Lee and M. D. Rosen, Phys. Rev. Lett. 42, 236 (1979).
19. H. A. Kramers, Philos. Mag. 271, 836 (1923).
20. L. D. Landau and E. M. Lifshitz, Classical Theory of Fields, 2nd Ed., Pergamon Press, Oxford, 1962.

21. W. J. Karzas and R. Latter, Astrophysical Journal Suppl. no. 55, Vol. VI, p. 167 (1961); P. J. Brussard and H. C. Van de Hulst, Rev. Mod. Phys., 34, 507 (1962); I. P. Grant, Mon. Not. R. Astron. Soc. 118, 352 (1958).
22. M. Lamoureux and R. H. Pratt, Radiative Properties of Hot Dense Matter, p. 241, Ed. by J. Davis et al., World Scientific, Singapore (1985).
23. V. I. Kogan and A. B. Kukushkin, Soviet Physics JETP 60, 665 (1984).
24. I. J. Feng and R. H. Pratt, unpublished preprint; C. M. Lee and R. H. Pratt, Phys. Rev. A12, 707 (1975).
25. J. P. Cox and R. T. Giuli, Principles of Stellar Structure, Vol. 1, Gordon and Breach, New York, 1968.
26. J. Dawson and C. Oberman, Physics of Fluids 6, 394 (1963).
27. B. F. Rozsnyai, J.Q.S.R.T. 22, 337 (1979).
28. J. M. Green, R. and D. Associates unpublished report RDA-TR-108600-003 (1980).
29. M. Lamoureux, I. J. Feng, R. H. Pratt and H. K. Tseng, J.Q.S.R.T. 27, 227 (1982).
30. L. Kim, R. H. Pratt and H. K. Tseng, Phys. Rev. A32, 1693 (1985).
31. S. Ichimaru, Basic Principles of Plasma Physics, W. A. Benjamin, Inc., Reading, Mass., 1973.
32. H. Mayer, Los Alamos Scientific Laboratory report LA-647 (unpublished) 1947.
33. W. Lokke and W. Grasberger, Lawrence Livermore National Laboratory report UCRL-52276 (1977).
34. G. B. Zimmerman and R. M. More, J.Q.S.R.T. 23, 517 (1980).
35. R. M. More, J.Q.S.R.T. 27, 345 (1982).
36. R. M. More, Lawrence Livermore National Laboratory report UCRL-93926, January (1986).
37. W. Huebner, M. F. Argo and L. D. Ohlsen, J.Q.S.R.T. 19, 93 (1978).

38. H. A. Bethe and E. E. Salpeter, Quantum Mechanics of One- and Two-Electron Atoms, Academic Press, Inc., New York, 1957.

39. S. J. Rose, Rutherford-Appleton Laboratory, unpublished preprint RL-82-114, "The Effect of Relativity on the Oscillator Strengths of Hydrogen-Like Ions," Dec. 1982.

40. O. Benka and R. Watson, Phys. Rev. A29, 2255 (1984).

41. International Workshop on Atomic Physics for Ion Driven Fusion, Journal de Physique, Colloque No. 8, Tome 44, Suppl. an FASC.II (1983).

42. C. Deutsch, Ann. Phys. Fr. 11, 1986.

43. J. A. Harte and R. M. More, Lawrence Livermore National Laboratory, unpublished report UCRL-50021-82, "Laser Program Annual Report," p. 3-66.

44. S. Skupsky, Phys. Rev. Lett. 44, 1760 (1980).

45. H. A. Bethe and J. Ashkin, Experimental Nuclear Physics, p. 166, J. Wiley and Sons, New York, 1953.

46. H. D. Betz, "Charge Equilibration of High-Velocity Ions in Matter," in Methods of Experimental Physics, ed. P. Richard, Academic Press, Vol. 17, p. 73 (1980).

47. E. Nardi, E. Peleg and Z. Zinamon, Phys. Fluids 21, 578 (1978); E. Nardi and Z. Zinamon, Phys. Rev. A18, 1246 (1978).

48. J. D. Jackson, Classical Electrodynamics, 2nd Ed., J. Wiley & Sons, Inc., New York, p. 724 (1975).

49. D. S. Bailey, Y. T. Lee and R. M. More, Journal de Physique, Colloque No. 8, p. 149 (1983).

50. E. Nardi, Z. Zinamon, Journal de Physique, Colloque No. 8, p. 93 (1983).

51. S. Skupsky, Phys. Rev. A16, 727 (1977).

52. N. Arista and W. Brandt, Phys. Rev. A23, 1898 (1981).

53. R. M. More, Lawrence Livermore National Laboratory, unpublished reports 84115 (1980), 84991 (1981) and 87147 (1982).

54. C. Deutsch, G. Maynard and H. Minoo, Laser Interaction

and Related Plasma Phenomena, Ed. by H. Hora and G. Miley, p. 1029, Plenum Publishing Corp. (1984).

55. E. J. McGuire, PRA26, 125 (1982); E. J. McGuire, J. M. Peek and L. C. Pitchford, Phys. Rev. A26, 1318 (1982); E. J. McGuire, unpublished.
56. H. Anderson and J. Ziegler, Stopping and Ranges of Ions in Matter, Vol. 3, Pergamon Press, New York (1977).
57. F. Bloch, Ann. Phys. (5), 16, 287 (1933).
58. J. M. Peek, Phys. Rev. A26, 1030 (1982).
59. T. A. Melhorn, J. Appl. Phys. 52, 6522 (1981).
60. The inhomogeneous electron gas method is described in Ref. 53 as well as a number of more recent papers.
61. W. K. Chu and D. Powers, Phys. Rev. 187, 478 (1969); Phys. Rev. B4, 10 (1971).
62. L. D. Landau, JETP 7, 203 (1937). The careful reader will observe there is a numerical mistake in the final result of this paper.
63. P. Hagelstein, Physics of Short Wavelength Laser Design, Lawrence Livermore National Laboratory, UCRL-53100 (1981).

INERTIAL CONFINEMENT FUSION

J. Meyer-ter-Vehn

Max-Planck-Institut für Quantenoptik
D-8046 Garching, FRG

In these notes, we discuss inertially confined thermonuclear fusion obtained by means of spherically imploded deuterium-tritium fuel. The emphasis is on the 'inner part' of ICF physics, on the implosion dynamics, central fuel ignition, and energy gain, rather than on the problems of beam/target interaction. Section 1 summarizes the basic concept of ICF and recent achievements on fuel compression and briefly indicates advantages and disadvantages of different driver beams. The key parameters for energy gain are discussed in Section 2 in terms of a model, and some recent results on the gas dynamics of central collapse, final compression and formation of the ignition spark are presented in Section 3.

1.
CONCEPT AND ACHIEVEMENTS

1.1 Inertial Confinement

The goal of inertial confinement fusion (ICF) is to achieve controlled thermonuclear burn in a deuterium-tritium plasma for power generation. The fusion reaction

$$D + T \rightarrow \alpha + n + 17.6\ \text{MeV} \qquad (1.1)$$

has a specific energy release of

$$q_{DT} = 3.37 \cdot 10^{11} \text{ J/g} \tag{1.2}$$

The plasma density n and the confinement time τ envisioned for ICF are very different from those for magnetic confinement fusion (11 orders of magnitude):

confinement	$n(cm^{-3})$	$\tau(sec)$	$n \cdot \tau (sec/cm^3)$
magnetic	10^{14}	10	10^{15}
inertial	10^{25}	10^{-10}	10^{15}

In both cases, however, the product $n \cdot \tau$ has to satisfy the Lawson criterion ($n \cdot \tau \geqq 10^{15}$ sec/cm^3) which is set by the DT-reaction rate $\langle\sigma v\rangle$. Instead of $n \cdot \tau$, the product $\rho \cdot R$ is more commonly used as the confinement parameter in ICF; a spherical configuration of the burning DT fuel is assumed with a mass density ρ and an initial radius R. The confinement time is given by the time of rarefaction $\tau = R/c_s$ which spreads with the sound velocity c_s. The fuel fraction burned during this time is obtained as

$$\phi = \rho R \,/(H_B + \rho R) \tag{1.3}$$

with

$$H_B = 4c_s(m_D + m_T)/\langle\sigma v\rangle \tag{1.4}$$

Eq. (1.3) takes into account burn depletion, i.e. $\phi \to 1$ for $\rho R \gg H_B$. The burn parameter is approximately

$$H_B \cong 7 \text{ g/cm}^2 \tag{1.5}$$

for DT burning at 20-50 keV. For 30 % burn-up, one therefore needs $\rho R \cong 3$ g/cm^2. In the case of 1 mg fuel, this requires a fuel density of 300 g/cm^3 or 1500 times compression of solid DT, and it leads to a thermonuclear energy release of 100 MJ. Since the micro-explosion has to be contained in a reactor cavity, probably not more than 10 mg of DT fuel can be considered. This means that high fuel compression is an essential condition to make ICF work.

The key idea of how to achieve such high fuel compression was first published by Nuckolls and coworkers[1] in 1972. They proposed spherical implosion of microspheres driven by high power laser beams. Time-shaped laser pulses were required to achieve isentropic compression. This concept was then refined by considering multi-layered hollow spheres with the innermost layer made of cryogenic DT[2,3]. Instead of lasers also ion beams and thermal x-rays have been considered as drivers. The simulation of a heavy ion beam driven implosion will be shown explicitly in Section 1.3 to make the standard scheme more transparent. The essential point is that the driver energy is not used to heat the fuel to fusion conditions directly, but to convert it first into kinetic energy of an imploding shell which is then transformed into internal energy of highly compressed fuel after central collapse.

1.2 Central ignition and propagating burn

Beside compression, the second condition is to heat the fuel to the ignition temperature of $T_s \geqq 5$ keV. The specific energy of a plasma with equal amounts of D and T ions at a temperature of 5 keV is

$$\varepsilon_s = 4 \cdot 3/2 \cdot T_s/(m_D + m_T) \cong 600 \text{ MJ/g} \qquad (1.6)$$

From this, a naive energy gain can be calculated by dividing the maximum energy output q_{DT} by the minimum energy investment ε_s. One obtains from Eqs. (1.2) and (1.6)

$$G_{naive} = q_{DT}/\varepsilon_s \cong 500$$

This gain is strongly reduced by the various losses occuring in the process of driving the DT fuel to ignition. It turns out that it is too low for a power reactor.

The way out of this difficulty is to ignite only a small portion of the fuel (the spark region) and then let the thermonuclear energy of the spark propagate and ignite a larger fuel reservoir. This concept is known as central ignition and propagating burn. It is now generally accepted as a necessary step for ICF. A detailed analysis of the gain which can be obtained on this basis will be given in section 2.

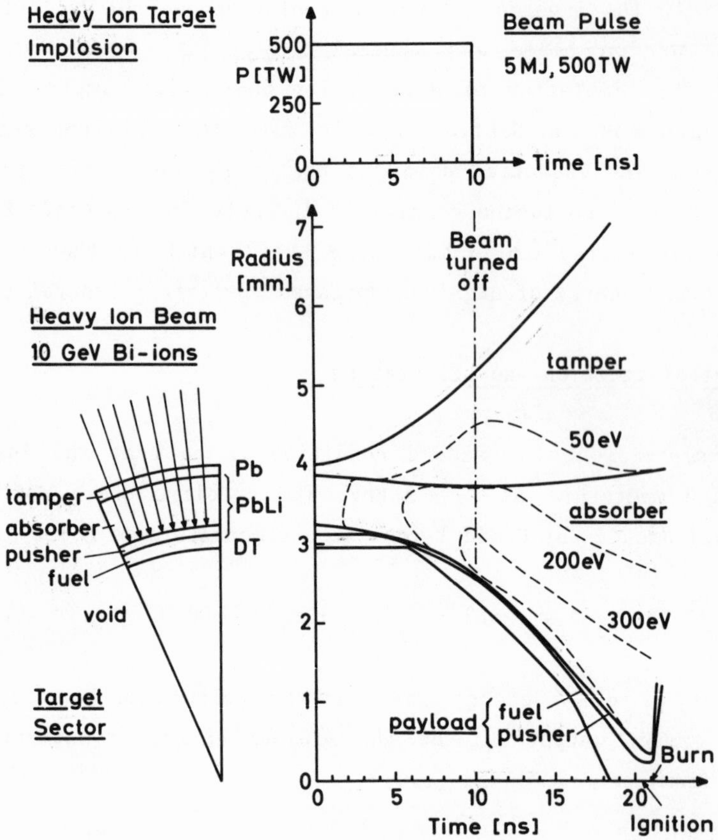

Fig. 1 Numerical simulation of a multilayered, single-shell target implosion driven by a heavy ion beam. A target sector is shown at the left, the power pulse at the top. The solid lines in the radius-time diagram show trajectories of interfaces between the different layers; the broken lines are temperature contours. More details are given in the text.

1.3 An Implosion Simulation

The numerical simulation of a hollow-sphere target implosion driven by a heavy ion beam[4] is shown in Fig. 1. It was performed as part of the HIBALL reactor study[5] and is based on a standard hydrodynamic code. A detailed description is found in Ref. 4. Here, it is reproduced as an example to display some typical features of the target dynamics.

A 5 MJ, 500 TW beam pulse is spherically incident on the target sphere (see target sector at the left side of Fig. 1) and heats the outer absorber layer. The radius-time diagram shows the explosive response of the deposition layer pushing at its inner front the payload towards the centre. The payload consists of an initially cryogenic DT layer (4 mg) and a pusher layer (20 mg) screening the fuel from the hot deposition region. The solid lines mark the interfaces between the different layers. The fuel reaches an implosion velocity of about $3 \cdot 10^7$ cm/s. At 18.3 ns, its inner front arrives at the centre (void closure). The pusher then further compresses the fuel, and ignition conditions are reached at 20.4 ns.

Profiles of density, temperature and pressure at different times are shown in Fig. 2. The beam deposition layer is characterized by temperatures of 200-300 eV and a pressure of 100 Mbar. The density profiles show the payload as a thin high-density layer at the inner front. This layer is Rayleigh-Taylor unstable, and its survival during implosion is one of the most severe problems for ICF. However, we shall not further discuss this topic here. Most dramatic changes are observed in Fig. 2 after void closure when the central values of the hydrodynamic variables jump by several orders of magnitude to truly astrophysical levels. It is the final convergence which leads to fuel ignition and burn and decides about the energy gain.

The target design presented here is untypical in some features. The box-shaped beam pulse creates a strong first shock which causes considerable shock-preheat in the fuel. The implosion could be driven much more isentropically with a gradually increasing beam pulse. However, the advantage of the box pulse is increased hydro-efficiency (8 %) and hydro-stability[4]. The energy gain of the present run is 165.

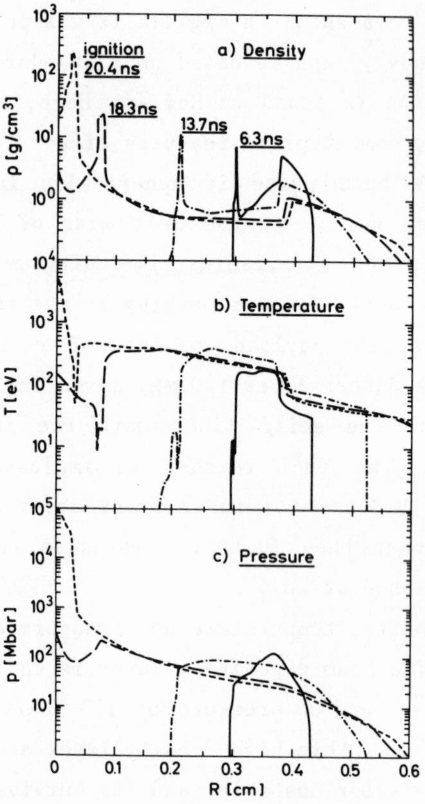

Fig. 2 Spatial profiles of density, temperature and pressure at different times for the target implosion shown in Fig. 1.

1.4 ICF Achievements, Drivers

A synopsis of experimental results from different magnetic confinement and inertial confinement fusion facilities [(6)] is shown in Fig. 3. Record results for ICF have been obtained with the 20 beam Livermore SHIVA laser (maximum pulse energy 10 kJ) and more recently with the 12 beam, 30 kJ GEKKO XII laser at Osaka. In Fig. 3, results from explosive pusher experiments (implosion of glass microballoons filled with DT gas) as well as from ablatively driven implosions are shown. High temperatures (up to 10 keV) and corresponding high neutrons yields (number of fusion neutrons $\leq 4 \cdot 10^{10}$) were reached in the explosive pusher experiments, whereas high confinement $n \cdot \tau$ at relatively low temperature was obtained with the ablatively driven implosions. In these latter experiments, the method of indirect drive was used in which the laser shines through holes into a cavity and the actual pellet is imploded inside this cavity driven by thermal x-rays and plasma pressure (so-called cannon-ball-target). One of the advantages of this method is an improved implosion symmetry. A record compression of 100 times cryogenic DT density was obtained with SHIVA.(7)

The dotted areas in Fig. 3 are envisaged with larger facilities which come into operation in the next years. The 100 kJ NOVA facility at Livermore which is now being completed is hoped to achieve 500 to 1000 times solid DT density with the formation of a hot spot for ignition. It is seen from Fig. 3 that the development of inertial confinement fusion and of magnetic confinement fusion is on a comparable level, and it is expected that ignition and burn is achieved on both ways in the near future.

Lasers are certainly the drivers for demonstrating the feasibility of ICF. They can be designed for high power at relatively low cost, and laser beams are easy to transport and to focus. One has to choose low wavelength ($\lambda < 1$ µm) to ensure sufficient beam/target coupling. On the other hand, heavy ion beams are promising candidates for a reactor driver(5) since they have very good efficiency (20-30 % of net energy can be converted into beam energy) and allow for a high repetition rate (10-20 Hz). High driver efficiency and repetition rate are key requirements for power reactors. Existing lasers are poor in both these points. A third driver option are light ion beams which can be produced in pulsed power diodes at a sufficient power level and at low cost when compared with heavy ion accelerators. However, these beams are difficult to transport and to focus.

At the moment, heavy ion beams appear as the best driver candidate for power production(7). This may change in the future with further development of short wavelength lasers (e.g. KrF) and free electron lasers.

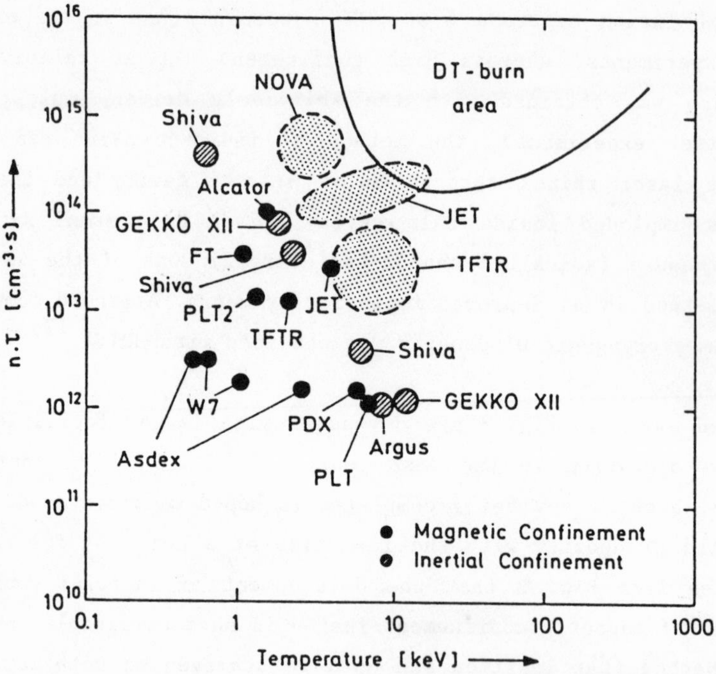

Fig. 3 Lawson diagram showing the product of density n and confinement time τ versus the temperature achieved in different magnetic and inertial fusion facilities. Full dots refer to tokamaks and stellerators. Hatched dots indicate results from laser fusion experiments in Livermore and Osaka. The dotted areas are envisaged with extended facilities in the near future. The figure is taken from Ref. 6.

2. A MODEL FOR ENERGY GAIN

In this section, a simple model for ICF energy gain is presented, based on the concept of central ignition and propagating burn. The model was first proposed by Kidder(8) and then extended by Bodner(9), assuming uniform density of the fuel at ignition. It was later pointed out by the present author(10) that this model becomes much more realistic and leads to a consistent interpretation of code calculations when considering an isobaric (rather than isochoric) ignition configuration. Approximately isobaric configurations are observed in code calculations and are also found in the selfsimilar solutions for shell implosions discussed in Section 3. Thermonuclear burn calculations starting with an isobaric fuel have been performed before by Guskov et al.(11) and also by Atzeni and Caruso[12] who derived a closed-form ignition criterion and introduced it into the present gain model. A general discussion and some technical improvements of this model were given by Rosen et al.[13] In the following, an attempt is made to combine these different developments.

2.1 The ignition spark and its radius R_s

It is assumed that the fuel configuration at ignition consists of a hot central region (the ignition spark) and an outer region of highly compressed fuel (the cold fuel) as shown in Fig. 4. In a schematic way, this model describes the fuel at the time of maximum compression when the shock emerging from the centre after shell collapse has completely passed the fuel and most of the imploding kinetic energy has been converted into internal energy. At this time of stagnation, the pressure is almost uniform over the fuel.

Within the model, the spark region is described as an ideal hydrogen plasma

$$p = 2\ \rho_s T_s/\mu_{DT} \tag{2.1}$$

with the pressure p, the density ρ_s and the temperature T_s. The average DT atomic weight is $\mu_{DT} \cong 4.17 \cdot 10^{-24}$g. Multiplying Eq. (2.1) with the spark radius R_s, one finds that the product

$$p \cdot R_s = 2\ H_s T_s/\mu_{DT} \tag{2.2}$$

is completely determined by the fusion physics. Often used values for the ignition temperature and the 'ρR' of the spark are

$$T_s = 5\ \text{keV} \quad , \quad H_s = \rho_s R_s = 0.4\ \text{g/cm}^2 \tag{2.3}$$

The ignition criterion, derived by Atzeni and Caruso[14] and plotted in Fig. 5, shows that it is in particular the product $H_s \cdot T_s$ occuring in Eq. (2.2) which determines ignition. The criterion can be expressed in the form

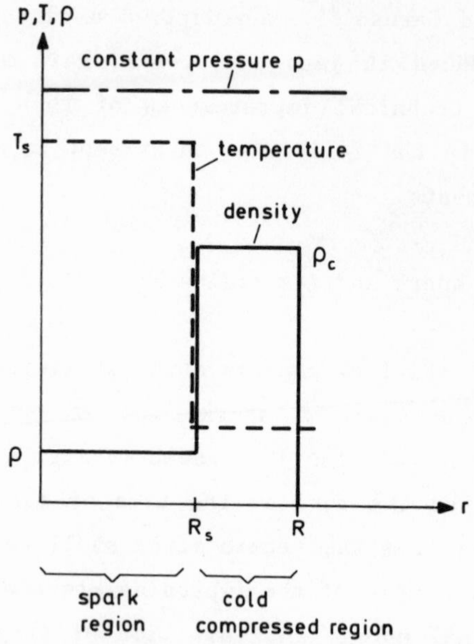

Fig. 4 Profiles for pressure, temperature and density of the model ignition configuration.

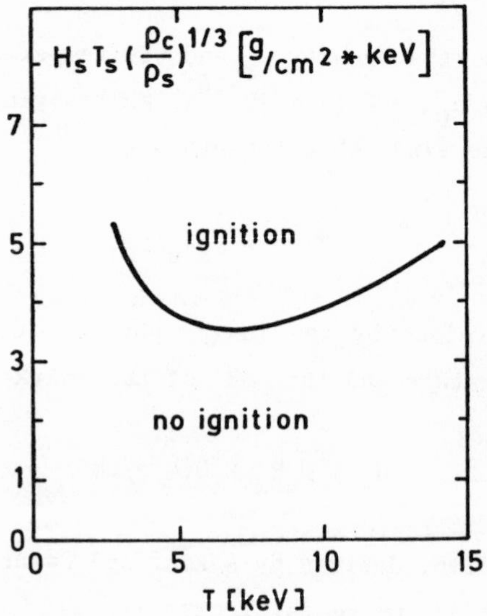

Fig. 5 The ignition threshold derived by Atzeni and Caruso[12].

$$H_sT_s(\rho_c/\rho_s)^{1/3} \geqq 5\cdot 10^{-16}\text{keV}\cdot\text{g/cm}^2\cdot T_s^{31/12}/\langle\sigma v\rangle^{5/6} \geqq 4\ \text{keV}\cdot\text{g/cm}^2 \qquad (2.4)$$

for a broad temperature region $5 \leqq T_s \leqq 10$ keV and $\langle\sigma v\rangle$ in cgs units. Since the density ratio between the cold compressed fuel and the spark is typically $\rho_c/\rho_s \cong 8$, as we shall see in the following, one finds from Eq.(2.4) $H_s\cdot T_s \geqq 2$ keV$\cdot$ g/cm^2, consistent with the values (2.3).

For a given R_s, one obtains

$$\rho_s = H_s/R_s \quad , \; M_s = (4\pi/3)R_s^{\,3}\rho_s \quad , \quad E_s = 2(3T_s/2)M_s/\mu_{DT} \quad (2.5)$$

where M_s and E_s are mass and internal energy of the spark, respectively. Apparently, the spark region is completely determined by specifying its radius R_s which is chosen as a free parameter of the model in the following. The minimum R_s that can be achieved is constrained by the degree of implosion symmetry and is therefore a most significant parameter for characterizing ICF target implosions.

2.2 The cold fuel and its isentrope α

The region of highly compressed, low entropy fuel is described as a degenerate electron gas with the pressure

$$p_c = 2\ \alpha n_e \varepsilon_F/5 = 2.3 \cdot 10^{12}\ \alpha\rho_c^{\,5/3} \qquad (2.6)$$

and the internal energy

$$E_c = 3/5\cdot\alpha\varepsilon_F\ M_c/\mu_{DT} = 3.5 \cdot 10^{12}\ \alpha M_c\rho_c^{\,2/3} \qquad (2.7)$$

in cgs units. The isentrope parameter α denotes the deviation from complete degeneracy and measures the entropy. Obviously, one has $\alpha \geqq 1$. The Fermi energy is given by $\varepsilon_F = \hbar^2(3\pi^2 n_e)^{2/3}/(2\ m_e)$ with the electron mass m_e and the electron density $n_e = \rho_c/\mu_{DT}$. In the model, the pressure p_c is equal to that of the spark p, the density of the highly compressed region ρ_c is obtained from Eq. (2.6) and its mass M_c from Eq. (2.7), assuming that E_c is known.

2.3 Hydrodynamic efficiency η and gain

By means of the target implosion, only a small fraction η of the energy of the driving beam pulse E_{Beam} is transferred into internal energy of the fuel at ignition:

$$E_F = \eta\, E_{Beam} \tag{2.8}$$

The parameter η is called hydrodynamic efficiency. Typical values are $0.05 < \eta < 0.15$, depending on the target design and the type of driver. Ignition occurs when the available fuel energy is just sufficient to form the spark: $E_s = E_F$. If $E_F > E_s$, the additional energy is used to build up the cold fuel reservoir: $E_c = E_F - E_s$. Further, it is noticed that the relation

$$E_F = 3pV_F/2 = 2\pi p R_F^{\,3}$$

holds for the combined fuel region and may be used to calculate its volume V_F and its radius R_F.

The target energy gain is defined as the ratio of thermonuclear energy release E_{TN} and absorbed beam energy:

$$G(E_{Beam}) = E_{TN}/E_{Beam} \tag{2.9}$$

From Eqs. (1.2) and (1.3), one has

$$E_{TN} = q_{DT} \cdot M_F \cdot \phi \tag{2.10}$$

with the fuel mass $M_F = M_s + M_c$, and the fractional burn-up $\phi = H_F/(H_B + H_F)$ with the burn parameter (1.5) $H_B \cong 7\ \mathrm{g/cm^2}$ and the confinement parameter

$$H_F = \rho_s R_s + \rho_c (R_F - R_s) \tag{2.11}$$

The value of H_B corresponds to a freely expanding DT sphere. In actual target implosions as the one discussed in Section 1.3, pusher material tamps this expansion and increases the burned fuel fraction. On the other hand, asymmetry effects will degrade the burn so that the chosen H_B may effectively describe a realistic situation.

The equations (2.1) to (2.11) form a closed set to determine the energy gain $G(E_{Beam}; \eta,\alpha,R_s)$ as a function of E_{Beam}. These gain curves are of fundamental importance for the assessment of inertial confinement fusion. In the following, we discuss their dependence on hydro-efficiency η, cold fuel isentrope α, and spark radius R_s. The fixed parameters are $T_s = 5$ keV, $H_s = 0.4$ g/cm^2, and $H_B = 7$ g/cm^2, implying $p \cdot R_s \cong 15$ Tbar·µm.

2.4 Gain Curves $G(E_{Beam}; \eta,\alpha,R_s)$

An example for gain curves generated by this model is given in Fig. 6. Curves for different hydroefficiencies and isentropes are shown. The spark radius is kept fixed at $R_s = 100$ µm, corresponding to a pressure of $p = 0.15$ Tbar. Varying the hydrodynamic efficiency results in a parallel shift of the curves; in the double logarithmic plot, this follows from the definition (2.9) and the scaling relation $G(E) = \eta G(\eta E)$. The curves start from their ignition point

$$E^{ign}_{Beam} = (4\pi H_s T_s/\mu_{DT}) \cdot R_s^2/\eta \qquad (2.12)$$

$$G^{ign} = q_{DT}/\varepsilon_s \cdot \eta \cdot H_s/(H_B+H_s) \qquad (2.13)$$

located in Fig. 6 on the thin broken line. The gain (2.13) at ignition is the naive gain $q_{DT}/\varepsilon_s \cong 500$ multiplied by the hydro-efficiency and the burn rate of the spark. This ignition gain is of order unity. When more beam energy is available to build up the region of highly compressed fuel around the spark, propagating burn occurs. This leads to steeply rising gain due to a sharp increase in ρR and the fraction of burned fuel. For much higher energies, the gain curves level off because of burn depletion and go asymptotically like

$$G(E_B \to \infty) = \text{const.} \cdot \eta \cdot p^{-2/5} \cdot \alpha^{-3/5}$$

High isentrope parameters α caused by pre-heat during implosion degrade

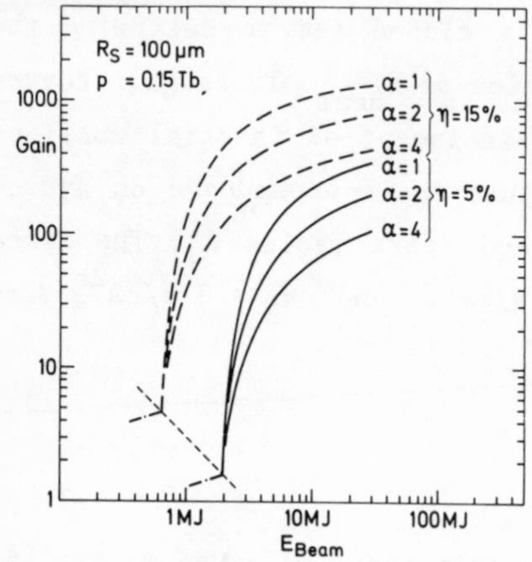

Fig. 6 Gain curves for different hydrodynamic efficiencies η and isentrope parameters α in the range relevant to reactor targets. Spark radius R_s and pressure p are kept fixed. Ignition points for different η are located on the thin dashed line.

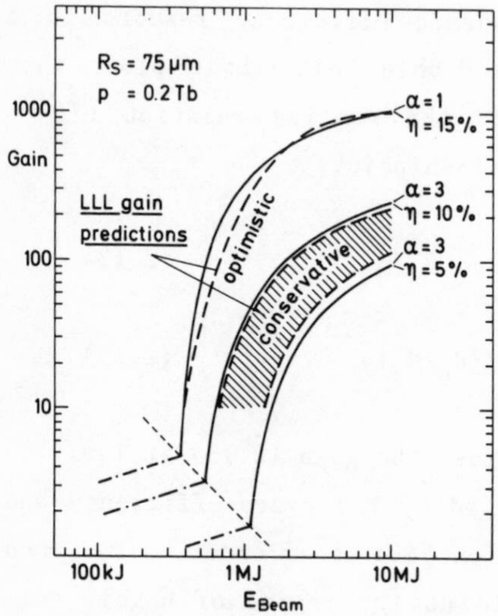

Fig. 7 Fit of the present gain model to the 'conservative' and 'optimistic' gain predictions obtained at Livermore on the basis of extensive code simulations. The LLL results correspond to a spark radius R_s = 75 μm and a central pressure p = 0.2 Tbar; the 'conservative' band is well reproduced with an isentrope α = 3 and an hydroefficiency ranging between 5 % < η < 10 % whereas the 'optimistic' curve corresponds to α = 1 and η = 15 %.

the high-gain region. The $\alpha^{-3/5}$ scaling is clearly seen in Fig. 6. Increasing α by a factor of two leads to about 35 % reduction in gain. Also high pressure p corresponding to small spark radii is unfavourable in the high-gain region.

In Fig. 7 the Livermore gain predictions for reactor targets based on extensive code calculations are compared with the present model. The results as published by the Livermore group[(15)] are separated into 'conservative' and 'optimistic' gain curves. It is found that these predictions are well reproduced with a spark radius $R_s \cong 75$ µm or, equivalently, a pressure of p = 0.2 Tbar. Assuming a reactor-size single-shell pellet with an initial fuel radius R_0 between 1.5 mm and 3 mm, the spark radius found above corresponds to a convergence ratio $\varepsilon = R_s/R_0$ between 5 % and 2.5 %, respectively. This seems to be a reasonable assumption concerning the implosion symmetry which may be achieved in the future. The 'conservative gain band' is spanned by hydrodynamic efficiency ranging between 5 % < η < 10 % and corresponds to a cold fuel isentrope $\alpha \cong 3$. These values may be called conservative in comparison with the 'optimistic gain curve' which is obtained with $\alpha = 1$ and $\eta = 15$ % and is close to the physical limits. It should be noticed that the fit shown in Fig. 7 determines the parameters rather uniquely. It appears that the present model analysis leads to a simple, quantitative, physical picture of the Livermore results. It may help to judge these results which play such a central role in the current discussion on inertial confinement fusion.

Figure 8 displays the decisive role of the spark radius R_s on expected gain and ignition energy. It is the limits on R_s which drive up the requirements on beam energy. Hydrodynamic efficiency and isentrope

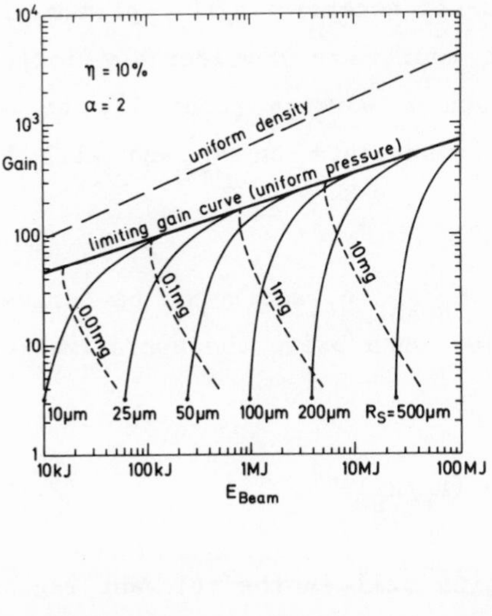

Fig. 8 Gain curves for fixed isentrope α and hydroefficiency η and increasing spark radius R_s. Lines of equal fuel mass M_F (short dashes) are labelled by the mass value. Limiting gain curves are shown for the present model assuming uniform pressure (thick solid line) as well as for the Kidder-Bodner model assuming uniform density (long dashed line).

have been fixed at $\eta = 10\ \%$ and $\alpha = 2$. Gain 60 could be reached with $E_{Beam} = 100$ kJ would it be possible to make a 10 μm, 1.5 Tbar spark at this low pre-heat level. Apparently, there are severe problems in achieving such values because of fuel preheat and insufficient implosion symmetry. At present, breakeven is expected at a level of 100 kJ laser beam energy. An option for high-gain reactor targets is located in the region

$$G \cong 100$$
$$1\ \text{MJ} < E_{Beam} < 10\ \text{MJ}$$
$$50\ \mu\text{m} < R_{Spark} < 200\ \mu\text{m}$$
$$1\ \text{mg} < M_{Fuel} < 10\ \text{mg}$$

The upper limit is given by the requirements of containing the micro-explosion in a reactor.

2.5 Optimal Gain

An interesting feature of the gain model is that there exists a limiting gain curve which determines the maximum gain at a given level of beam energy. In fact, G is a function of R_s at fixed E_{Beam} and, of course, at fixed η and α has a maximum G* which defines the limiting gain curve and appears in Fig. 8 as the envelope of the individual gain curves with different R_s. This maximum gain is not necessarily the optimum. It turns out that the admissable spark radii are considerably larger already at somewhat reduced gain. Such a working point may be of advantage in view of the symmetry constraints on R_s and will be investigated further below.

The scaling of the limiting gain with E_{Beam}, η, and α can be derived analytically from Eqs. (2.9) and (2.10) when using the approximation for

$$\phi = H_f/(H_B + H_F) \cong 0.5 \cdot (H_F/H_B)^{1/2} \qquad (2.14)$$

as suggested in ref. 13. It holds quite well in the relevant region $0.3 < \phi < 0.7$. From $\partial G(E_B;\eta,\alpha,R_s)/\partial R_s = 0$, one finds the limiting gain

$$G^* \cong 6700\ \eta(\eta E_B/\alpha^3)^{0.3}\cdot(\beta\gamma)^{-0.4}\cdot\delta^{0.5} \tag{2.15}$$

It is obtained at a fixed ratio of spark/fuel radii and energies

$$R_s^*/R_F^* \cong 0.34 \quad , \qquad E_s^*/E_F^* = (R_s^*/R_F^*)^3 \cong 0.04 \tag{2.16}$$

Apparently, only 4 % of the energy has to be used for ignition at maximum gain. Furthermore, one obtains

$$R_s^* \cong 68\ \mu m \cdot (\eta E_B/\beta\gamma)^{1/2}, \qquad p^* \cong 0.23\ \text{Tbar} \cdot (\beta\gamma/\eta E_B)^{1/2} \tag{2.17}$$

and for the densities

$$\rho_s^* \cong 59\ g/cm^3\cdot(\beta^3\gamma/\eta E_B)^{1/2}, \qquad \rho_c^* \cong 1025\ g/cm^3\cdot(\beta\gamma)^{0.9}/(\eta E_B\alpha^2)^{0.3} \tag{2.18}$$

In Eqs. (2.15)-(2.18), the beam energy E_B is taken in units of MJ, and $\beta = H_s/0.4\ g/cm^2$, $\gamma = T_s/5$ keV, $\delta = (7\ g/cm^2)/H_B$.

A better working point of ICF targets is found when looking at gain curves with constant fuel mass (broken lines in Fig. 8). Such a curve is shown for $M_F = 4$ mg, $\eta = 0.1$, and $\alpha = 2$ in Fig. 9. Volume ignition

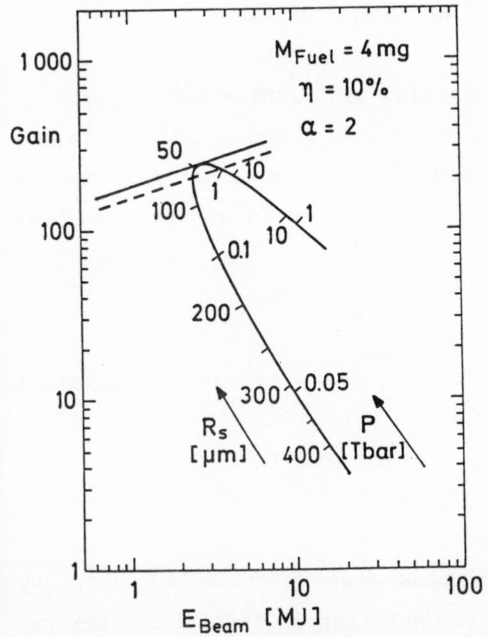

Fig. 9 Gain curve for fixed fuel mass M_F and fixed η and α. Radii R_s and pressures p are marked on the gain curve. The solid straight line gives the limiting gain, the broken straight line the optimum gain.

of this amount of fuel would require $E_{Beam} = 22$ MJ and would give only a low gain $G \cong 3$, although at moderate compression ($R_F = R_s =$ 450 μm, p = 0.035 Gbar). Figure 9 shows that the 4 mg fuel can be ignited and burned with much less beam energy and higher gain when driving it to higher pressure and smaller radius. It is seen that the minimum beam energy is 2.4 MJ with $G^+ = 200$ and $R_s^+ = 67$ μm; we define this extremum as the optimal working point. For different beam energies, this optimal gain is located on the broken straight line in Fig. 9. Compressing further leads to the maximum gain discussed before and located on the solid straight line. For $M_F = 4$ mg, it is reached at $E_B = 2.7$ MJ with $G^* = 245$ and $R_s^* = 35$ μm.

Calculating the optimal gain analytically from $\partial E_B/\partial R_s = 0$ for fixed values of η, α, and M_F, gives

$$G^+ \cong 5700\ \eta(\eta E_B^+/\alpha^3)^{0.3} \cdot \beta^{-2/5} \cdot \gamma^{-4/15} \cdot \delta^{0.5} \qquad (2.19)$$

with the optimal beam energy

$$E_B^+ \cong 0.054\ \mathrm{MJ} \cdot (\alpha\beta)^{1/2} \cdot \gamma^{1/3} \cdot M_F^{5/6}/\eta \qquad (2.20)$$

in MJ and the fuel mass M_F in mg. In addition, one finds

$$R_s^+ \cong 137\ \mu\mathrm{m}\ (\eta E_B^+/\beta)^{1/2}, \quad \rho_c^+ = 675\ \mathrm{g/cm^3} \cdot (\beta^3\gamma^2/\eta E_B \alpha^2)^{0.3}$$

and $R_s^+/R_F^+ = 0.57$, $E_s^+/E_F^+ = (R_s^+/R_F^+)^3 \cong 0.18$. Of particular interest is the comparison between 'maximum' and 'optimal' gain at the same beam energy:

$$G^+/G^* \cong 0.85, \qquad R_s^+/R_s^* \cong 2.0$$

Although the gain is reduced by only 15 %, the spark radius is larger by a factor 2 for the 'optimal' case.

2.6 Polarized-Fuel Scaling

Spin-polarizing the DT fuel may increase the reaction rate $\langle\sigma v\rangle$ by 50 %, corresponding to $\delta = 1.5$ with the notation of Sect. 2.5. This way of reducing the fusion thresholds has recently attracted considerable attention. The present model provides an immediate answer to how the gain and the driver energy scale with δ.(13)

Increasing $\langle\sigma v\rangle \sim \delta$ leads to an increase of the burn rate $\phi \cong 0.5(H_F/H_B)^{0.5} \sim \delta^{1/2}$ and to a decrease of the ignition threshold (2.4)

$$H_s T_s \sim \beta\gamma \sim \delta^{-5/6}$$

Accordingly, the gain (2.15) scales like

$$G^* \sim \eta(\eta E_B/\alpha^3)^{3/10} \cdot \delta^{1/3} \cdot \delta^{1/2} \sim \delta^{5/6}$$

Thus if we are satisfied with a particular gain G^*, the required energy E_B scales as $\delta^{-25/9}$. Polarized fuel with $\delta = 1.5$ would thus reduce the required driver energy by a factor of 3.

2.7 Gain Model Conclusions

We have identified three key parameters which determine the energy gain of ICF targets. These are

(1) the hydrodynamic efficiency which describes implicitly the conversion of beam energy into internal fuel energy at ignition;

(2) the isentrope parameter which measures the preheat entropy deposited in the cold fuel part during implosion by shocks and other transport processes (e.g. hot electrons and radiation);

(3) the spark radius which is intimately related to the implosion symmetry, a key problem in target performance.

The model provides important scaling relations which govern the change of gain when varying the various fuel parameters. It gives a quantitative and physically transparent interpretation of the Livermore gain curves which currently play a central role in the assessment of inertial confinement fusion. It may therefore serve as a very useful interface between detailed target calculations and general parameter studies needed for the design of power reactors.

3.
CENTRAL IMPLOSION DYNAMICS

The question discussed in this section is how the ICF ignition configuration can be formed in a single-shell implosion, how the terabar pressure of the fuel is achieved after central collapse and how the innermost part of the fuel is driven to much higher entropy than the rest of the fuel to form the spark. Although one needs numerical simulation for a quantitative answer, specific qualitative insight into the singular gas dynamics of entropy shaping and spherical collapse can be obtained from similarity solutions of Guderley's type.[16,17] This is the approach followed below. It will be valid only to the extent that actual implosions behave in the selfsimilar manner postulated here. It is our experience, however, that the general conclusions derived are indeed born out approximately by code calculations. It is our belief that these solutions represent some universal features, although there exists no rigorous proof at present.

Some results of 'selfsimilar analysis' relevant to ICF target implosions are summarized. A more systematic presentation is found in ref. 18. In Sect. 3.1, the general form of similarity solutions of the gasdynamical equations and some important consequences are outlined. The application to shock propagation through density gradients[19] is discussed in Sect. 3.2 and the problem of spherically collapsing shells and central reflection in Sect. 3.3.

3.1 Selfsimilar Gas Dynamics

The equations of ideal gas dynamics

$$\begin{aligned} &\partial\rho/\partial t + r^{-(n-1)}\partial/\partial r(r^{n-1}\rho u) = 0 \\ &\partial u/\partial t + u\partial u/\partial r + \rho^{-1}\partial p/\partial r = 0 \qquad (3.1) \\ &(\partial/\partial t + u\partial/\partial r)(p/\rho^{\gamma}) = 0 \end{aligned}$$

express the conservation of mass, momentum and entropy, respectively; n is the dimension parameter (n=1 plane, n=2 cylindrical, n=3 spherical) and γ the adiabatic exponent of the gas. Dissipative processes, e.g. heat conduction etc., are neglected in Eqs. (3.1) and entropy is therefore a constant of motion for each gas element except when a shock passes. The similarity ansatz[16] for the density $\rho(r,t)$, the velocity $u(r,t)$ and the sound velocity $c(r,t)$ (defined by $c^2=\gamma p/\rho$) has the form

$$u(r,t) = \alpha r/t\cdot U(\xi),\ c(r,t) = \alpha r/t\cdot C(\xi),\ \rho(r,t) = r^{\kappa}G(\xi) \qquad (3.2)$$

where the similarity coordinate $\xi = r/|t|^{\alpha}$. The similarity exponent α and the density exponent κ are considered to be free parameters. In the case of spherically convergent waves, ansatz (3.2) is motivated by the observation[16] that the inner front of the wave moves on a path $R_F(t) = \xi_F(-t)^{\alpha}$ with constant ξ_F for times close to central collapse ($t = 0$) and that the shock emerging from the centre after collapse has a trajectory $R_s(t) = \xi_s t^{\alpha}$ with constant ξ_s. An example for these trajectories is shown in Fig. 10.

At collapse time $t = 0$, the solutions (3.2) reduce to simple power laws

$$u(r,t=0) = u_0 r^{-\lambda}, \qquad c(r,t=0) = c_0 r^{-\lambda}$$

$$\rho(r,t=0) = \rho_0 r^{\kappa}, \qquad p(r,t=0) = p_0 r^{\kappa-2\lambda} \qquad (3.3)$$

$$A(r,t=0) = p/\rho^{\gamma} = A_0 r^{-\varepsilon} \quad \text{(entropy function)}$$

with $\lambda = (1/\alpha-1)$, and the amplitudes u_0, c_0, ρ_0, etc. are obtained from Eqs. (3.2) with $|t| = (r/\xi)^{1/\alpha}$ in the limit $\xi \to \infty$. The implosion is physically described by the uniform Mach number M_0 of the flow and the entropy exponent ε

$$M_0 = u_0/c_0 \quad , \quad \varepsilon = \kappa(\gamma-1) + 2\lambda \qquad (3.4)$$

at collapse which can be used as free parameters instead of α and κ. It will be shown that final compression is governed by M_0, whereas spark formation is controlled by ε.

Another important consequence of the ansatz (3.2) involves the trajectories $R(t,a)$ of gas elements. The Lagrangian coordinate $a = R(t=0,a)$ labels each element by its position at time $t = 0$. From $dR/dt = u(R,t) = \alpha R/t \cdot U(\xi)$ and $R = \xi|t|^{\alpha}$ one obtains

$$d(\ell nR)/d(\ell n\xi) = U(\xi)/(U(\xi)-1) \qquad (3.5)$$

where $R(\xi,a)$ is now interpreted as a function of ξ, and time follows from $|t| = (R/\xi)^{1/\alpha}$. From Eq. (3.5) it is obvious that ratios $R(\xi_2,a)/R(\xi_1,a)$ for a given gas element on different ξ lines do not depend on a, and it is easily seen that the same holds for ratios of densities, pressures and other physical variables of the given element. Choosing $\xi_1=\infty$ (t=0 axis, see Fig. 10) and $\xi_2 = \xi_s$ (reflected shock), this means that the ratios $\rho_s(a)/\rho_0(a)$, $p_s(a)/p_0(a)$, etc. are independent of a. Here $\rho_0(a)$ is the density of element a at $t = 0$ and $\rho_s(a)$ its density when the reflected shock has just passed ($p_0(a)$, $p_s(a)$, etc. defined analogously). One arrives at the important result that the final compression ratio is the same for all gas elements. Even more important: ratios across the reflected shock front are constant. This means that the reflected shock has constant strength and that it adds the same amount of entropy to each imploding gas element. Therefore, the gas dynamics after shell collapse do not change the entropy profile, and entropy shaping to form the spark has to be accomplished during implosion.

It is a remarkable feature of the gasdynamical equations (3.1) that they allow similarity solutions of the type (3.2). A recent review on this problem, the derivation of the basic formulas and various applications to ICF are found in Ref. 18. Inserting the ansatz (3.2) into Eqs. (3.1), leads to one ordinary differential equation of the form

$$dU/dC = f(U,C)/g(U,C)$$

which is expressed exclusively in terms of polynomials $f(U,C)$, $g(U,C)$ of the reduced velocities U and C. The functions $U(\xi)$ and $C(\xi)$ are then obtained in form of integrals. An algebraic integral exists for

$$G(\xi) = K\cdot(\xi^{1/\alpha}C(\xi))^{\mu(n+\kappa)/\beta} \cdot (1-U(\xi))^{(\kappa+\mu\lambda)/\beta} \qquad (3.6)$$

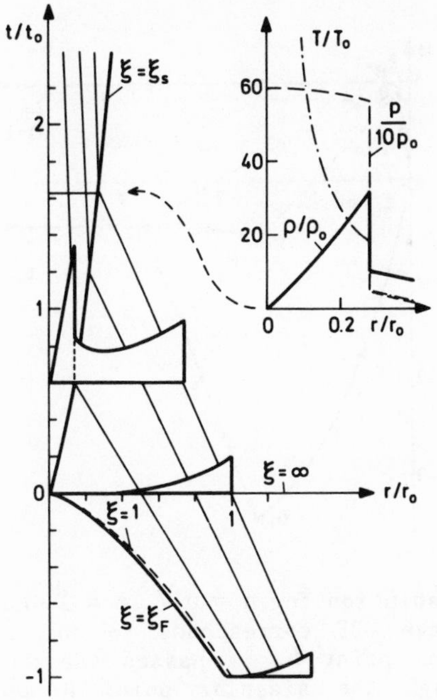

Fig. 10 Selfsimilar solution for $\alpha = 0.7$, $\kappa = 3$ and $\gamma = 5/3$ in the r-t diagram. Several ξ-lines are shown representing the path of the shell's inner surface ($\xi_F = 0.963$), the limiting characteristic ($\xi_B = 1$, broken line), the r-axis at $t = 0$ ($\xi = \infty$) and the path of the reflected shock ($\xi_S = 0.198$). Trajectories of three gas elements a = 0.5, 0.8, 1.0 are also given. In addition, density profiles at $t/t_o = -1, 0, 0.6$ have been inserted with a cut-off at gas element a = 1 where one may image a spherical piston pushing the shell inwards. The reflected shock is clearly seen in the density profile at $t/t_o = 0.6$. It would hit the piston at $t_S/t_o = 1.65$ and $R_S/r_o = 0.281$. Profiles of density ρ, pressure p and temperature T at this time are shown separately in the upper right corner. The values ρ_o, p_o, T_o refer to element a = 1 at $t = 0$.

with a constant K and $\mu = 2/(\gamma-1)$, $\beta = n-\mu\lambda$. In general, the integrals have to be found numerically. The important case of a non-isentropic, hollow shell implosion is presented in Fig. 10; the integration curve in the U,C plane is shown in Fig. 11. Details of these figures are explained in the captions. This solution is close to the situation of actual single-shell implosions considered for ICF. Its physical significance will be further discussed in Sect. 3.3.

As a general rule[18], physical solutions are allowed to cross the line U + C = 1 in Fig. 11 only in singular points like B. This condition strongly reduces the number of admissable solutions, and, in some

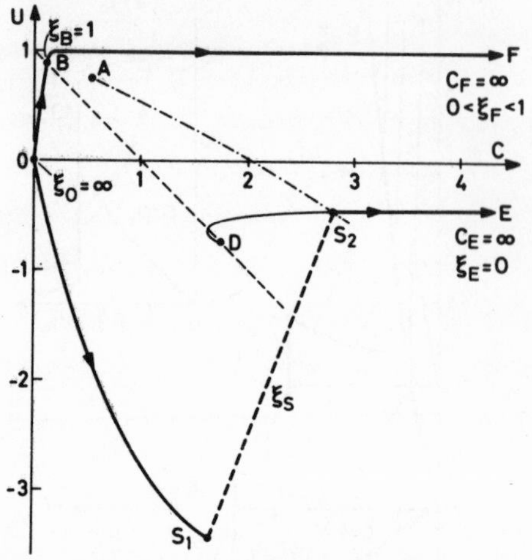

Fig. 11 Selfsimilar solution for $\alpha = 0.7$, $\kappa = 3$ and $\gamma = 5/3$ in the U-C plane. The solid curve OBF corresponds to the imploding gas shell ($t < 0$). Starting from point 0, it passes the sonic line $U + C = 1$ (small dashes) through the singular point B and proceeds to the singular point F at $C = \infty$ which describes the inner front ($\xi = \xi_F$) of the imploding shell. The arrows give the direction of decreasing ξ. The curve OS_1S_2E corresponds to the gas after void closure ($t > 0$). The branch OS_1 describes the outer part of the gas ($\xi_S \leq \xi \leq \infty$) which is still imploding. Near point 0, where $\xi \to \infty$, the solutions for $t < 0$ and $t > 0$ smoothly fit (U changes sign since t does!). Gas states reached from OS_1 by shocks are located on line AS_2; point A ($U = 3/4$, $C = \sqrt{5}/4$) can be reached from a constant gas by a strong shock as required in Guderley's problem. Shock fitting to the inner solution S_2E ($0 \leq \xi \leq \xi_S$) is achieved along the line S_1S_2. Point E at $C = \infty$ describes the implosion center ($\xi_E = 0$). Actually, points E and F represent the same node point at infinity which is approached by DS_2E as a solitary solution and by OBF as one of a bundle of solutions along $U = 1$.

cases, a solution is only found for a fixed value of the similarity exponent α. A problem of this type is discussed in the next section.

3.2 Entropy Shaping by Shocks during Implosion

In order to shape the hot central spark for ignition, it is required that the innermost part of the fuel is driven to high entropy, whereas the surrounding fuel has to be compressed more or less isentropically. In single-shell targets, this entropy shaping has to be accomplished during implosion before void closure, as we have learned in Sect. 3.2.

For this purpose, one has to take advantage of non-linear shock dynamics. At least two shocks have to pass the shell. This is explained in Fig. 12. A first shock is always created by beam switch-on. It passes the shell raising the entropy almost uniformly and converting the fuel into a plasma of relatively low temperature (Fig. 12a). After shock arrival at the inner surface, a rarefaction wave runs back into the shell (Fig. 12b). In this phase, the second shock has to be launched; it is running through the sloping density profile of the rarefaction wave (Fig. 12c) generating an entropy profile which is diverging

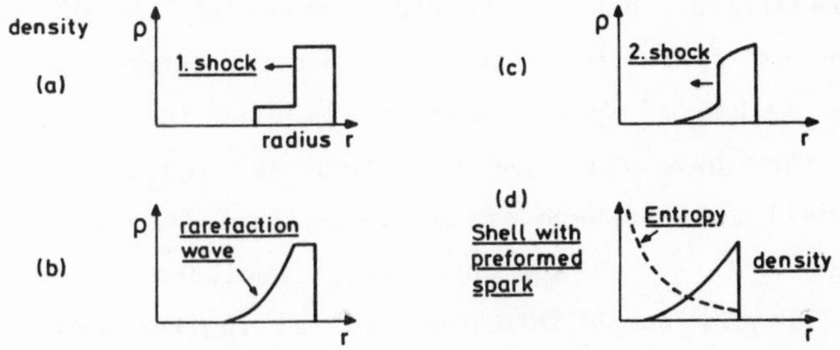

Fig. 12 The mechanism of entropy shaping by two shocks in a single-shell implosion.

towards the inner surface and preforms the spark. Proper shaping of the beam pulse allows one to tailor the spark region by changing strength and timing of the two shocks. It should be mentioned that even the box-shaped pulse, used for our example of target design in Sect. 1.4, leads to a two shock implosion. These shocks show up in Fig. 1 as kinks of the inner fuel boundary at 6 ns and 16 ns (compare the discussion in Sect. 1.4).

A shock passing through a layer in which the density falls to zero is characterized by a steep increase of shock strength and entropy generation close to the surface. This problem was studied for shock waves emerging at the surface of a star.[17] For an initially isentropic, plane layer at rest and a density profile $\rho(x) \sim x^k$ vanishing at the surface $x = 0$, it leads to a similarity solution described by Eqs. (3.1)-(3.6) with $n = 1$. This solution also describes approximately the situation described above with a shock running into a rarefaction wave during shell implosion. The shock trajectory $X_F(t) \sim |t|^\alpha$ is given by fixed values of the similarity parameter α. For a $\gamma = 5/3$ gas, some calculated values[19] are $\alpha = 0.590$, 0.696, 0.816 for $\kappa = 13/4$, 2, and 1, respectively. At time $t = 0$, when the shock has just passed the layer, the spatial profiles of the hydrodynamic variables are given by Eq. (3.3), in particular, the density $\rho(x) \sim x^\kappa$ and the entropy $A(x) \sim x^{-\varepsilon}$ with $\varepsilon = \kappa(\gamma-1) + 2(1/\alpha-1)$. The important point is that the entropy diverges for $x \to 0$ and that this entropy distribution over the gas elements is conserved for times $t > 0$ unless another shock passes.

This description of entropy shaping in single-shell implosions is highly idealized. But it provides physical insight into the basic mechanism of spark formation. We mention here that another, very effective method of spark formation is used in so-called double-shell targets. They have the same structure as single-shell targets except that a small solid sphere of cryogenic fuel (the ignitor) is added in the centre of the void. The outer shell imploded isentropically to high velocity finally bumps into the central ignitor sphere which is then strongly shock-heated and forms the spark.

3.3 Final Compression and Spark Build-up

In this section, we describe what happens to non-isentropic, spherically imploding gas shells of the type shown in Fig. 12d when they arrive in the centre. Numerical simulations are typically very poor in bringing out the details of the gas dynamics occuring at the point of spherical convergence. The solutions discussed here represent a generalization of Guderley's fundamental work on spherically converging shocks in a uniform gas.[16] They follow from Eqs. (3.1)-(3.6) with $n = 3$. An example is shown in Figs. 10 and 11. At time $t = 0$, when the inner front of the shell arrives in the centre, the spatial profiles of the hydrodynamic variables have the form (3.3) of power laws. The solutions form a two parameter family conveniently characterized by the Mach number M_o and the entropy exponent ε as discussed in Sect. 3.1. It should be noticed that the similarity exponent α is not fixed in the present case, but may vary within certain limits.[18]

Explicit solution for imploding shells have been obtained by integrating Eq. (3.3) numerically.[18] Results for the implosion Mach number M_0, the entropy exponent ε, the trajectories of the incoming fuel front ξ_F and the reflected shock front ξ_s, as well as for the final density and pressure increase, ρ_s/ρ_0 and p_s/p_0, are listed in Table 2 for different values of the similarity exponent α and the density exponent κ. It was shown in Sect. 3.1 that the final compression ratios ρ_s/ρ_0 and p_s/p_0 between the times t = 0 (void closure) and $t = t_s$ (the reflected shock has just passed) are the same for each gas element. The numerical results show that these ratios depend on α and κ almost exclusively through the Mach number M_0, but only weakly through ε. The approximate relations

$$\rho_s/\rho_0 \cong 2.4 \cdot M_0^{3/2}, \qquad p_s/p_0 \cong 3.6 \cdot M_0^3 \tag{3.7}$$

Table 2. Results for Mach number M_0, entropy exponent ε, inner front position ξ_F, reflected shock position ξ_S, ρ_S/ρ_0 and p_S/p_0, as functions of similarity exponent α and density exponent κ for $\gamma = 5/3$.

α	κ	M_0	ε	ξ_F	ξ_S	ρ_S/ρ_0	p_S/p_0
0.60	2	2.61	0	0.902	0.320	9.86	59
	3	3.66	2/3	0.935	0.244	15.8	148
	4	4.68	4/3	0.953	0.199	22.7	303
0.75	2	5.26	2/3	0.961	0.229	29.0	498
	3	6.65	4/3	0.972	0.185	41.5	1024
	4	8.08	6/3	0.979	0.155	56.2	1896
0.90	2	9.36	10/9	0.988	0.209	65.5	3050
	3	11.6	16/9	0.992	0.172	90.4	5972
	4	13.9	22/9	0.994	0.146	119	10780

are obtained for the range covered by Table 2. With M_o = 3-5 obtained from typical target simulations, one finds density jumps $\rho_s/\rho_o \cong 10\text{-}25$ and pressure jumps $p_s/p_o \cong 100\text{-}500$ after void closure.

The state of the gas in the centre behind the reflected shock is described by the line S_2E in the U,C plane of Fig. 11. Point E at $C = \infty$ corresponds to an infinite temperature at the centre $r = 0$. This line also generates the famous Taylor-Sedov solution for a strong point explosion, although with different parameters ($\alpha = 2/5$, $\kappa = 0$). As shown in Ref. 18, it is well approximated by

$$U(\xi) \cong -(\kappa-2\lambda)/n\gamma \quad , \quad C(\xi)/C_s \cong (\xi_s/\xi)^{(1+n\varepsilon/2\nu)} \tag{3.8}$$

with $\lambda = (1/\alpha-1)$ and $\nu = (n\gamma+\kappa-2\lambda)$. Inserting these expressions into Eqs. (3.2) and (3.6) and setting $n = 3$, one obtains the following approximate relations for the gas near to the centre

$$\begin{aligned} u(r,t) &\cong -\,\alpha(\kappa-2\lambda)/3\gamma \cdot r/t \\ T(r,t) &\sim r^{-3\varepsilon/\nu} \cdot t^{-\alpha(3\varepsilon/\nu-2\lambda)} \\ p(r,t) &\sim r^{o} \cdot t^{\alpha(\kappa-2\lambda)} \\ \rho(r,t) &\sim r^{3\varepsilon/\nu} \cdot t^{\alpha(\kappa-3\varepsilon/\nu)} \end{aligned} \tag{3.9}$$

These are very important relations since they describe the spatial profiles and the time evolution of the hydrodynamic variables in the central region behind the reflected shock where DT ignition and burn has to be achieved. Most important: the space-time dependence is given in terms of the parameters of the imploding gas, and these analytic relations therefore connect the state of gas before and after void closure.

The relations (3.9) hold exactly for $\kappa = 2\lambda$. In this case, the incoming gas is isobaric before void closure (see Eq. (3.3)), and it is at rest and has a uniform, time-independent pressure after void closure (see Eq. (3.9)). The gas in the centre is further contracting or expanding depending on wether $(\kappa-2\lambda) > 0$ or $(\kappa-2\lambda) < 0$, respectively. In both cases, the pressure is approximately uniform ($p \sim r^{o}$). This isobaric pressure distribution in the centre behind the reflected shock is a key ingredient of the gain model described in Sect. 2. It is obtained here in a rather general way.

Another remarkable result concerns the temperature distribution near the centre. It is seen that $T \sim r^{-3\varepsilon/\nu}$ diverges for $r \to 0$ provided that $\varepsilon > 0$ (ν is always positive for cases of interest). This means that the parameter $\varepsilon = \kappa(\gamma-1)+2\lambda$ which describes the entropy distribution of the imploding gas ($A(r,t=0) \sim r^{-\varepsilon}$, compare Eq.(3.3)) also controls the formation of the temperature spike in the centre (spark). In order to form the spark after void closure, one has to impress an entropy distribution to the imploding shell which increases towards the inner front. A way to do this has been discussed in Sect. 3.2. It is interesting to note that it is not the reflected shock which generates the spark. On the contrary, the reflected shock has constant strength and only adds a constant amount of entropy to each gas element without changing the entropy distribution. Compare the discussion following Eq. (3.5).

3.4 Selfsimilar Implosion Conclusions

The selfsimilar solution for an imploding non-isentropic gas shell described above has provided abundant new insight in the gas dynamics of final compression and spark formation after shell collapse.

In single-shell implosions, most of the compression and heating leading to DT ignition and burn occurs during these final moments after void closure. We have seen that the final compression is governed by the implosion Mach number which measures the ratio between kinetic and internal energy of the incoming shell. It should be realized that a considerable increase of density and pressure also occurs during implosion before void closure. This part of the total compression is also described within the selfsimilar analysis and is related to so-called cumulative solutions (e.g. Kidder's solution for homogenous isentropic compression[20]). Due to limited space, we could not discuss these solutions here. A detailed description is found in ref. 18.

Important results derived above concern the gas state in the centre behind the reflected shock. The hydrodynamical variables as functions of radius and time are obtained approximately in form of power laws in which the exponents are determined by the gas state before void closure. As a general result, the pressure is found to be uniform in the central region, which is a key assumption for the gain model in Sect. 2, whereas the temperature diverges $\sim r^{-3\varepsilon/\nu}$ for $r \to 0$ provided that the entropy distribution does so $A \sim r^{-\varepsilon}$ before void closure. Therefore, spark formation is controlled by the entropy parameter ε. We have described how $\varepsilon > 0$ can be obtained by shock tailoring during implosion. The selfsimilar analysis reveals that the final shock reflected from the centre after void closure has constant strength and does not contribute to the spark shaping.

Comparing typical code simulations of target implosions with the similarity solutions, one finds that most of the results derived here are indeed substantiated, at least qualitatively. A difficulty with numerical schemes based on finite differences is that they are rather inaccurate in describing singular flows like the one occuring in the implosion centre. Frequently it is hard to distinguish between true physics and numerical noise in the code calculations. This hampers a detailed comparison. But more global features like the almost constant implosion velocity of individual gas elements (compare Fig. 10) and the uniform Mach number characteristic for the similarity solution in the final implosion phase are observed in the numerical simulations. It is found that the similarity analysis brings out fundamental features of central implosions which are difficult to discover by other methods. It therefore represents an important piece in the basic theory of inertial confinement fusion.

REFERENCES

1. J. Nuckolls, L. Wood, A. Thiessen, and G. Zimmermann, Nature 239, 139 (1972).
2. R.J. Mason and R.L. Morse, Phys. Fluids 18, 814 (1975).
3. Yu. V. Afanasev, N.G. Basov, P.O. Volosevich, E.G. Gamalii, O.N. Krokhin, S.P. Kurdyumov, E.I. Levanov, V.B. Rosanov, A.A. Samarskii, and A.N. Tikhonov, JETP Letters 21, 68 (1975).
4. N. Metzler and J. Meyer-ter-Vehn, Laser and Particle Beams 2, 27 (1984).
5. D. Böhne, I. Hofmann, G. Kessler, G.L. Kulcinski, J. Meyer-ter-Vehn, U. von Möllendorf, G.A. Moses, R.W. Müller, I.N. Sviatoslowsky, D.K. Sze and W. Vogelsang, Nuclear Engineering and Design 73, 195 (1982).
6. S. Witkowski, Nucl. Fusion 25, 224, (1985).
7. J. Nuckolls, "The feasibility of inertial confinement fusion", Physics Today, p. 24, Sept. 1982.
8. R.E. Kidder, Nucl. Fusion 19, 223 (1976).
9. S. Bodner, 'Critical Elements of High Gain Laser Fusion', NRL Memorandum Report 4453 (Jan. 21, 1981).
10. J. Meyer-ter-Vehn, Nucl. Fusion 22, 561 (1982).
11. S.Yu. Guskov, O.N. Krokhin and V.B. Rozanov, Nucl. Fusion 16, 957 (1976).
12. S. Atzeni and A. Caruso, Nucl. Fusion 23, 1092 (1983).
13. M.D. Rosen, J.D. Lindl and A.R. Thiessen, 'Simple Models of High-Gain Targets-Comparison and Generalization', Lawrence Livermore Laboratory, Laser Program Annual Report 1983, p. 3-5.
14. S. Atzeni and A. Caruso, Phys. Lett. A85, 345 (1981).
15. J.H. Nuckolls, in Lawrence Livermore Laboratory, Laser Program Annual Report 1979, Vol. 2, p. 3-2.
16. G. Guderley, Luftfahrt-Forschung 19, 302 (1942).
17. Ya. Z. Zeldovich and Yu. P. Raizer, Physics of Shock Waves and High-Temperature Hydrodynamic Phenomena, Academic Press, London, 1967.
18. J. Meyer-ter-Vehn and C. Schalk, Z. Naturforschung 37a, 955 (1982).
19. A. Sakurai, Commun. Pure Appl. Math. 13, 353 (1960).
20. R.E. Kidder, Nucl. Fusion 14, 53 (1974), and 16, 3 (1976).

PLASMA SIMULATION USING FLUID AND PARTICLE MODELS

R G Evans

Rutherford Appleton Laboratory

1
INTRODUCTION

With the enormous increase in the power of electronic computers since 1960 it is now possible to solve numerically the partial differential equations describing quite complex physical situations. In some cases eg the aerodynamics of civil and military aeroplanes the numerical simulations are so cost effective that they act as a spur to further computer development.

In plasma physics there are essentially two distinct approaches to simulation. The first is the MHD fluid model of the plasma(1), which uses macroscopic variables such as pressure, density and magnetic field and is able to follow quite long time scales in the plasma. The second method concentrates on the microscopic description of the plasma as a collection of mutually interacting particles(2) and gives a detailed picture of waves and particle distribution functions.

This simple distinction is blurred in practice by recent developments in particle models(3) which average out some of the high frequency waves and follow MHD time scales, and also by the use of fluid methods(4) to solve the Vlasov equation as a 6D fluid equation in phase space.

To make these lectures moderately self contained an introduction to numerical methods will be given first, followed by a description of fluid and particle methods and a brief review of their applications to real problems.

2
FINITE DIFFERENCE EQUATIONS

The real equations describing a physical system of interest can be thought of in the most general form as

$$\frac{\partial F}{\partial t} = f\left(\underline{F}, \underline{x}, \frac{\partial F}{\partial x}, \frac{\partial^2 F}{\partial x^2}, \quad \ldots \right)$$

and the function f need not be linear. Two simple examples serve to indicate the basic features of finite difference equations in one dimension:

a) Heat conduction

$$\frac{\partial T}{\partial t} = \sigma \frac{\partial^2 T}{\partial x^2} + S(x,t) \tag{1}$$

b) particle motion in a potential field

$$\frac{\partial v}{\partial t} = -\nabla\phi(x)$$
$$\frac{\partial x}{\partial t} = v \tag{2}$$

Since computers cannot handle continuous functions and derivatives, $\underline{F}$ must be represented on some finite number of spatial grid points $\underline{x}_i$ and at some discrete time levels t^n as $\underline{F}_i^n$ The task of numerical analysis is to find an accurate and computationally feasible approximation to the derivatives and to solve the resulting (possibly) non-linear algebraic equations(5).

2.1 Centred Differences

Consider the approximation to $\partial F/\partial x$ on the one dimensional grid $x_i = i\Delta x$. The simplest approximations are

$$\left(\frac{\partial F}{\partial x}\right)_i = (F_{i+1} - F_i)/\Delta x$$

or

$$\left(\frac{\partial F}{\partial x}\right)_i = (F_i - F_{i-1})/\Delta x$$

Both of these are weak approximations in the sense that the error is $O(\Delta x)$ and tends to zero only linearly in Δx. On the other hand the

Taylor series expansion of F shows that

$$\left(\frac{\partial F}{\partial x}\right)_i = \frac{F_{i+1} - F_{i-1}}{2\ \Delta x} + \frac{x^2}{3}\left(\frac{\partial^3 F}{\partial x^3}\right)$$

which is a much better approximation. On the other hand it is slightly inconvenient since it spans three mesh points and it is often convenient to think of a split mesh with some quantities defined half a mesh interval out of phase, ie

$$\left(\frac{\partial F}{\partial x}\right)_{i+\frac{1}{2}} = \frac{F_{i+1} - F_i}{\Delta x} + O(\Delta x^2)$$

This is known as centred differencing and gives the best accuracy for a given amount of computational effort. Even order derivatives are much easier to handle since

$$\frac{\partial^2 F}{\partial x^2} = \frac{\partial}{\partial x}\left(\frac{\partial F}{\partial x}\right)$$

and the two meshes come back into phase. The usual approximation is

$$\frac{\partial^2 F}{\partial x^2} = \frac{F_{i+1} - 2F_i + F_{i-1}}{\Delta x^2} \qquad (3)$$

2.2 Leapfrog

If the idea of central differences is applied to the particle equations of motion (2) then we obtain the natural splitting onto integer and half integer time steps

$$\frac{v^{n+1} - v^n}{\Delta t} = \left(\frac{\partial v}{\partial t}\right)^{n+\frac{1}{2}} = -\nabla\phi(x^{n+\frac{1}{2}})$$

$$\frac{x^{n+\frac{1}{2}} - x^{n-\frac{1}{2}}}{\Delta t} = \left(\frac{\partial x}{\partial t}\right)^n = v^n$$

If $x^{n-\frac{1}{2}}$, v^n are known this gives two simple equations for $x^{n+\frac{1}{2}}$, v^{n+1} and effectively advances the system by one step Δt.

$$x^{n+\frac{1}{2}} = x^{n-\frac{1}{2}} + v^n\ \Delta t$$

$$v^{n+1} = v^n - \nabla\phi\ (x^{n+\frac{1}{2}})\ \Delta t$$

Alernatively advancing two variables on a split mesh is known as 'leapfrog' and is a very simple and accurate method if it can be applied. A minor problem is that the initial conditions on x and v must be set up one half time step apart.

2.3 Implicit and Explicit Differencing

This time we take the diffusion equation(1) as an example. Take a uniform mesh in x and t, $x_i = i\,\Delta x$, $t^n = n\Delta t$, then the left hand side becomes

$$\frac{T_i^{n+1} - T_i^{n}}{\Delta t} = \left(\frac{\partial T}{\partial t}\right)^{n+\frac{1}{2}}$$

The right hand side cannot be evaluated exactly at $t^{n+\frac{1}{2}}$ so as a first try we evaluate it at t^n using (3). This gives an explicit equation

$$T_i^{n+1} = T_i^n + \frac{\sigma\Delta t}{\Delta x^2}\left(T_{i+1}^n - 2T_i^n + T_{i-1}^n\right) + S_i^n\,\Delta t \tag{4}$$

Since all the quantities on the right hand side are known this is simple and fast but will have an error of $O(\Delta t)$ as noted earlier. If the problem needs to be solved for some time interval (o, t) then reducing the time step Δt does not necessarily help since the number of time steps varies as $^1/\Delta t$ and the accumulated error does not change. Equation (4) hints at another problem with explicit differencing: if

$$\sigma\Delta t/\Delta x^2 = \tfrac{1}{2}$$

then the right hand side of (4) has no dependence on T_i^n, and if

$\sigma\,\Delta t/\Delta x^2 > \frac{1}{2}$ it actually has a negative dependence on T_i^n. This causes the solution to lose "memory" of its previous state and to generate increasingly large errors, ie it is unstable.

The second possibility is to evaluate the right hand side of (1) at the new time step t^{n+1}, this gives a matrix equation for T^{n+1} which by convention is written

$$-A_i\,T_{i+1}^{n+1} + B_i\,T_i^{n+1} - C_i\,T_{i-1}^{n+1} = D_i \tag{5}$$

where

$$A_i = \sigma\Delta t/\Delta x^2$$

$$B_i = 1 + 2\,\sigma\,\Delta t/\Delta x^2$$

$$C_i = \sigma\Delta t/\Delta x^2$$

$$D_i = S_i^{n+1}\,\Delta t$$

This is considerably more difficult to solve than (4) and also has error $O(\Delta t)$. What makes it attractive in some situations is its behaviour for large time steps. If we write the implicit form of (1) as

$$\frac{T_i^{n+1} - T_i^n}{\Delta t} = (\sigma\frac{\partial^2 T}{\partial x^2} + S)^{n+1}$$

then as $\Delta t \to \infty$ the implicit solution tends towards the solution of

$$\sigma \frac{\partial^2 T}{\partial x^2} + S = 0$$

ie the equilibrium solution. If the equations of interest have an equilibrium then implicit differencing will always approach the equilibrium, independently of the value of Δt. In this sense implicit differencing is unconditionally stable. However the rate at which the system approaches equilibrium will not be reproduced in the implicit solution unless Δt is sufficiently small.

The third choice on forming the right hand side of (1) is to approximate

$$(\frac{\partial^2 T}{\partial x^2})^{n+\frac{1}{2}} = \tfrac{1}{2}(\frac{\partial^2 T^n}{\partial x^2} + \frac{\partial^2 T^{n+1}}{\partial x^2})$$

ie it is the mean of the explicit and implicit forms. Taylor analysis shows that this has an error $O(\Delta t^2)$ and is the most accurate representation of the original differential equation using just adjacent mesh points. Again this gives rise to a matrix equation of the form (5) with coefficients

$$A_i = \sigma\Delta t/2\Delta x^2$$

$$B_i = 1 + \sigma\Delta t/\Delta x^2$$

$$C_i = \sigma\Delta t/2\Delta x^2$$

$$D_i = T_i^n + \frac{\sigma\Delta t}{2\Delta x^2}(T_{i+1}^n - 2T_i^n + T_{i-1}^n) + \frac{\Delta t}{2}(S_i^n + S_i^{n+1})$$

This form of differencing, centred in space and time is known as the Crank Nicholson method[6].

2.4 Time step constraints

The condition $\sigma\Delta t / \Delta x^2 < \frac{1}{2}$ is suggested by the above, very simple analysis to represent a stability limit for the solution of the diffusion equation. This can be proved vigorously by Fourier analysing the temperature distribution T_i and requiring that all frequency components decay in time[7]. A very similar result can be deduced for the solution of the wave equation

$$\frac{\partial^2 f}{\partial t^2} = c^2 \frac{\partial^2 f}{\partial x^2}$$

This has the following straightforward finite difference representation:

$$\frac{f_i^{n+1} - 2f_i^n + f_i^{n-1}}{\Delta t^2} = c^2 \frac{f_{i+1}^n - 2f_i^n + f_{i-1}^n}{\Delta x^2}$$

or rearranging

$$f_i^{n+1} = c^2 \Delta t^2/\Delta x^2 \; (f_{i+1}^n - f_{i-1}^n) - f_i^{n-1} + 2(1 - \frac{c^2 \Delta t^2}{\Delta x^2}) f_i^n$$

This time we see that if $c\,\Delta t / \Delta x > 1$ then f_i^{n+1} has a "negative" dependence on f_i^n and again a Fourier analysis will show that this gives rise to an instability.

Each of these constraints expresses the idea that information in the real system should not propagate more than one cell Δx in one time step Δt. In the case of the wave equation this is obvious, and in the case of the diffusion equation it represents the diffusive "velocity" of a feature with size Δx, ie $v_{eff} = \sigma/\Delta x$. These are particular forms of the Courant Lewy Friedrichs conditions(8) for stability of finite difference equations.

2.5 Conservation

The equations used to describe a physical situation should reflect the conservation properties eg energy, momentum, of that system. Since the finite difference equations are an approximation to the real equations they may or may not reflect this conservation law. As an example take the heat diffusion equation (1) with S = 0, then energy is conserved and $\int T\,dx$ = const, and in the numerical representation on a uniform mesh we would like ΣT_i = const

For the constant coefficient case described above equations (4) and (5) may be added together to show that this is in fact true. However it is only true if σ is constant, and for instance in a plasma σ is a strong function of temperature. In this case writing the Laplacian of T in the form

$$\nabla^2 T = \frac{T_{i+1} - 2T_i + T_{i-1}}{\Delta x^2}$$

does not give rise to conservation of energy. Conservation is recovered by writing the original equation as

$$\frac{\partial T}{\partial t} = - \text{div } \underline{Q}, \quad \underline{Q} = - \sigma \underline{\nabla T}$$

ie

$$Q_{i+\frac{1}{2}} = - \frac{\sigma_{i+\frac{1}{2}}}{\Delta x} (T_{i+1} - T_i)$$

$$Q_{i-\frac{1}{2}} = - \frac{\sigma_{i-\frac{1}{2}}}{\Delta x} (T_i - T_{i-1})$$

and

$$\frac{T_i^{n+1} - T_i^n}{\Delta t} = \frac{1}{\Delta x^2} \cdot (\sigma_{i+\frac{1}{2}} T_{i+1} - (\sigma_{i+\frac{1}{2}} + \sigma_{i-\frac{1}{2}})T_i + \sigma_{i-\frac{1}{2}} T_{i-1})^n$$

In this way both implicit and explicit differencing conserve energy exactly.

A different application of conservation laws occurs in the particle equation of motion in a potential field. We have then $\frac{1}{2}v^2 + \phi = \phi_o =$ const Leapfrog does not conserve energy exactly, since

$$\tfrac{1}{2}(v^{n+1})^2 - \tfrac{1}{2}(v^n)^2 = v^n \nabla\Phi(x^{n+\frac{1}{2}}) \Delta t + (\nabla\Phi \Delta t)^2$$

The first term represents the change in potential energy and there is an error of $O(\Delta t^2)$. Exact conservation is only possible if v and x are defined at the same time steps, but the solution is then more laborious:

$$x^{n+1} = x^n + v^{n+\frac{1}{2}} \Delta t = x^n + (v^n + v^{n+1}) \Delta t/2$$

$$v^{n+1} = \sqrt{2\phi_o - 2\phi(x^{n+1})}$$

These two equations now have to be solved iteratively.

Exact conservation can be written into most numerical methods, but at the expense of greater computational effort; whether or not this is worthwhile depends on the precise application.

2.6 Matrix Equations

The first matrix equation that we have met is the simple tridiagonal equation, written out in full matrix form as:

$$\begin{vmatrix} B_1 & -A_1 & 0 & \cdot & \cdot & \cdot & & \\ -C_2 & B_2 & -A_2 & \cdot & \cdot & \cdot & & \\ 0 & -C_3 & B_3 & \cdot & \cdot & \cdot & & \\ & & & & & & & \\ & & & \cdot & -C_{n-1} & B_{n-1} & -A_{n-1} \\ & & & & 0 & -C_n & B_n \end{vmatrix} \cdot \begin{vmatrix} U_1 \\ U_2 \\ U_3 \\ \\ U_{n-1} \\ U_n \end{vmatrix} = \begin{vmatrix} D_1 \\ D_2 \\ D_3 \\ \\ D_{n-1} \\ D_n \end{vmatrix}$$

Apart from the leading diagonal and the two adjacent diagonals all the N x N elements are zero. This form of the matrix allows the simple Gauss elimination (ie row manipulation) to proceed without generating any new non-zero elements. (If the whole matrix filled up it would lead to enormous storage problems in the computer). A particularly compact and fast form of Gauss method is given by the sequence

$$E_o = F_o = 0 : E_j = \frac{A_j}{B_j - C_j E_{j-1}} \qquad : F_j = \frac{D_j + C_j F_{j-1}}{B_j - C_j E_{j-1}}$$

$$U_N = F_N \quad : \quad U_j = E_j \, U_{j+1} + F_j$$

In two dimensions the heat flow equation gives rise to an equation of the form

$$-A_{ij} \, U_{i-1,j} - B_{ij} \, U_{i,j-1} + C_{ij} U_{ij} - D_{ij} U_{i+1,j} - E_{ij} U_{i,j+1} = F_{ij}$$

with $A = \sigma \Delta t / 2\Delta x^2$

$B = \sigma \Delta t / 2\Delta y^2$

In order to lay this out as a matrix equation we take the row vectors in sequence $U_{1i}, U_{2i}, \ldots\ldots U_{Ni}$ $(i = 1 \ldots M)$
and assemble them as one long row vector

$$U = (U_{11}, U_{21} \ldots U_{M1}, U_{12} \ldots U_{M2} \ \ldots\ldots \ U_{1N}, U_{2N} \ldots U_{MN})$$

this gives the matrix equation

$$
\begin{pmatrix}
C_{11} & -D_{11} & 0 \cdots & & & -E_{11} & 0 & 0 \cdots & & & & \\
-A_{21} & C_{21} & -D_{21} & & & 0 & -E_{21} & 0 \cdots & & & & \\
0 & -A_{31} & C_{31} & & & & & & & & & \\
 & & & & -D_{M-1,1} & & & & & & & \\
 & & & -A_{M1} & C_{M1} & & & & & -E_{M1} & & \\
-B_{11} & 0 & 0 & & & C_{12} & -D_{12} & 0 & & & & \\
0 & -B_{21} & & & & -A_{22} & C_{22} & -D_{22} & 0 & & & \\
 & & & & -B_{M1} & & & & & C_{M2} & & \\
 & & & & & & & & & & & -D_{M-1,N} \\
 & & & & & & & & & & -A_{MN} & C_{MN}
\end{pmatrix}
\begin{pmatrix}
U_{11} \\ U_{21} \\ \\ \\ U_{M1} \\ U_{12} \\ U_{22} \\ U_{M2} \\ U_{M-1,N} \\ U_{MN}
\end{pmatrix}
=
\begin{pmatrix}
F_{11} \\ F_{21} \\ \\ \\ F_{M1} \\ F_{12} \\ F_{22} \\ F_{M2} \\ F_{M-1,N} \\ F_{MN}
\end{pmatrix}
$$

This has five non-zero diagonals but they are no longer adjacent. Gauss elimination will fill up most of the MNxMN matrix and cannot be stored in the computer. Most methods of solution of these "sparse matrix" equations are approximate and two fairly simple ones will be described here.

2.7 Alternate Direction Implicit (ADI)

ADI will be described in general terms involving the Laplacian operator to reduce the problems of notation. The esssence of ADI is to split the finite difference representation as

$$\nabla^2 U = \frac{U_{i,j+1}^{(n)} - 2U_{ij}^{(n)} + U_{i,j-1}^{(n)}}{\Delta y^2} + \frac{U_{i+1,j}^{(n+1)} - 2U_{ij}^{(n+1)} + U_{i-1,j}^{(n+1)}}{\Delta x^2}$$

((n) is the order of iteration, not the time step)

The sequence of iteration is as follows

Take trial solution $U_{ij}^{(o)}$ (could be previous timestep)
Construct tridiagonal equation in $U_{i,j-1}, U_{ij}, U_{i,j+1}$
solve for $U_{ij}^{(1)}$

reverse (n) and (n+1) in representation of ∇^2

Construct tridiagonal equation in $U_{i-1,j}, U_{ij}, U_{i+1,j}$
solve for $U_{ij}^{(2)}$
repeat until convergence is achieved.

ADI was very popular for heat flow problems in two dimensions but has been superceded in many applications.

2.8 Successive Over Relaxation (SOR)

This is another iterative method which is simple to code and applies to any pattern of sparsity provided the matrix is diagonally dominant, ie the diagonal element is much greater than the sum of all other elements in a row or column.

The basic relaxation method writes the matrix equation to give u_{ij} alone on the left hand side, eg the heat flow equation

$$T_{ij} = \frac{1}{C_{ij}} \left(A_{ij} T_{i-1,j} + B_{ij} T_{i,j-1} + D_{ij} T_{i+1,j} + E_{ij} T_{i,j+1} + F_{ij}\right)$$

Given any trial solution $T_{ij}^{(0)}$ substitute this on the right hand side to give $T_{ij}^{(1)}$, and repeat until the iterations converge.

Over relaxation is a minor variation on this, if we write the sequence of operations for relaxation as $T_{ij}^{(n+1)} = T_{ij}^{(n)} + \delta_{ij}^{(n)}$

then the extension to $T_{ij}^{(n+1)} = T_{ij}^{(n)} + \alpha\, \delta_{ij}^{(n)}$ results in over relaxation if $\alpha > 1$ and under relaxation if $\alpha < 1$. Prescriptions exist for calculating the optimum value of α for a given set of coefficients but this frequently results in large computational overheads. It is frequently more efficient to obtain a "good" constant value of α (typically $1.1 < \alpha < 1.3$) by trial, and then to use this for repeated solutions.

SOR is not the most efficient method on scalar computers(11) but it vectorises on nearly all array processing or pipelined super computers and is most valuable in these applications.

3
FLUID CODES

The fluid approximation to a plasma assumes that it is adequately described by the thermodynamic quantities density, pressure, energy and velocity. The details of the particle distribution function are lost, and it is assumed to be Maxwellian in order to calculate various transport coefficients. The system of equations describing the hydrodynamic behaviour of the plasma splits naturally into two parts both physically and numerically.

a) Fluid motion, described by the Navier Stokes equation. In the absence of viscosity it has no dissipation apart from shock fronts.

b) Energy transport by laser irradiation, thermal conduction and possibly by thermal radiation. This is strongly dissipative as the system seeks to remove the imposed temperature gradients.

The equations describing the fluid behaviour are written using the Lagrangian derivative $\frac{Df}{Dt} = \frac{\partial f}{\partial t} + (\underline{v}.\underline{\nabla})\, f$

$$\frac{D\rho}{Dt} = -\,\rho \text{ div } \underline{v} \qquad \frac{D\underline{v}}{Dt} = -\,\frac{1}{\rho}\,\underline{\nabla}\, p$$

$$\frac{DU}{Dt} + p\,\frac{DV}{Dt} = \text{div}\,(-\kappa\,\underline{\nabla} T) + S$$

Where U is the specific internal energy and $V = 1/\rho$ is the specific volume.

Fluid codes split into two classes according as to whether the computational mesh is fixed in the fluid (Lagrangian) or fixed in the laboratory frame (Eulerian). In one space dimension the Lagrangian scheme is numerically more straightforward and will be described first.

3.1 The 1-D Lagrangian scheme of Richtmeyer and von Neumann[12]

In the Lagrangian description on a moving mesh x_i the Lagrangian derivative is simply the partial derivative

$$\frac{Df}{Dt} = \frac{\partial f}{\partial t}$$

In the presence of non-uniform fluid motion the mesh will stretch or compress and the space between two mesh points will always contain the same amount of fluid ie dM_i = const.

The motion of the cell boundaries is given by

$$x^{n+1}_{i+\frac{1}{2}} - x^{n}_{i+\frac{1}{2}} = v^{n+\frac{1}{2}}_{i+\frac{1}{2}}\ \Delta t$$

and the cell volumes by $dV_i = x_{i+\frac{1}{2}} - x_{i-\frac{1}{2}}$

where we are anticipating splitting the time and space meshes. Since the mesh is no longer uniform the splitting does not necessarily give second order accuracy but is the most useful representation without incurring heavy additional computation.

The thermodynamic variables are defined at the cell centres, and give rise to the following momentum and energy equations

$$v_{i+\frac{1}{2}}^{n+\frac{1}{2}} - v_{i+\frac{1}{2}}^{n-\frac{1}{2}} = \frac{\Delta t}{2} \frac{(p_{i+1}^{n} - p_{i}^{n})}{(dM_{i} + dM_{i+1})}$$

$$U_i^{n+1} - U_i^n + \tfrac{1}{2}(p_i^n + p_i^{n+1})(V_i^{n+1} - V_i^n) = (Q_{i+\frac{1}{2}} - Q_{i-\frac{1}{2}} + S_i^{n+\frac{1}{2}})\,\Delta t$$

where $$Q_{i+\frac{1}{2}} = -\tfrac{1}{2}\ \frac{(\kappa_i + \kappa_{i+1})\ (T_{i+1} - T_i)}{\tfrac{1}{2}(x_{i+3/2} - x_{i-\frac{1}{2}})}$$

The time index on Q has been suppressed, but assuming that the usual Crank Nicholson scheme is used the average of Q^n and Q^{n+1} would be inserted above. In order to solve these equations some relationship between U, p and T is required, ie an equation of state. In sophisticated fluid codes the equation of state is quite complex but for introductory purposes a perfect gas equation of state can be assumed ie $U = 3/2\ n\ k\ T$

$p = n\ k\ T$, where $n = \rho/M_i$ and M_i is the ion mass.

The sequence of solution is as follows:

assume $v^{n-\frac{1}{2}}$, x^n, T^n, ρ^n are known

a) Calculate $v^{n+\frac{1}{2}}$

b) Calculate x^{n+1}, V^{n+1}, ρ^{n+1}

c) Knowing V^{n+1} and V^n solve the matrix equation for T^{n+1}

Time step constrainsts are set by the Courant Lewy Friedrichs (CLF) condition, which will be different on different parts of the mesh; the smallest value must always be used.

This basic 1D Lagrangian code is simple and reasonably accurate but as it stands will not handle shock waves because the fluid motion is not dissipative in the numerical scheme. Shock fronts produce strong oscillations of density and temperature. This is normally handled by introducing a fictitious pressure, known as artificial viscosity, near the shock front. It is applied only when the fluid is compressing ie $\frac{\partial v}{\partial x} < 0$; its form is not crucial and is normally written as

$$P = \begin{cases} a\,\Delta x^2\,\rho\,(\partial v/\partial x)^2 & \text{if } \partial v/\partial x < 0 \\ 0 & \text{if } \partial v/\partial x > 0 \end{cases}$$

The constant 'a' determines how many mesh points will be taken up by the shock front.

3.2 Additions to the basic code

1) Ion and Electron temperatures

In laser heated targets the electrons are heated directly by the laser while the ions gain energy through electron ion collisions giving $T_e > T_i$ in the absorbing region. Near shock fronts the ions are preferentially heated giving $T_i > T_e$.
This may be handled by having separate energy equations for electrons and ions. The laser source term appears only in the electron equation, while the work done by artificial viscosity PdV appears only in the ion equation. The two temperatures are coupled together by

$$\frac{\partial}{\partial t}(T_e - T_i) = (T_e - T_i)/\tau_{ei}, \text{ where } \tau_{ei} \text{ is}$$

the electron ion equipartition time.

2) Electron heat flux limit

The Spitzer(13) diffusion theory $Q_o = -\kappa\underline{\nabla}T$ breaks down when the electron mean free path is comparable to the temperature scale length. This requires a solution of the electron Fokker Planck equation(14) but is frequently approximated by a harmonic mean "flux limit"

$$1/Q = 1/Q_o + 1/Q_s$$

where

$$Q_o = -\kappa \underline{\nabla} T \text{ and } Q_s = \tfrac{1}{2} f n_e m_e v_e^3$$

The constant f is frequently adjusted to attempt to fit experimental data(15,16).

3) Multi group electron Transport

Laser light in plasmas frequently gives rise to strongly non-Maxwellian distributions. These may be approximated as a Maxwellian plus a non-thermal tail, and the tail is divided into discrete energy bins. Each energy bin is described by a diffusion equation with sources and sinks and an electric field must be generated to preserve quasi-neutrality(17).

4) Radiation Transport

Particularly in high Z materials an appreciable part of the energy transport is by X-ray photons. The sources and sinks of the radiation require a detailed calculation of atomic level populations in the plasma and in some cases this must be carried out in step with the hydrodynamic simulation(18), while in other cases it may be possible to precalculate tables of source function and opacity. The transport of the photons is sometimes by multi group diffusion although this is a poor approximation. X-ray photons are emitted and absorbed rather than scattered and hardly ever obey a diffusion equation.

3.3 Eulerian Methods

In the Eulerian formulation the energy transport equation is solved in the same way as for the Lagrangian description but the motion of the Lagrangian mesh points is replaced by advection of the fluid on the fixed Eulerian grid. As the simplest possible example of this consider a fixed density profile being carried along by a uniform velocity field v, then

$$\frac{\partial \rho}{\partial t} + v\frac{\partial \rho}{\partial x} = 0$$

and the explicit centred difference form is

$$\rho_i^{n+1} - \rho_i^n = -\frac{v\Delta t}{2\Delta x}(\rho_{i+1}^n - \rho_{i-1}^n)$$

A simple calculation with a square wave density profile shows that this is extremely unstable, as is shown in Fig 1(a) for Courant number $c = v\Delta t/\Delta x = 0.2$. Taking smaller time steps does not help since the method is first order in Δt. Fig 1(b) shows the calculation for $c = 0.05$ at the same times. Using either the "Crank Nicholson" or fully implicit differencing (Fig 1c) does not help greatly and the severe distortion is due to high spatial frequencies propagating more slowly than low frequencies giving rise to phase errors. Paradoxically a scheme which is only first order in Δx actually gives "better" results. This is known as "donor cell" or upstream differencing and takes the x derivative on the upstream side of the cell under consideration

$$\rho_i^{n+1} - \rho_i^n = \frac{-v\Delta t}{\Delta x} \times \begin{matrix} (\rho_i^n - \rho_{i-1}^n) \text{ if } v > 0 \\ (\rho_{i+1}^n - \rho_i^n) \text{ if } v < o \end{matrix}$$

Subject to a Courant condition $c < \frac{1}{2}$ this scheme is stable and maintains positivity (which is physically very important). Fig 1(d) shows the calculation for donor cell differencing for $c = 0.2$ and the main defect is a diffusive smoothing of the steep sides of the original profile.

Many more accurate schemes have been developed eg those of Lax-Wendroff[20] and Fromm[21] but all symmetrical difference schemes suffer from phase errors and do not in general maintain positivity. The numerical diffusion coefficients of any particular scheme can be calculated and a corresponding "anti-diffusion" can be provided to sharpen up the steep boundaries. However anti-diffusion is inherently unstable and can give rise to rapid high frequency fluctuations. The

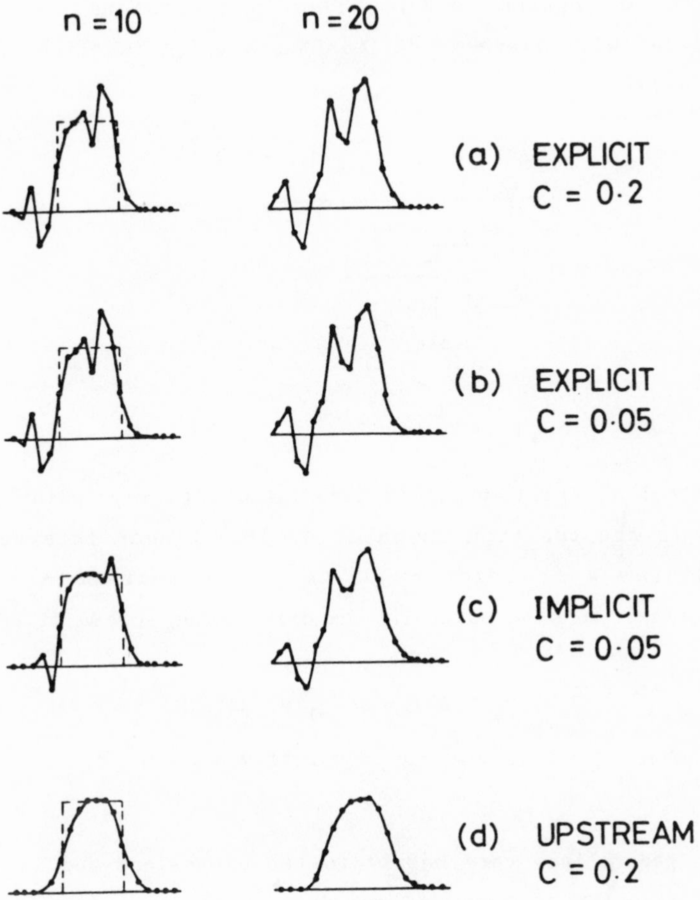

Fig 1 Advection of a square density profile (shown dotted) using different numerical schemes

Flux Corrected Transport scheme of Boris and Book[22] uses anti-diffusion but applies a non-linear clipping to the anti-diffusive fluxes to prevent any new maxima or minima appearing in the solution. FCT can be applied to any advection technique but it is often associated with a scheme called SHASTA which is slightly better than donor cell.

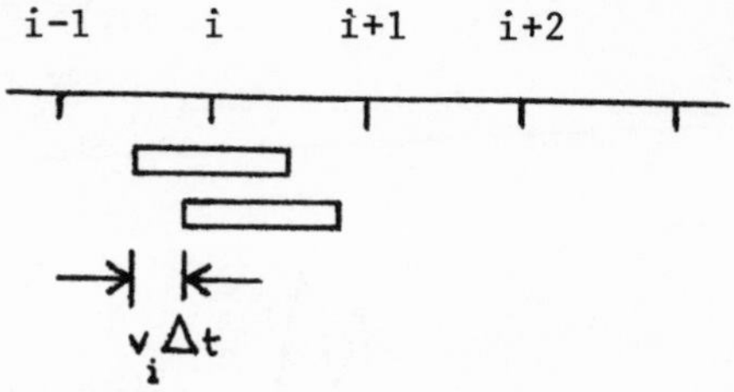

The parcel of fluid initially associated with mesh point i, at time t moves to the position shown at t $+\Delta t$. Linear interpolation then associates a fraction $v_i \Delta t/\Delta x$ to mesh point (i + 1) and $(1-v_i\Delta t/\Delta x)$ to mesh point i. The diferencing scheme of SHASTA is then

$$\rho_i^{n+1} = \rho_i^n (1-v_i\Delta t/\Delta x) + \alpha\rho_{i-1}^n (v_{i-1}\Delta t/\Delta x) + (1-\alpha)\rho_{i+1}^n (v_{i+1}\Delta t/\Delta x)$$

where $\alpha = 1$ if $v > 0$; $\alpha = 0$ if $v < 0$

SHASTA generalises very easily to two dimensions and is an extremely useful algorithm. Since FCT requires different computations depending on the sign of v, and whether or not clipping of the anti diffusive fluxes occurs it may not vectorise very well on certain machines. Van Leer's[23] extension of Fromm's method is more amenable to vectorisation and has a similar performance to FCT.

SHASTA is a simple mapping of mass from one mesh to another and works equally well if the Eulerian mesh moves between successive timesteps. This enables the mesh in an Eulerian code to adapt itself to the evolution of the system and in one dimension to behave rather like a Lagrangian code, with the advantage that the non-linear nature of FCT removes the need for an artificial viscosity. Moreover Eulerian codes can be conservative to within rounding error if the conserved quantities are advected directly and in this order:

$$\begin{aligned}&\rho\\&\rho v\\&p + \tfrac{1}{2}\rho v^2\\&U + pV + \tfrac{1}{2}v^2\end{aligned}$$

3.4 Multi Dimensional Codes

In one space dimension the usual argument in favour of Lagrangian codes is that since the mesh points follow the fluid motion the mesh always resolves regions of interest. An additional advantage is that material boundaries are always associated with the same mesh point so that multi-material codes are quite straightforward to implement.

In more than one space dimension the problem with the Lagrangian description is that the mesh will not remain orthogonal[24]. The representation of gradient operators is then very complicated if the basic requirements of symmetry, and heat flow parallel to temperature gradients are to be maintained. Typically at least nine adjacent grid points must be considered in the heat flow equation giving rather lengthy matrix equations. The extreme case of mesh distortion is when a cell can turn inside out without its volume going to zero. Such pathological situations can never be handled correctly and most 2D Lagrangian schemes allow some fluid slip through the cell boundaries to prevent this occurring.

On the other hand Eulerian codes do not always place the mesh points where they are needed, even with adaptive rezoning and are very difficult to use for multi-material calculations. However the orthogonal mesh can always be relied on to give accurate differencing and the average fluid motion can be followed with confidence.

3.5 MHD Codes

In two or more dimensions ∇n_e and ∇T_e are not necessarily parallel and give rise to[25]

$$\frac{\partial B}{\partial t} = \frac{k}{cn_e}\,(\underline{\nabla n}_e \times \underline{\nabla T}_e)$$

The equation for the evolution of B also include the advection of B by the electrons, the curvature terms and resistive diffusion. The presence of a non-zero magnetic field makes an accurate hydrodynamic description of the plasma vastly more difficult since the following additional effects must be considered(26,27)

a) The magnetic energy appears in the energy equation

b) Since $\frac{\partial \underline{B}}{\partial t} = \text{curl } E$, E is non-zero and so is $\underline{j}$

This gives Ohmic heating $\underline{E}.\underline{j}$ and the Lorentz force $\underline{j}\times\underline{B}$

c) The thermal conductivity is different parallel to and across $\underline{B}$

d) The heat fluxes and currents are related by the thermoelectric, Nernst, Hall and Righi Leduc terms.

MHD effects in the low density plasma can be important in changing density and temperature profiles and it has been suggested that magnetic fields at densities above critical density can modify the ablation flow and the fluid instabilities(28).

4

PARTICLE IN CELL (PIC) CODES

The essential feature of particle simulation(2) is that the plasma is described as a large number of discrete particles moving in their self-consistent electromagnetic fields. The typical time scale of the simulation is the plasma period and in general PIC codes are unsuitable for description of the plasma on a hydrodynamic time scale. On the other hand PIC methods give details of wave motions in the plasma and of the particle distribution function. Ions and electrons are normally simulated as two distinct species but for brevity the discussion here will only be of the electron equations of motion.

The method of particle codes is to repeatedly solve the equation of motion

$$\frac{d}{dt}(\gamma\underline{v}) = \frac{e}{m}(\underline{E} + \frac{1}{c}\underline{v} \times \underline{B}) \quad : \gamma = \sqrt{1-v^2/c^2}$$

and Maxwell's equations

$$\frac{\partial \underline{B}}{\partial t} = - c \text{ curl } \underline{E} \qquad \frac{\partial \underline{E}}{\partial t} = c \text{ curl } \underline{B} - 4\pi \underline{J}$$

The remaining two Maxwell equations div E = $4\pi\rho$ and div B = 0 serve mainly as initial conditions and consistency checks. The system is closed by the definition of charge and current density

$$\rho = \sum_j e\,\delta(\underline{r}_j - \underline{r}) - \sum_j Ze\delta(\underline{r}_j - \underline{r})$$

$$\underline{J} = \sum_j e\underline{v}\delta(\underline{r}_j - \underline{r}) - \sum_j Ze\underline{v}\,\delta(\underline{r}_j - \underline{r})$$

where the δ functions represent general shape functions for the particles.

The full 3D equations are rarely solved, although they can be written down easily, since the computational demands in terms of number of particles are excessive. The following subsets are useful, and are obtained by setting various derivatives and constant field quantities to zero.

$2\frac{1}{2}$D:	$\partial/\partial z = 0$:	$(B_x \quad B_y \quad B_z)$	$(E_x \quad E_y \quad E_z)$
		$(x \quad y \quad 0)$	$(v_x \quad v_y \quad v_z)$
2D :	$\partial/\partial z = 0$:	$(B_x \quad B_y \quad B_z)$	$(E_x \quad E_y \quad 0)$
		$(x \quad y \quad 0)$	$(v_x \quad v_y \quad 0)$
$1\frac{1}{2}$D:	$\partial/\partial z = 0$:	$(0 \quad 0 \quad B_z)$	$(E_x \quad E_y \quad 0)$
	$\partial/\partial y = 0$	$(x \quad 0 \quad 0)$	$(v_x \quad v_y \quad 0)$
1D :	$\partial/\partial z = 0$:	$(0 \quad 0 \quad 0)$	$(E_x \quad 0 \quad 0)$
	$\partial/\partial y = 0$	$(x \quad 0 \quad 0)$	$(v_x \quad 0 \quad 0)$

As the number of space and velocity dimensions increases, so the required number of simulation particles increases in order to populate the particle phase space at a reasonable density. More than 10^6 particles may be necessary in the 5 dimensional phase space of the $2\frac{1}{2}$ D codes.

The purely 1D problem is electrostatic since no transverse fields are allowed and in this case E_x is most conveniently obtained from Poisson's equation. Historically the first particle codes used direct force calculations which are feasible because of a result due to Dawson(29). In one dimension the particles are sheets of charge in the (y,z) plane, area mass density λ, area charge density σ, so that $\sigma/\lambda = e/m$. They are considered to be embedded in a continuous positive

background of density ρ and initially have position $x_i = i\Delta x$. For neutrality $\rho = \sigma/\Delta x$. In between charge sheets $\partial E/\partial x=4\pi\rho$, while on each sheet there is a jump $(E_+ - E_-) = 4\pi\sigma$.
Suppose that all the particles to the left of the origin are unmoved, but those to the right are each moved by a displacement ξ_i . Then at particle 1;

$$E_- = -2\pi\sigma + \frac{4\pi\sigma}{\Delta x}(\Delta x + \xi_1)$$

$$E_+ = E_- - 4\pi\sigma$$

Force at particle 1 $= \frac{1}{2}(E_- + E_+) = \frac{4\pi\sigma^2}{\Delta x}\xi_1$

At particle 2

$$E_- = -2\pi\sigma + \frac{4\pi\sigma}{\Delta x}\xi_1 + \frac{4\pi\sigma}{\Delta x}(\Delta x + \xi_2 - \xi_1)$$

and force on particle 2 $= \frac{4\pi\sigma^2}{\Delta x}\xi_2$

This holds for all particles provided that the charge sheets do not cross, and does not require $\xi << \Delta x$ This obviously gives the equation of motion

$$\lambda\ddot{\xi} = \frac{4\pi\sigma^2}{\Delta x}\xi \quad : \quad \omega_p^2 = 4\pi\frac{e}{m}\frac{\sigma}{\Delta x}$$

ie $\sigma/\Delta x$ is the effective volume charge density as would be expected.
This direct calculation of force is possible only in one dimension and more generally it is necessary to calculate the field quantities $\underline{E}$ and $\underline{B}$ on a discrete mesh and to interpolate the fields to the position of each particle. To advance the field quantities in time requires a knowledge of ρ and J and these are formed on the grid points by means of the shape function δ . The two simplest functions for δ are

1) Nearest grid point, $\delta_i = \begin{cases} 1 \text{ if } |x - x_i| < \frac{1}{2}\Delta x \\ 0 \text{ if } |x - x_i| > \frac{1}{2}\Delta x \end{cases}$

2) Cloud in cell, or linear interpolation

$$\delta_i = 1 - |x - x_i|/\Delta x$$

Higher order interpolations are rarely used since the extra

computational cost produces little reward. It is important that the same weighting function be used to form ρ and J as is used to interpolate $\underline{E}$ and $\underline{B}$, otherwise the particles will move under the influence of their own charge.

4.1 The 1D Electrostatic Model

The field mesh, labelled i is uniform with spacing Δx, while particles indexed j have arbitrary positions and velocities on the mesh. Poisson's equation gives

$$E_{i+\frac{1}{2}} = E_{i-\frac{1}{2}} + 4\pi\rho_i \Delta x$$

This is directly integrated to give E_i subject to the appropriate boundary conditions. In order to save computation with little loss of accuracy $E_{i+\frac{1}{2}}$ is averaged to the same mesh as ρ_i

$$E_i = \tfrac{1}{2}(E_{i-\frac{1}{2}} + E_{i+\frac{1}{2}})$$

Each of the particles is advanced in velocity and position

$$v_j^{n+\frac{1}{2}} = v_j^{n-\frac{1}{2}} + \frac{e}{m} E_j^n \Delta t \quad (E_j \text{ is interpolated})$$

$$x_j^{n+1} = x_j^n + v_j^{n+\frac{1}{2}} \Delta t$$

finally calculate ρ_i^{n+1} using the new particle positions x_j^{n+1}.

The 1D code is trivially simple, but shows up a number of interesting problems.

a) Timestep The difference scheme can be Fourier analysed[30] to show that stability requires $\omega_p \Delta t < 2$ while retaining reasonable accuracy for the plasma wave frequency requires $\omega_p \Delta t < 0.2$

b) Aliasing High frequency modes are not represented correctly on the mesh and modes k_x and $k_x + 2\pi/\Delta x$ cannot be distinguished. This problem can be alleviated by arranging for high spatial frequencies to be Landau damped. Using the time step constraint $\omega_p \Delta t < 2$ and Landau damping for $k\lambda_D > 0.2$ shows that the mesh spacing Δx should be $\Delta x < \lambda_D$ and enables the time step constraint to be re-expressed as $\Delta t < \Delta x/v_e$ where v_e is the electron thermal velocity.

c) <u>Boundary Conditions</u>

The boundary condition on E is simply that the total potential difference across the plasma is equal to the "applied" potential (the capacitor model). For an infinitely periodic system this potential difference is zero. On the other hand only periodic boundary conditions can be consistently formulated for the particles and for a finite system the following methods can be used

(i) Reflect the particles at the boundary,

(ii) Reflect, but choose v_x from a random distribution

(iii) Reflect with random velocity, but wait for a random time before reflection.

d) <u>Initialisation</u>

Usually it is desired to initialise the particles with some density distribution and a Maxwellian velocity distribution. Unfortunately few computers have Gaussian random number generators and the following trick will be found useful.

Suppose n is randomly distributed on (0,1)

$$\text{then } p(n)\,dn = dn$$

If f(n) is a continuous function of n then

$$p(f)\,df = p(n)\,dn, \text{ ie } p(f) = {}^{dn}/df$$

The Gaussian form e^{-v} is not analytically integrable, but $f = 2ve^{-v}$ integrates to give $n = 1-e^{-f}$, and

$$f(n) = \sqrt{-\ln(1-n)}$$

This gives a Maxwellian distribution in two dimensions in v and so if n_1, n_2 are two random numbers

$f(n_1)\sin(2\pi n_2)$ and $f(n_1)\cos(2\pi n_2)$ give the required velocity distribution.

e) <u>Quiet Start</u>

With relatively few particles per cell compared to a real plasma the noise in a PIC code, ie fluctuations from mean quantities can be quite large. Initial noise can be suppressed by forcing $\langle u \rangle = o$ and $\langle u^2 \rangle = kT$ in each cell.

4.2 $1\frac{1}{2}$ D Electromagnetic Code

Since Maxwell's equations are consistent with Special Relativity it is convenient to use the relativistic equations of motion

$$\frac{d}{dt}(\gamma \underline{v}) = \frac{e}{m}(\underline{E} + \frac{v}{c} \times \underline{B})$$

The particular subset of Maxwell's equations reduces to

$$\frac{\partial B_z}{\partial t} = -c\frac{\partial E_y}{\partial x} \qquad \frac{\partial E_y}{\partial t} = c\frac{\partial B_z}{\partial x} - 4\pi J_y$$

and E_x is purely electrostatic, ie derives from Poisson's equation. Since these equations represent electromagnetic waves with velocity c, there will obviously be a time step constraint $\Delta t < \Delta x/c$

A centred difference form of these equations gives

$$BZ_{i+\frac{1}{2}}^{n+\frac{1}{2}} - BZ_{i+\frac{1}{2}}^{n-\frac{1}{2}} = -\frac{c\Delta t}{\Delta x}(EY_{i+1}^{n} - EY_{i}^{n})$$

$$EY_{i}^{n+1} - EY_{i}^{n} = \frac{c\Delta t}{\Delta x}(BZ_{i+\frac{1}{2}}^{n+\frac{1}{2}} - BZ_{i+\frac{1}{2}}^{n+\frac{1}{2}}) - 4\pi\Delta t J_{i}^{n+\frac{1}{2}}$$

In the centred difference form EY and BZ are not known at the same time so it is necessary to advance BZ in two steps, $BZ^{n-\frac{1}{2}} \rightarrow BZ^{n} \rightarrow BZ^{n+\frac{1}{2}}$ with the particle motion being updated at t^n. Since there are far fewer mesh points than particles this is a small overhead, also BZ is interpolated in space to give $BZ_i = \frac{1}{2}(BZ_{i-\frac{1}{2}} + BZ_{i+\frac{1}{2}})$

The particle momentum is updated as $u_j = \gamma v_j$ to avoid problems as $v \rightarrow c$

$$u_j^{n+\frac{1}{2}} = u_j^{n-\frac{1}{2}} + \frac{e\Delta t}{m}(\underline{E} + \frac{v}{c} \times \underline{B})^n$$

$$x_j^{n+1} = x_j^{n} + v_j^{n+\frac{1}{2}}\Delta t \quad : \quad v = \frac{u}{\sqrt{1 + u^2/c^2}}$$

To form $J^{n+\frac{1}{2}}$ we need to know $v^{n+\frac{1}{2}}$ and $x^{n+\frac{1}{2}}$, in practice this is done by forming $\underline{J}$ half before updating x, and half after updating x, ie with a new set of interpolation constants. The charge density is formed using x^{n+1} and used in Poisson's equation to give E^{n+1}. The form of the Lorentz force prevents the use of leap frog and the following scheme is found to work well and is reversible in time.

$$(1) \quad u^- = u^{n-\frac{1}{2}} + \tfrac{1}{2}\frac{e}{m}E\Delta t$$

$$(2) \quad u^- \rightarrow u^+ \quad \text{due to } \underline{v} \times \underline{B}$$

$$(3) \quad u^{n+\frac{1}{2}} = u^+ + \tfrac{1}{2}\frac{e}{m}E\Delta t$$

Step (2) is a pure rotation in velocity space by an angle $\theta = \frac{eB\Delta t}{\gamma mc}$

Buneman(32) gives the recipe

$$u'_x = u^-_x + u^-_y t$$
$$u^+_y = u^-_y - u'_x s$$
$$u^+_x = u'_x - u^+_y t$$

where $t = \tan\theta/2$, $s = 2t/(1 + t^2)$

4.3 Boundary Conditions in $1\frac{1}{2}$D

The particle boundary conditions can be as in a 1D code but if periodic boundary conditions are used for the fields then there is no way to introduce a laser or other disturbance. If the mesh extends from EY_1 to EY_N, then $BZ_{\frac{1}{2}}$ and $BZ_{N+\frac{1}{2}}$ are needed to update EY. The requisite boundary terms are found by requiring that electromagnetic waves generated within the mesh should escape without reflection. At the right hand boundary waves propagating to the right have EY = BZ

ie

$$EY_N^n + EY_N^{n+1} = BZ_{N-\frac{1}{2}}^{n+\frac{1}{2}} + BZ_{N+\frac{1}{2}}^{n+\frac{1}{2}}$$

where the last term is the boundary value.

Maxwell's equation gives

$$EY_N^{n+1} - EY_N^n = \frac{c\Delta t}{\Delta x}\left(BZ_{N+\frac{1}{2}}^{n+\frac{1}{2}} - BZ_{N-\frac{1}{2}}^{n+\frac{1}{2}}\right)$$

eliminating EY_N^{n+1} we obtain

$$BZ_{N+\frac{1}{2}}^{n+\frac{1}{2}} = \frac{2\,EY_N^n - (1 - c\Delta t/\Delta x)\,BZ_{N-\frac{1}{2}}^{n+\frac{1}{2}}}{(1 + c\Delta t/\Delta x)}$$

with a similar result being obtained for the left boundary with EY=−BZ A laser wave can be introduced by adding the field of the laser to BZ at the boundary.

4.4 $2\frac{1}{2}$ DCode(31,33)

The $2\frac{1}{2}$ D code is similar in principle to the $1\frac{1}{2}$ D code with a staggered mesh

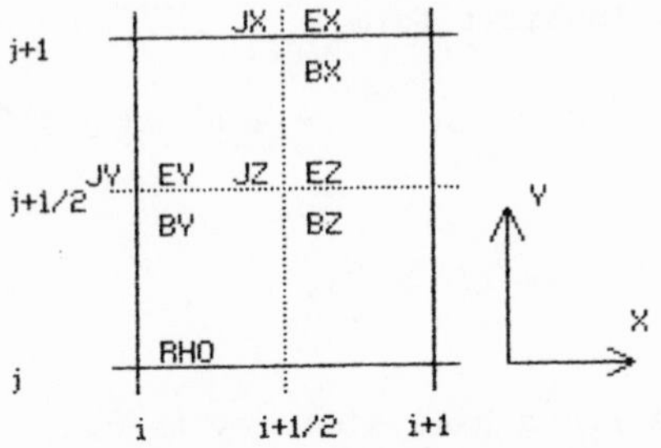

The field and particle equations are solved as for $1\frac{1}{2}$D except that all three components of $\underline{E}$ are electromagnetic. The difference scheme does not preserve div $\underline{E} = 4\pi\rho$ exactly so a correction is made by solving Poisson's equation for the correction field div $\underline{E}^c = 4\pi\rho$ -div $\underline{E}^n$

In $2\frac{1}{2}$ D the boundary conditions are much more difficult to implement since waves at arbitrary angles to the mesh lines must be transmitted(34).

4.5 Implicit Particle Codes

The normal explicit differencing in particle codes gives the time step constraint $\omega_p \Delta t < 2$ and limits the time scales that can be followed. In some cases, eg the study of electron energy transport, the plasma oscillation period is of no interest and longer time steps are desirable. Since implicit methods are generally stable it is of interest to examine whether they can be applied to particle simulations(3).

The equations of motion are written

$$u^{n+\frac{1}{2}} = u^{n-\frac{1}{2}} + \frac{e}{m} E^* \Delta t$$

$$x^{n+1} = x^n + u^{n+\frac{1}{2}} \Delta t$$

and we would like E^* to be implicit. A prediction can be made for E^{n+1} by taking the 'fluid' average of the momentum equation

$$j^{n+\frac{1}{2}} = j^{n-\frac{1}{2}} - \frac{1}{m}(\partial P/\partial x - enE^*)\Delta t$$

j is the particle flux, P the pressure, $P = \Sigma\, m\, u^2$, and n the particle number density.

The continuity equation gives $\qquad n^{n+1} = n^n - \frac{\partial}{\partial x}(j^{n+\frac{1}{2}})\,\Delta t$

and Poisson's equation gives

$$\frac{\partial}{\partial x}(E^{n+1}) = 4\pi e n^{n+1}$$

Thus if we take the fully implicit form $E^* = E^{n+1}$ we obtain

$$E^{n+1} = 4\pi \left(\int^x e n^n dx - e\Delta t j^{n-\frac{1}{2}} + \frac{e\Delta t^2}{m}\frac{\partial P}{\partial x} - \frac{\Delta t^2 n e^2}{m} E^{n+1}\right)$$

ie

$$E^{n+1}(1 + \omega_p^2 \Delta t^2) = 4\pi \left(\int^x e n^n dx - e\Delta t j^{n-\frac{1}{2}} + \frac{e\Delta t^2}{m}\frac{\partial P}{\partial x}\right)$$

All the quantities on the right hand side may be calculated as moments of known particle positions and velocities, giving a fully implicit form of E. The method may be extended to higher dimensions but the algebra is lengthy(35). Implicit PIC codes are of greatest use for the study of transport effects in E and B fields. If it is desired to study collisional effects then Monte Carlo methods(36) may be used to randomly deflect the particles in addition to the self-consistent fields. Such random deflections should however conserve energy and momentum.

5
APPLICATIONS

5.1 Fluid Codes

Fluid simulation is intimately connected with experimental observations of laser produced plasmas since most observable quantities, eg temperature and density profiles are macroscopic quantities. Fluid codes have been used to simulate experiments since the early days of laser fusion and the discrepancies have usually been attributed to breakdown of the classical fluid approximations(15,16), particularly in the realm of electron thermal transport. This has led to more sophisticated codes designed to look in detail at energy transport(14) but lacking the generality needed to model experimental data.

Fluid codes are very useful for extrapolating a modest factor beyond experimental data using "prescriptions" which (hopefully with some physical basis) have proved satisfactory to explain current observations. This was done for instance[37] to estimate the increase in ablation pressure by going to shorter laser wavelengths and is frequently used to estimate the parameter ranges of interest in planned experiments.

In some cases fluid codes have been used to predict phenomena in advance of experiments being feasible, particularly in the field of hydrodynamic instabilities. Many computer simulations of the Rayleigh Taylor instability[38-41] were published before the first experimental data became available[42].

MHD models have been used to simulate magnetic field generation but different numerical models are inconsistent and it is likely that deficiencies in the electron transport model are most noticeable in this case. Magnetic fields at densities above critical (where they are not observeable) have been proposed as a contributor to the anomalously low heat flow[28] and may also play a part in the Rayleigh Taylor instability since they have similar source terms.

5.2 PIC Codes

If fluid codes are the tools of the experimentor then it can fairly be said that particle codes are the tools of the theorist[33]. In most cases the analysis of plasma instability is limited to the linear regime while the practical importance eg the generation of fast particles is a result of the non-linear regime. Particle codes provide the means of passing from the linear to the non-linear regime accurately, verifying linear growth rates and observing the non-linear behaviour[43].

As an example consider the growth of a plasma wave in the "beat wave accelerator". This wave has a phase velocity almost exactly equal to c, and its non-linear behaviour is complicated by relativistic effects since

$$\delta n/n = v_o/v_{ph} = v_o/c$$

PIC simulation of this process shows that the wave becomes strongly anharmonic at large amplitude and the electrons acquire a drift in the direction of wave propagation

PIC codes have been used to simulate all the major instabilities of laser plasma interaction, ie stimulated Raman scattering(33), stimulated Brillouin scattering(33), filamentation(45) and two plasmon decay(43). In most cases the results cannot immediately be compared with experiment since experimental conditions are rarely known with sufficient accuracy. Hybrid or collisional PIC models have been used to look at long time scale phenomena such as the lateral spreading of hot electrons in magnetic fields and have made a major contribution to understanding in this area(35).

REFERENCES

1) K A Bruckner and S Jorna Rev Mod Phys 46, 325 (1974)
2) O Buneman Phys Rev 115, 503 (1959)
3) R J Mason J Comp Phys 41, 233 (1981)
4) C Z Cheng and G Knorr J Comp Phys 22, 330 (1976)
5) R D Richtmeyer and K W Morton "Difference Methods for Initial Value Problems" Wiley 1967
6) J Crank and P Nicholson Proc Camb Phil Soc vol 43, 50 (1947)
7) See for instance Ref (5) p 8 ff
8) R Courant K O Friedrichs H Lewy Math Ann 100, 32 (1928)
9) D W Peaceman and H H Rachford J Soc Industrial Appl Math 4, 277 (1955)
10) D N Allen "Relaxation Methods" McGraw Hill, New York (1954)
11) D W Kershaw J Comp Phys 26, 43 (1978)
12) J von Neumann R D Richtmeyer J Appl Phys 21, 232 (1950)
13) L Spitzer "The Physics of Fully Ionised Gases" (Interscience New York 1956)
14) A R Bell R G Evans D J Nicholas Phys Rev Lett 46, 4, 243 (1981)
15) M D Rosen et al Phys Fluid 22, 10, 2020 (1979)
16) W C Mead et al Phys Fluid 26, 8, 2316 (1983)
17) R A Haas et al Phys Fluid 20, 322 (1977)
18) D Duston and J Davis Phys Rev A 23, 2602 (1981)
19) G J Pert (Proceedings of the 20th SUSSP, St Andrews 1979 publ SUSSP 1980)
20) D P Lax and B Wendroff Comm Pure Appl Math 13, 217 (1960)
21) J E Fromm J Comp Phys 3, 176 (1968)
22) J P Boris and D L Book J Comp Phys 11, 38, (1973) also Methods in Computational Physics Vol 16 p85 Academic Press New York 1976
23) B van Leer J Comp Phys 23, 276 (1977)
24) W D Schultz "Methods in Computational Physics" Vol 3 p196 Academic Press New York (1963)
25) D A Tidman and R A Shanny Phys Fluid 17, 1207 (1974)
26) R S Craxton and M G Haines Phys Rev Lett 35, 1336 (1975)
27) G J Pert J Comp Phys 43, 1, 111 (1981)

28) M H Emery J H Gardner J P Boris paper B-III-3, 10th Int Conf Plasma Physics and Controlled Fusion IAEA, London, 1984, to be published IAEA 1985

29) J M Dawson Phys Rev 118, 381 (1960)

30) J Denavit and W L Kruer Lawrence Livermore Laboratory preprint UCRL 84293 (1980) Comments in Plasma Physics 6, 1, 35 (1980)

31) The 1 D and 2 D models are taken from A B Langdon and B F Lasinski Methods in Computational Physics 16, 327 (1976)

32) O Buneman J Comp Phys 12, 124 (1973)

33) D W Forslund J M Kindel E L Lindman Phys Fluid 18, 1017, (1975)

34) E L Lindman J Comp Phys 18, 66 (1973)

35) D W Forsland and J U Brackbill Phys Rev Lett 48, 1614 (1982)

36) R J Mason J Comp Phys 41 233 (1981)

37) R G Evans A R Bell and B J MacGowan J Phys D 15, 711 (1982)

38) J D Lindl and W C Mead Phys Rev Lett 34, 1273 (1975)

39) R L McRory L Montierth R L Morse C P Verdun Phys Rev Lett 46, 336 (1981)

40) M H Emery J H Gardner and J P Boris Phys Rev Lett 48, 677 (1982)

41) R G Evans A J Bennett and G J Pert Phys Rev Lett 49, 1639 (1982)

42) A J Cole J D Kilkenny R G Evans C J Hooker P T Rumsby M H Key Nature (Phys Sci) 299, 329 (1982)

43) A B Langdon B F Lasinkski and W L Kruer Phys Rev Lett 43, 133 (1979)

44) R Bingham R A Cairns R G Evans Rutherford Lab Report RAL 84-122 (1984)

45) A B Langdon and B F Lasinski Phys Rev Lett 34, 934 (1975)

RARE GAS HALIDE LASERS

F O'Neill

Rutherford Appleton Laboratory
Chilton, Didcot, Oxon OX11 OQX, England

1.
INTRODUCTION

Optical fluorescent emission at ultraviolet wavelengths was first observed from rare gas halide (RGH) excimer molecules in 1974 by Golde and Thrush[(1)] at Cambridge University and independantly by Velazco and Setser[(2)] at Kansas State University. At both laboratories these observations were made in flowing afterglow equipment during experiments which were designed to measure the rate of reaction of rare gas metastable atoms with various quencher molecules. In the work of Golde and Thrush it was observed that the rare gas halide molecule $ArCl^*$ was formed in the following reaction (the * symbol denotes a species in an electronically excited state)

$$Ar^* + Cl_2 \xrightarrow{k_1} ArCl^* + Cl; \qquad k_1 = 7.1 \times 10^{-10}\ cm^3\ s^{-1} \tag{1.1}$$

where the presence of the $ArCl^*$ excimer was determined from the broad-band UV chemiluminescence

$$ArCl^* \rightarrow Ar + Cl + h\nu (175nm) \tag{1.2}$$

The work of Velazco and Setser concentrated on observing Xe^*

quenching reactions with halogenated molecules and, again by recording UV fluorescence spectra, they concluded that the xenon halides XeF^*, $XeCl^*$, $XeBr^*$, and XeI^* were being formed. These workers also made the crucial observation that the rare gas halide excimer molecules were potential UV laser candidates. The molecular energy level diagram of a rare gas halide molecule is shown in Fig 1 and the reasons why these molecules were considered useful as laser candidates were,

(1) rare gas halide molecule, RX, is formed with high quantum efficiency in reactions such as (1.1)

(2) molecules are formed in strongly bound excited states and the fluorescence transition to the ground state is highly allowed

(3) UV fluorescent transition terminates on a dissociative (or weakly bound) state giving fast emptying of the lower laser level

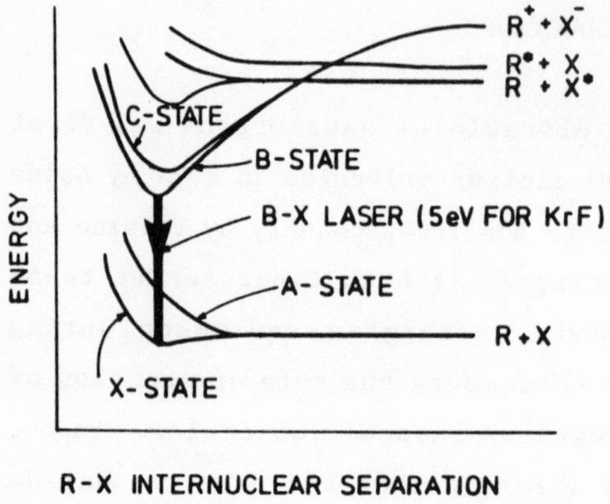

Fig. 1 Schematic potential energy diagram showing the electronic levels of a rare gas monohalide molecule RX. R = rare gas atom, X = halogen atom

Experiments to demonstrate a working rare gas halide laser started at a number of laboratories in the USA in late 1974 and within a few months Searles and Hart(3) at the Naval Research Laboratory in Washington DC using e-beam pumping achieved laser action at 282nm on the $XeBr^*$ molecule. This initial discovery was quickly followed by the achievement of laser action on KrF^* (249 nm) and $XeCl^*$ (308 nm) by Ewing and Brau(4) and on XeF^* (351 nm) and ArF^* (193 nm) by other workers(5,6).

The wavelengths of the rare gas halide lasers are shown in Fig 2

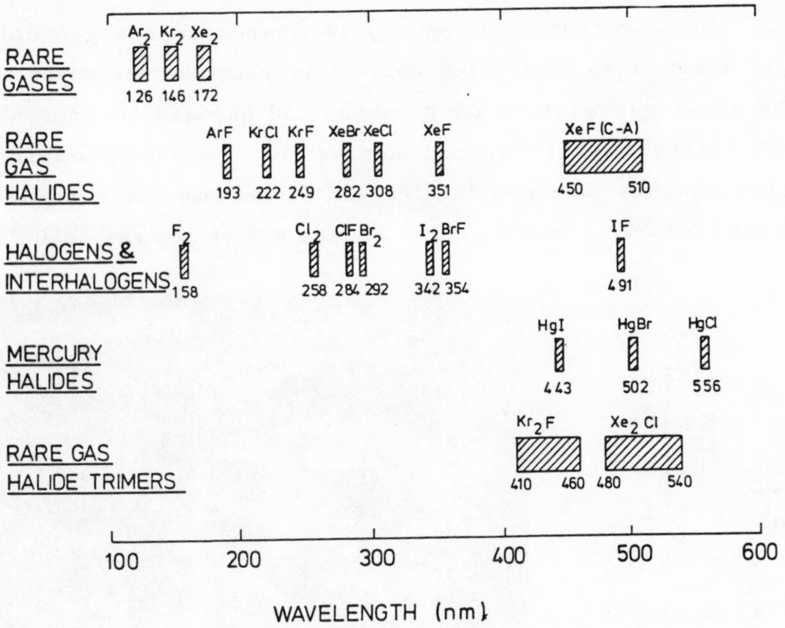

Fig.2 Wavelengths of excimer lasers

along with the wavelengths of other excimer lasers. It is worth noting that laser emission has been observed on two transitions in XeF^* viz. the normal B-X laser at 351 nm and a longer wavelength C-A laser at approximately 480 nm. It can also be seen from Fig 2 that laser emission has been demonstrated on the rare gas halide trimers Kr_2F^* and Xe_2Cl^* in addition to the normal dimer laser transitions.

For laser-plasma interaction studies, which is the topic of this summer school, it is necessary to achieve focussed laser powers > 10^{13} W/cm^2 on target in spots ~ 100 μm diameter and this requires the use of lasers of very high output power of $>10^9$ W with nanosecond pulse durations. One of the highest power rare gas halide lasers to date is a discharge excited system(7) which produces 40mJ pulses of 10 picoseconds duration. In order to obtain high power nanosecond pulses however it is necessary to use electron-beam pumping as shown schematically in Fig 3. In this apparatus a high current pulse of electrons is accelerated to high voltage in a vacuum diode and is fired into the laser gas cell through a thin (~ 25 μm) metal foil window. The electron beam energy is dumped in the gas cell by multiple small angle scattering collisions with the gas constituents and the laser gas mixture of rare gas and halogen becomes highly ionised. After a complicated series of ion-ion recombination reactions rare gas halide molecules are formed and laser oscillation is obtained between mirrors placed on each end of the gas cell.

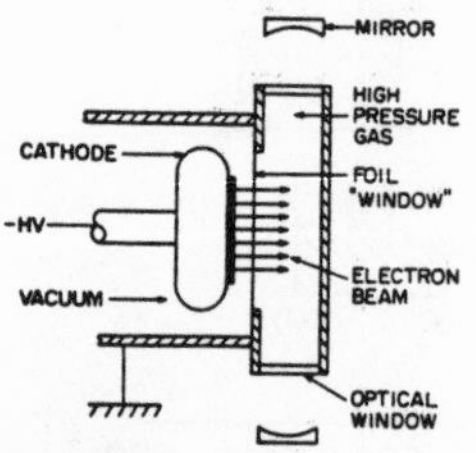

Fig. 3 Schematic diagram of a laser pumped by a high current electron beam[8]

Fig. 4 200J, 50 ns KrF laser, Sprite

The electron beam is usually fired into the laser cell through the side walls, ie transverse pumping, and single sided[3], two sided[9] and four sided[10] pumping geometries have been employed. KrF which lases at 249 nm is the most efficient of the rare gas halide lasers and at the Rutherford Appleton Laboratory (RAL) Central Laser Facility we have spent a number of years developing high power KrF systems. These lecture notes therefore concentrate strongly on the description of high power e-beam-pumped KrF lasers but many aspects of this technology could apply equally well to other RGH lasers. The largest KrF laser at RAL is code-named Sprite[10] and this system is now routinely used for laser-matter interaction experiments. The laser which is seen in Fig 4 produces 200J, 50ns pulses in a 265 mm diameter annular beam from an unstable resonator cavity. For laser-plasma interaction experiments a peak focussed power on target of 5×10^{13} W/cm^2 is achieved in a spot size of 40 μm diameter.

KrF laser development is also being vigorously pursued in other countries notably the USA, Canada, Japan, Holland and the USSR. Many of these programs are aimed at developing KrF lasers for application to inertial confinement fusion (ICF) and the largest system in operation is the Aurora laser at the Los Alamos National Laboratory (LANL) in the USA. This laser, which is shown in Fig 5, produces 10 kJ pulses of 600 ns duration in a beam of dimensions 1m x 1m. LANL are now designing an even bigger laser for fusion experiments code named Polaris with a design energy of 100 kJ.

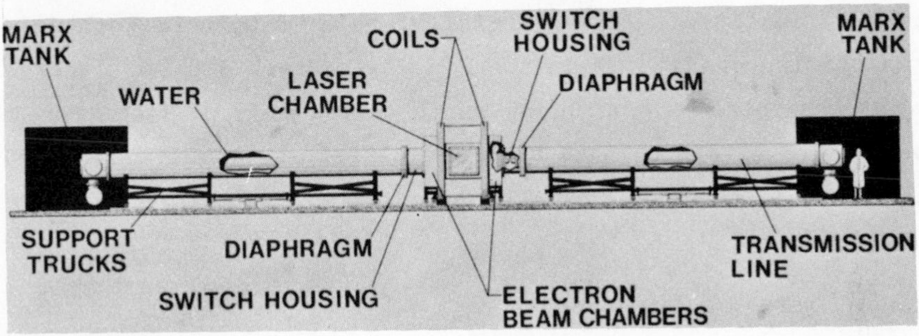

Fig. 5 Artists impression of 10kJ, 600 ns Aurora KrF laser

The KrF laser is in many respects the ideal candidate as a laser fusion driver because of the following characteristics.

(1) High overall operating efficiency of > 5% is feasible

(2) Short operating wavelength of 249 nm gives efficient coupling to the target plasma

(3) Gaseous laser medium can be cooled when operating at high repetition rate

(4) Laser system can be constructed from robust low grade technology

The KrF laser does however have one major, still unsolved, technical drawback viz. the gain medium is not a storage medium. The pump time of a typical e-beam system is in the range 50-500 ns which is considerably longer than the 2 ns upper state lifetime of the KrF molecule. The laser therefore operates quasi-CW where the laser output pulse has the same duration as the pumping pulse. Thus if say a 5 ns KrF pulse is amplified in a 50 ns amplifier it will extract only a maximum of 10% of the available inversion (5 ns out of 50 ns) and the pulse intensity will be no greater than could be achieved by simply operating the amplifier as a laser oscillator with mirrors.

In this respect the KrF laser compares unfavourably with for example the 1.06 μm Nd:glass laser where the upper state laser level has a long lifetime of 350 μs. With Nd:glass lasers very high output powers of > 1TW are achieved in a nanosecond pulse by first building up stored energy in the active medium by pumping with long pulse low intensity (10 MW) flashlamps and then extracting this energy by amplifying a short 1ns laser pulse in a series of amplifiers. Very high power gains of $>10^8$ are easily achieved by this method over normal lasing conditions in the Nd: glass laser medium. There is however considerable work going on to devise ways of increasing the power of KrF lasers to make them useful for ICF applications and I will return to this topic at a later point.

2.

BASIC PRINCIPLES OF OPERATION OF E-BEAM-PUMPED KrF LASERS

2.1 Spectroscopy

The principles of operation of the KrF excimer laser can be understood by referring to Fig 6 which is a detailed potential energy level diagram of the KrF molecule. In keeping with common notation the lowest energy state is labelled the X-state and the higher levels are labelled the A-state, B-state, etc in order of increasing energy. The strong UV fluorescence observed from this molecule is due to the B-X transition and laser action occurs at the peak of this emission band at a wavelength of 248.6 nm (5 eV photon).

The B-state in KrF is strongly bound and it can be seen from Fig 6 that the formation of a B-state molecule in a collision between a krypton atom and a fluorine atom requires the combined excess energy of the reactants at infinite separation to exceed approximately 8eV ie

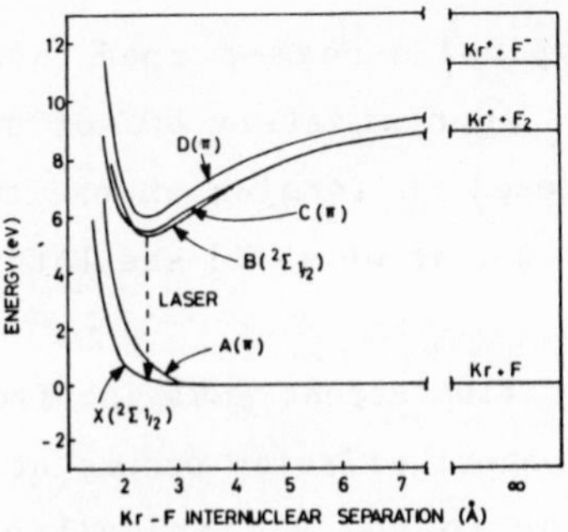

Fig. 6 Detailed potential energy level diagram of KrF[11]

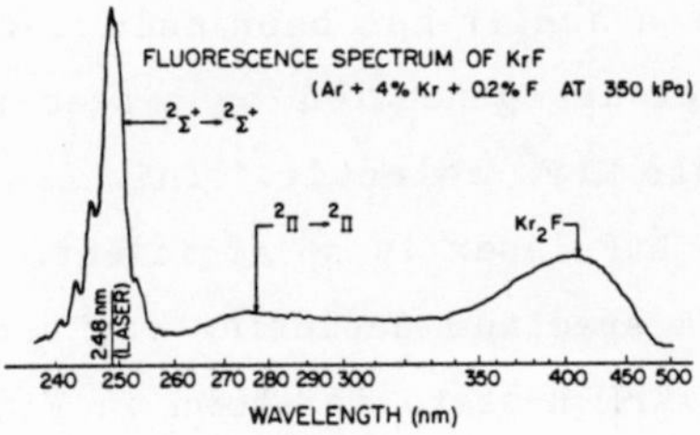

Fig. 7 Fluorescence spectrum of KrF[8]

you have to "pump" electronic energy into the molecule at 8eV in order to obtain laser output at 5eV. In KrF this excitation to the upper laser level can be achieved in two main ways.

$$Kr^+ + F^- \overset{k_3}{\rightarrow} KrF^*; \qquad k_3 = 3.0 \times 10^{-6}\ cm^3\ s^{-1} \tag{2.1}$$

$$Kr^* + RF \overset{k_4}{\rightarrow} KrF^* + R\ (R = F,\ NF_2,\ etc) \tag{2.2}$$

$$k_4 = 7.2 \times 10^{-10}\ cm^3\ S^{-1}\ (R{=}F)$$

In e-beam-pumped lasers the fluorine donor molecule, RF, used in reaction (2.2) is usually F_2 and the ionic (Kr^+, F^-) and excited state (Kr^*) primary species are produced by high energy electron collisions in the gas mixture.

Reaction (2.1) is known as the ion-channel of KrF^* formation and the reaction is exothermic by an energy $\Delta E(1)$ where

$$\Delta E(1) = \text{Ionisation potential of Kr} - \text{Electron affinity of F}$$
$$= 14.0\ eV - 3.4\ eV = 11.6\ eV$$

Reaction (2.2) is known as the metastable-channel of KrF^* formation where the exothermicity $\Delta E(2)$ is

$$\Delta E(2) = \text{Metastable energy} - F_2 \text{ dissociation energy}$$
$$= 9.9\ eV - 1.6\ eV = 8.3\ eV$$

The excess energy in both reactions which appears as internal excitation energy in the product, KrF^*, is therefore $> 8eV$ as is required to enter the B-state potential energy curve. Reactions (2.1) and (2.2) proceed extremely rapidly in typical e-beam-pumped laser media and it has been calculated[12] that approximately 80% of the reactants generated by e-beam-pumping succeed in forming an excited state KrF^* molecule. This is one reason why, as we shall see later, the KrF laser is so efficient.

A spectrum depicting the e-beam-excited fluorescent emission from the KrF B-state is shown in Fig 7[8]. Strongest emission occurs at a peak wavelength of 248.6nm corresponding to the B-X transition. Weaker broadband emissions are also observed at longer wavelengths and have been identified as KrF C-A emission (at 275 nm) and Kr_2F B-X emission (at 410 nm).

Efficient laser action in e-beam-pumped gas mixtures is easily obtained at the peak of the B-X band in KrF and some of the important parameters relevant to laser operation are shown below.

Typical laser mixture in Sprite	4 torr F_2/110 torr Kr/1120 torr Ar
Laser wavelength (λ_L)	248.6 nm
Photon energy ($h\nu_L$)	5 eV
Stimulated emission cross-section at 248.6 nm (σ_s)	2.6×10^{-16} cm^2
Natural lasing linewidth ($\Delta\bar{\nu}_L$)	50 - $100cm^{-1}$
B-state fluorescent lifetime (τ_f)	6.4 ns
Typical lifetime including collisional quenching (τ_{coll})	2ns
Saturation energy ($h\nu_L/\sigma_s$)	3×10^{-3} J/cm^2
Typical saturation intensity ($h\nu_L/\sigma_s\tau_{coll}$)	1.5×10^6 W/cm^2

From the above list of data we can see that the B-X transition is a very high gain one with characteristics similar to those observed in a tunable dye laser medium. In fact tunable, line-narrowed operation of KrF lasers has been achieved by using dispersive optics in the laser resonator[13].

2.2 Kinetic Processes in E-Beam-Pumped KrF Lasers

In the early work with e-beam-pumped KrF lasers[4] it was observed experimentally that the most intense laser emission was obtained from laser gas mixtures of Ar, Kr and F_2 where the partial pressure of Ar was > 90% of the total. One major effect of using such Ar-rich mixtures is that the KrF formation kinetics are considerably more complicated than was shown in equations (2.1) and (2.2). E-beam energy is mainly deposited in the laser gas by collisions between high voltage electrons, e_f, and Ar atoms.

$$e_f + Ar \rightarrow Ar^+ + e_s + e_f(-W\ eV) \qquad (2.3)$$

$$e_f + Ar \rightarrow Ar^* + e_f(-X\ eV) \qquad (2.4)$$

where Ar^* denotes a metastable Ar atom and e_s is a slow secondary electron. It has been found experimentally[14] that when a fast electron loses energy according to reactions (2.3) and (2.4) then 7 ion pairs and 2 metastables are produced for an energy loss by the fast electron of approximately 180 eV ie 20 eV of energy is expended per primary excited species (Ar^+, Ar^*) generated. The quantum efficiency of the laser, η_{quant}, is thus 5eV (laser photon energy) divided by 20eV equals 25%.

The secondary electron produced in reaction (2.3) very rapidly attaches to F_2 in the reaction

$$e_s + F_2 \xrightarrow{k_5} F^- + F; \qquad k_5 = 7.5 \times 10^{-9}\ cm^3\ s^{-1} \qquad (2.5)$$

The F_2 concentration in the laser gas is gradually depleted as a result of reaction (2.5) and is initially turned into F^- and F-atoms. After the lasing process is complete the F^- returns to the gas as an F-atom ($KrF^* \rightarrow Kr + F + h\nu$). F-atoms recombine to form F_2 on a 100 μs time scale so when an F_2 molecule is used up in a reaction (2.5) it is not able to contribute a second time to the lasing process in a typical pulse of < 500 ns duration. Thus sufficient F_2 "fuel" has to be included in the initial laser mix to last the full duration of the laser pulse. It is not possible to use a large excess concentration of F_2 because as we shall see later this molecule absorbs strongly at the laser wavelength. The F_2 "fuel" can however be reused for many laser shots provided the time between shots is greater than the recombination time of 100 μs.

As a result of reactions (2.3)-(2.5) we have generated the species necessary to form rare gas halide molecules. While the primary rare gas species are of argon rather than krypton it is found that KrF^* molecules are rapidly formed by the following reactions.

Ion Channel

$$Ar^+ + Ar + M \xrightarrow{k_6} Ar_2^+ + M \ (M = Ar, Kr) \qquad (2.6)$$

$$k_6 = 2.5 \times 10^{-31} \ cm^6 \ S^{-1}$$

$$Ar_2^+ + F^- \xrightarrow{k_7} ArF^* + Ar; \qquad k_7 = 1.5 \times 10^{-6} \ cm^3 \ S^{-1} \qquad (2.7)$$

Metastable Channel

$$Ar^* + F_2 \xrightarrow{k_8} ArF^* + F; \qquad k_8 = 7.5 \times 10^{-10} \ cm^3 \ S^{-1} \qquad (2.8)$$

Exchange Reaction

$$ArF^* + Kr \xrightarrow{k_9} KrF^* + Ar; \qquad k_9 = 3.0 \times 10^{-10} \ cm^3 \ s^{-1} \qquad (2.9)$$

We see therefore that the KrF^* molecules are formed by an indirect route in that ArF^* is formed first and then KrF^* is generated in exchange reactions (2.9) with Kr. The kinetic efficiency of this process is high and approximately 80% of the Ar^+ and Ar^* species initially produced by the e-beam are converted into KrF^*[(12)].

The above kinetic scheme has neglected contributions due to e-beam excitation of Kr which makes up about 10% of the partial pressure of a typical laser fill. When Kr reactions are included the situation becomes more complicated and a full kinetics diagram including Kr is shown in Fig 8. This full diagram includes for the first time the rare gas halide trimer molecule Kr_2F^* which is found to have a strong absorption band close to the KrF laser wavelength the presence of which, as is seen later, has important implications for the efficiency of the KrF laser.

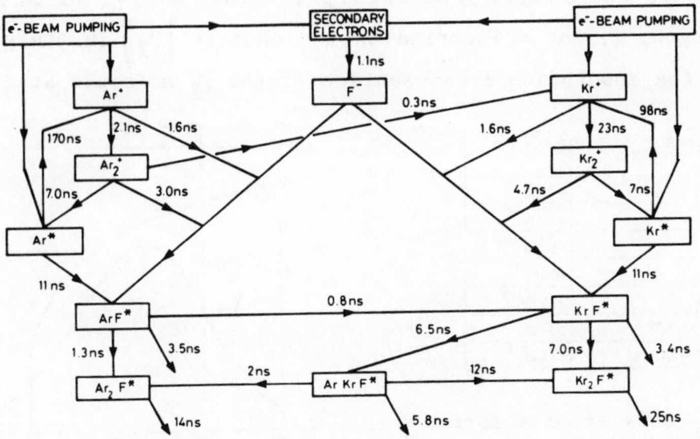

Fig. 8 Flow diagram for the main kinetic processes in e-beam-pumped KrF. The times quoted are lifetimes for each kinetic route as calculated for a laser gas mixture of 4 torr F_2, 125 torr Kr, 1371 torr Ar at an e-beam excitation rate of 1 MW/cm^3 assuming no laser action[15]

2.3 Absorbers in the KrF Gain Medium

Section 2.2 above has described how the energy of a high current, high voltage electron beam is deposited in the KrF laser gas and how this energy is converted to gain by the kinetic processes in Fig 8. E-beam pumping of the Ar, Kr, F_2 laser gas mixture not only results in the formation of the gain species KrF^* but also in the formation of a number of transient species which absorb strongly at the KrF laser wavelength of 249 nm. Details of the main absorber species are shown in Table 2.1 and UV absorption curves for F^- and F_2 are shown in Fig 9 as examples. Absorption of a KrF laser photon in F^- causes electron photodetachment[16] and in F_2 causes photodissociation[17]. F_2 absorption is not transient but is always present in the gain medium and its value is given by the equation,

$$A(F_2) = 1 - \exp\{-\sigma_a(F_2)\,[F_2]\,l\} \qquad (2.10)$$

where $A(F_2)$ is the single pass absorption loss at 249 nm in a laser cell of length, l, for a fluorine number density $[F_2]$ in the gas mix. $\sigma_a\,(F_2)$ is the absorption cross-section of the F_2 molecule at 249 nm.

REACTION	CROSS-SECTION (cm^2)
$Ar_2^+ + h\nu \rightarrow Ar^+ + Ar$	5.0×10^{-17}
$Kr_2^+ + h\nu \rightarrow Kr^+ + Kr$	1.0×10^{-18}
$F^- + h\nu \rightarrow F + e_s$	5.0×10^{-18}
$F_2 + h\nu \rightarrow 2F$	1.5×10^{-20}
$Ar^* + h\nu \rightarrow Ar^+ + e_s$	1.0×10^{-19}
$Kr^* + h\nu \rightarrow Kr^+ + e_s$	3.2×10^{-20}
$Kr_2F^* + h\nu \rightarrow Kr_2F^{**}$	3.8×10^{-18}
$Ar_2F^* + h\nu \rightarrow$ PRODUCTS	1.5×10^{-18}

Table 2.1 Details of absorber species in the e-beam-pumped KrF gain medium(15)

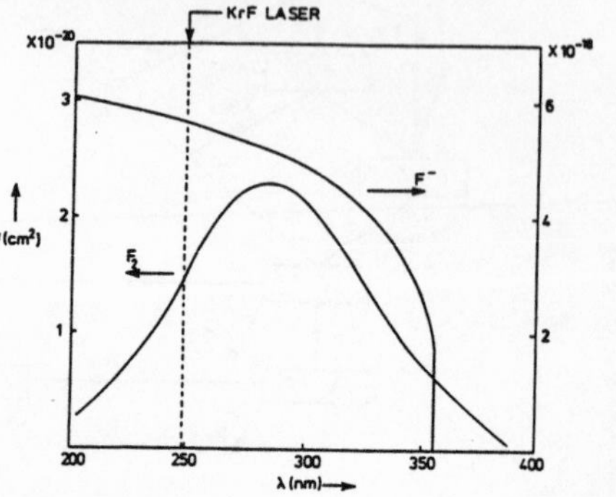

Fig. 9 Absorption versus wavelength for F^- *(16) and* F_2 *(17)*

Putting in values for the Sprite KrF laser of $\sigma_a(F_2) = 1.5 \times 10^{-20}$ cm^2, l = 150 cm (100 cm pumped length) and $[F_2] = 1.3 \times 10^{17}$ cm^{-3} (4 torr partial pressure) we obtain $A(F_2) = 0.25$ ie 25% loss/pass. While this is quite a large loss in the laser medium some of the other species listed in Table 2.1 contribute even greater losses as a result of a combination of a higher (number density x absorption cross-section) product. The major absorber is Kr_2F^* which is formed from KrF^* as follows,

$$KrF^* + Kr + M \overset{k_{10}}{\rightarrow} Kr_2F^* + M \quad (M = Kr, Ar) \qquad (2.11)$$

$$k_{10} = 6.5 \times 10^{-31}\ cm^6\ s^{-1}$$

The rate equation describing Kr_2F^* formation in the gain medium is,

$$\frac{d[Kr_2F^*]}{dt} = k_{10}[KrF^*][Kr][M] - \frac{[Kr_2F^*]}{\tau} \qquad (2.12)$$

where quantities in [ ] brackets denote number densities of those species and τ is the collision dominated life-time of Kr_2F^*. Under steady state conditions

$$\frac{d[Kr_2F^*]}{dt} = 0 \qquad (2.13)$$

$$\therefore [Kr_2F^*] = k_{10}\tau[KrF^*][Kr][M] \qquad (2.14)$$

For a typical KrF laser with a gain coefficient of approximately 20% cm^{-1} we have

$$[KrF^*] = 0.2/\sigma_s = 7.7 \times 10^{14}\ cm^{-3} \qquad (2.15)$$

Using values of τ = 25 ns, $[Kr] = 3.6 \times 10^{18}$ for 110 torr Kr mix, $[M] = 3.9 \times 10^{19}$ for a total mix pressure of 1200 torr, and substituting for $[KrF^*]$ from (2.15) gives

$$[Kr_2F^*] = 1.8 \times 10^{15}\ cm^{-3} \text{ ie } 0.05 \text{ torr}$$

The loss per cm due to this absorber is

$$\alpha(Kr_2F^*) = [Kr_2F^*]\,\sigma_a(Kr_2F^*) = 0.8\%\ cm^{-1} \qquad (2.16)$$

When the effects other absorber species (Table 2.1) are added the total loss is in the region of 2-3% cm^{-1} (18).

The effect of these non-saturable losses in the gain medium is to limit the efficiency of power extraction from the laser. The laser field in the cavity will grow until the saturable gain per cm is equal to the non-saturable loss per cm at which point the field intensity remains constant. The effect of losses can be analysed(19) by considering the extraction of laser power from the two pass amplifier shown in Fig 10. The performance of such an amplifier will be similar to a KrF laser oscillator which typically operates with approximately 90% output coupling.

In the two pass amplifier shown in Fig 10 a KrF laser beam of intensity I_{in} is injected into the right hand side. This beam propagates from right to left (I^-) through the amplifier growing in

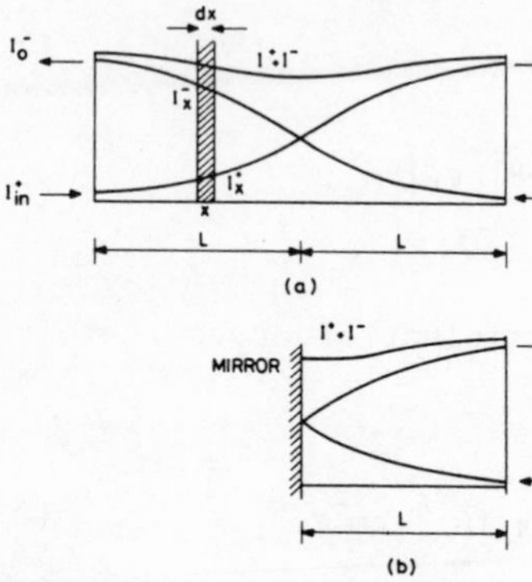

Fig. 10 (a) A bi-directional amplifier of length 2L having two identical inputs and outputs (b) The equivalent folded two pass amplifier of length L[19]

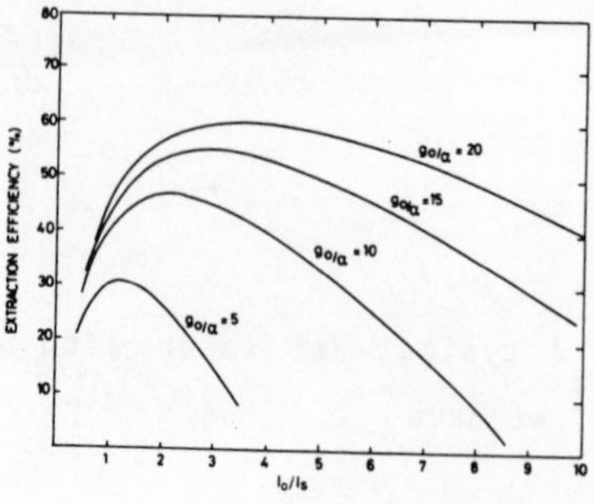

Fig. 11 Amplifier extraction efficiency[19]

intensity and is reflected off a mirror on the left side of the amplifier. The beam then travels from left to right (I^+) still growing in intensity until it leaves the right side of the amplifier having achieved an intensity I_o.

When a laser beam of intensity I propagates in a gain medium the beam intensity will increase as a function of distance according to the expression

$$\frac{dI}{dx} = gI \tag{2.17}$$

where g is the gain per unit length. As the beam intensity increases the gain will saturate according to the expression

$$g(I) = \frac{g_o}{1 + I/I_s} \tag{2.18}$$

where g_o is the small signal gain coefficient for I = 0 and I_s is the saturation intensity such that $g(I) = g_o/2$ when $I = I_s$.

In a laser medium with internal non-saturable loss α cm^{-1}, equation (2.18) becomes

$$g(I) = \frac{g_0}{1 + I/I_s} - \alpha \tag{2.19}$$

and equation (2.17) for a lossy medium is,

$$\frac{dI}{dx} = \left[\frac{g_0}{1 + I/I_s} - \alpha\right] I \tag{2.20}$$

Now let us consider a small slab of gain medium of thickness dx in the double pass amplifier shown in Fig 10. Power is being continually extracted from this elemental slab by light waves travelling from right to left (I^-) and left to right (I^+). Since the gain medium is simultaneously saturated by waves going in both directions we replace equation (2.20) by two equations as follows,

$$\frac{dI^+}{dx} = \left[\frac{g_0}{1 + (I^+ + I^-)/I_s} - \alpha\right] I^+ \tag{2.21}$$

$$\frac{dI^-}{dx} = \left[\frac{g_0}{1 + (I^+ + I^-)/I_s} - \alpha\right] I^- \tag{2.22}$$

The maximum intensity which can be extracted from the elemental slab is

$$I_{max} = g_0 dx\, I_s \tag{2.23}$$

and the extraction efficiency from this element is

$$\eta = \frac{(I^+_{x+dx} - I^+_x) + (I^-_x - I^-_{x+dx})}{g_0\, dx\, I_s} \tag{2.24}$$

Substituting from equations (2.21) and (2.22) gives,

$$\eta = \frac{1}{g_0 I_s}\left[\frac{dI^+}{dx} + \frac{dI^-}{dx}\right] \tag{2.25}$$

and substituting again from equation (2.20) gives,

$$\eta = \frac{1}{g_o I_s}\left[\frac{g_o}{1 + (I^+ + I^-)/I_s} - \alpha\right](I^+ + I^-) \qquad (2.26)$$

$$= \left[\frac{1}{1 + I_o/I_s} - \frac{\alpha}{g_o}\right]\frac{I_o}{I_s} \qquad (2.27)$$

where the approximation is made that

$$I_o = I^+ + I^- = \text{constant}$$

In this constant intensity approximation the extraction efficiency for each part of the amplifier depends only on the output intensity I_o and the gain to loss ratio g_o/α. The expression in equation (2.27) is plotted in Fig 11 for typical values of g_o/α. In e-beam-pumped Ar-rich mixtures it is found experimentally(18) that g_o/α is 15-20 thus from Fig 11 a maximum extraction efficiency ~50% is expected for the KrF laser. We can compare this theoretical value with experiment if we first define the intrinsic laser efficiency η_{Int} as,

$$\eta_{Int} = \eta_{Extr} \times \eta_{Kin} \times \eta_{Quant} \qquad (2.28)$$

where η_{Extr} = the extraction efficiency of the laser, η_{Kin} = the efficiency of the laser kinetics and η_{Quant} = the quantum efficiency. Putting in typical values we have,

$$\eta_{Extr} = 0.5^{(18,19)}; \quad \eta_{Kin} = 0.8^{(12)}; \quad \eta_{Quant} = 0.25$$

so we expect η_{Int} = 0.1 ie 10%. The intrinsic efficiency of an e-beam-pumped KrF can be measured from the quantity

$$\eta_{Int} = \frac{\text{Laser Output Energy}}{\text{Deposited E-Beam Energy}} \qquad (2.29)$$

and values of 10% have been observed experimentally(20) which is in exact agreement with the theoretical value expected from equation (2.28). It is interesting to note that an intrinsic efficiency of 25% is predicted if both the extraction efficiency and kinetic efficiency can be made to approach 100%.

2.4 Sprite - A 200J, 50ns KrF Laser

In this section we will see how the underlying principles of operation of KrF lasers presented in sections 2.1 - 2.3 above are incorporated into a working high power laser system. The technological details of the Sprite KrF laser at RAL are presented starting with the Sprite high voltage pulsed power supply shown schematically in Fig 12. The overall purpose of this power supply is to generate square high voltage pulses of 400-450 kV to apply to the four large area (100 cm x 10 cm) cold cathode e-beam diodes in Sprite. These field emission diodes generate the high current electron beams required to pump the laser gain medium.

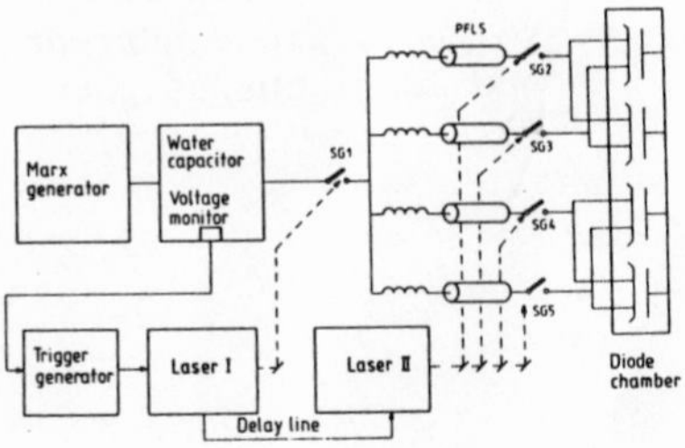

Fig. 12 Sprite pulsed power system[21]

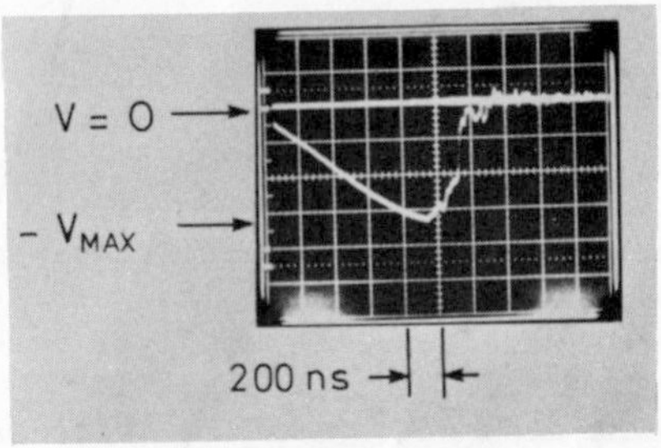

Fig. 13 Charging voltage waveform on Sprite water capacitor. V_{Max} = 625 kV for a charge voltage of 65 kV per stage on the Marx bank

The primary energy storage component in Sprite is a 12-stage Marx generator which stores 13 kJ of energy at a typical charge voltage of 65 kV (-ve) per stage. The generator is charged by a DC power supply and is discharged by a high voltage trigger when a laser pulse is required. When triggered the output pulse from the Marx-bank is applied to a 40 nF high voltage water capacitor which is pulse charged with a voltage waveform as shown in Fig 13. A voltage monitor on the water capacitor senses when peak volts has been achieved and at that time a low energy discharge excited KrF laser (Laser I) is fired and its beam is used to laser trigger an SF_6 spark gap, SG1. The water capacitor then discharges through SG1 to pulse charge four water-filled co-axial pulse forming lines (PFLs) of 5 Ω characteristic impedance. Typical charging waveforms on the PFLs are shown in Fig 14 where rapid charging is achieved because they are fed from the low inductance water capacitor.

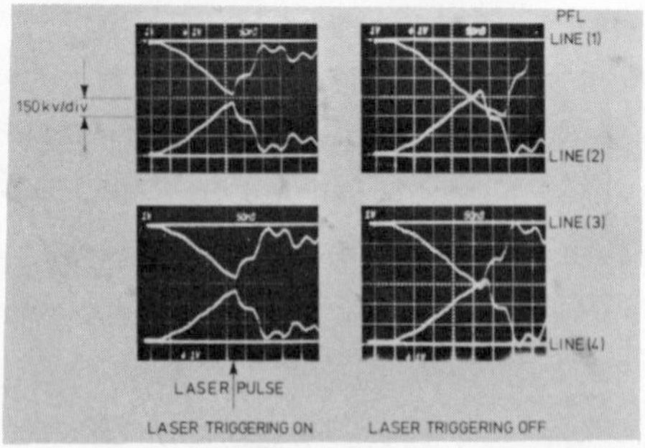

Fig. 14 Voltage waveforms of the four PFLS as measured by capacitance voltage monitors (lower traces of each pair inverted). The left-hand set shows the result of laser triggering the switches. The right-hand set shows the switches operating under self-break conditions (50 ns/div) [21]

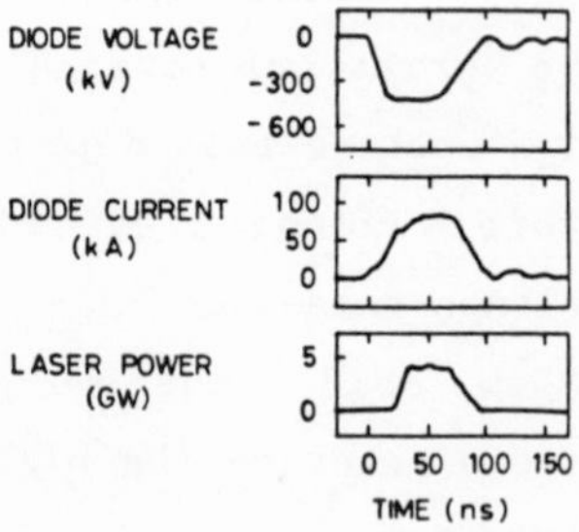

Fig. 15 Sprite diode voltage, diode current and laser output power verus time

When the voltages on the PFLs are maximum Laser II is fired and its beam is used to laser trigger four SF_6 output switches and the PFLs discharge into the e-beam diode loads. Each e-beam diode is impedance matched to the 5 Ω line impedance for maximum energy transfer and diode voltage and current pulses are shown in Fig 15.

The e-beam cathodes in Sprite are of the field emission type and consist of a large area array of spikes 5 mm apart punched in tantalum sheet. These spikes are required to give microscopic enhancement of the applied electric field to $> 10^7$ V/cm in order to obtain field emission of electrons from the spike tips. When the -ve high voltage pulses are applied to the cathodes from the PFLs there is strong field emission of electrons which starts from metallic whiskers[22]. Resistive heating causes these whiskers to quickly (1ns) explode thus covering the surface of the cathode with partially ionised metal vapor. This cathode plasma[23] provides the space charge limited electron current flow for the major part of the e-beam pulse. After generation the electron beam is accelerated across the diode and passes through a thin titanium foil window into the laser cell with a pumped length of 100 cm. Ionisation of the laser gas occurs and a 200J, 50 ns laser pulse is obtained with a time history as shown in Fig 15.

While the cold cathode field emission diodes are a robust and convenient way of obtaining high e-beam current densities the actual processes of generation are not completely understood even though the first observations of this phenomenon date back over 45 years[24,25]. In recent years the level of understanding of these cathodes has increased as a result of work on high power lasers, vacuum insulation and particle beams and there are a number of excellent papers devoted to this topic[23,26,27].

The current density in the e-beam-diode is described by the Child-Langmuir formula for space-charge-limited flow.

$$j_B = 2.33 \times 10^{-6}\ V^{3/2}/d(o)^2 \quad A/cm^2 \tag{2.30}$$

Here V is the applied voltage (volts) and d(o) is the anode-cathode spacing (cm). As described above this current is drawn from the cathode plasma which is formed by whisker explosions and as a result of this explosive process the plasma expands towards the anode during the pulse. Thus we can write,

$$d(t) = d(o) - ct \tag{2.31}$$

where c is the diode closure velocity which is found experimentally to be in the range 1-5 cm/ μs.

From equation (2.30) we can obtain the following expression for the diode impedance.

$$Z_D = \frac{d(0)^2}{2.33 \times 10^{-6} AV^{\frac{1}{2}}} \quad \text{ohms} \tag{2.32}$$

where A is the area of the diode in cm^2.

In the Sprite laser with V = 400kV, d(o) = 2.8 cm and A = (100 x 10)cm^2 we obtain from equations (2.30) - (2.32),

$$j_B = 75 \text{ A/cm}^2 \quad ; \qquad Z_D = 5.3\ \Omega$$

During the 60 ns voltage pulse the diode gap will close by 1-2mm.

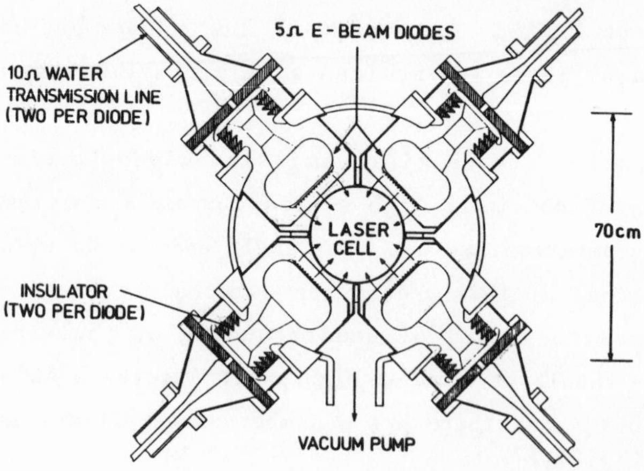

Fig. 16 Cross-sectional diagram of Sprite

A cross-sectional diagram of the Sprite laser is shown in Fig 16. This shows the structure of the four e-beam diodes surrounding the central laser cell. The e-beams in the diodes are accelerated across the cathode-anode gap to a voltage of 400 kV and pass first through the anode foil (12 μm Ti) which is supported in vacuum and then through the pressure foil (37 μm Ti) into the laser cell. A double foil arrangement is beneficial for two reasons. Firstly the use of a separate anode foil provides a well defined plane parallel electrical geometry in the diode and secondly the anode foil acts as an energy filter by stopping any spurious low voltage electrons from reaching the main pressure bearing foil. Electron emission at low voltage is typically generated in high current density filaments and if these reach the pressure foil they can cause localised melting and foil rupture.

The e-beam energy deposited in the laser cell is 2kJ which represents a conversion efficiency of 15% from the stored Marx energy of 13kJ. The e-beam operating voltage of 400 kV is chosen so that the electron range in the laser gas of pressure 1200 torr equals the diameter of the laser cell. Fig 17 shows a near field image of the laser beam from Sprite where it can be seen that the four-fold pumping symmetry is imprinted on the laser beam. This effect does not cause problems for target work however as it disappears in the far field ie in the focal plane of the focussing lens in the target chamber. The hole in the center of the beam is caused by the use of unstable resonator mirrors (Fig 18) and this hole also disappears in the far field.

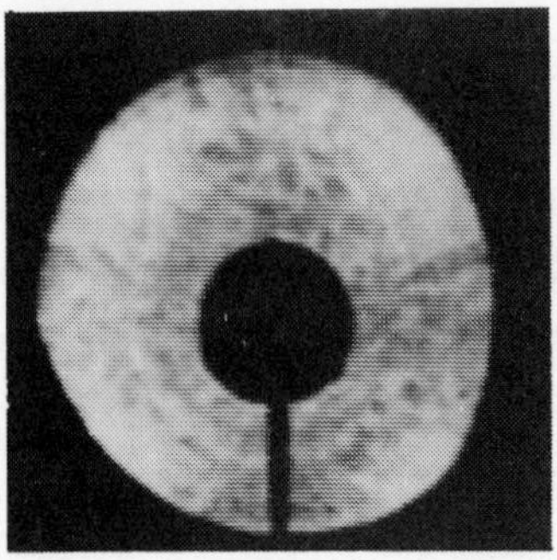

Fig. 17 Near field image of Sprite laser beam[28]

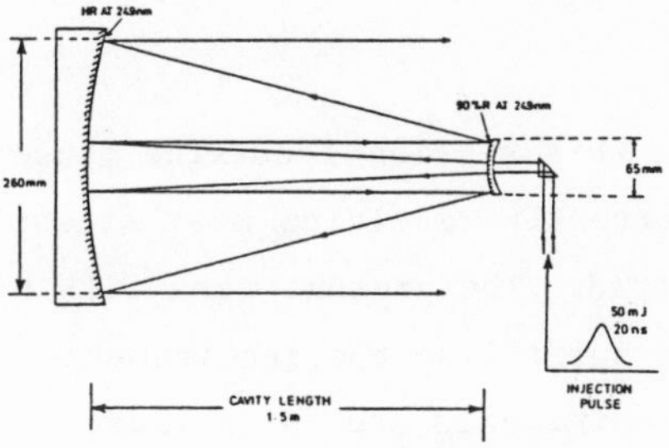

Fig. 18 Typical injection locked unstable resonator cavity. Injection pulse obtained from a low divergence discharge laser

The Sprite laser cell is closed off on each end with fused silica windows of 300 mm diameter and 30 mm thickness and the distance between the windows is 150 cm. The pumped length in the cell is 100 cm and the normal gas mixture employed is 4 torr F_2, 110 torr Kr plus 1120 torr Ar. Every effort is made to remove water vapour from the fill gases and the laser gas in the cell is circulated continuously through a gas processor and filter system which removes HF impurities and dust particles of diameter >0.5 μm. HF which is formed by the reaction of H_2O impurities with F_2 is particularly troublesome as it attacks SiO_2 and therefore can etch the laser windows thus reducing their optical transmission. Our experience however is that with the above precautions the laser cell and gas fill will typically last a full week of running at about 10-15 laser shots per day. Over this period the laser output energy will drop by only 10-20%.

By using an injection locked unstable resonator cavity as shown in Fig 18 the output from Sprite is typically 200J for a deposited e-beam energy of 2kJ which represents an intrinsic efficiency of 10%. The overall conversion efficiency from stored electrical energy in the Marx bank is 1.5%.

3.
CURRENT TOPICS IN KrF LASER RESEARCH

In this section I outline those areas of KrF laser research which are currently receiving most attention at various laboratories around the world. The present work with high energy KrF lasers is strongly aimed at fulfilling the requirements of a laser driver for ICF. These laser requirements are as follows,

(1) High output energy of 0.1 - 1.0 MJ
(2) Short output pulse of 10 - 20 ns
(3) Good beam quality
(4) Repetition rate ~10 Hz
(5) Gaseous medium to facilitate easy cooling
(6) Overall laser efficiency of 5 - 10%
(7) Short laser wavelength of 250nm - 500nm

At RAL our laser development work is not so strongly directed with only ICF in mind. We require lasers for a wide range of laser-plasma studies and of the above listed topics, numbers (3), (5) and (7) are most important. Our present interest is in developing a short wavelength laser that can provide high powers on target and that can be fired at a reasonable repetition rate to maximise the number of target shots per day.

3.1 Target experiments with the Sprite KrF Laser

A target irradiation facility based on the single beam Sprite laser is now operational at the RAL. The 249 nm laser beam is focussed onto target using an f:10 singlet aspheric focussing lens. The beam profile in the target plane has been measured and the details are shown in Figs 19 and 20. These measurements were taken by firstly attenuating the laser beam and then moving a knife edge through the laser spot on a shot by shot basis while measuring the unobscured fraction of the energy with a calorimeter placed behind the focal plane.

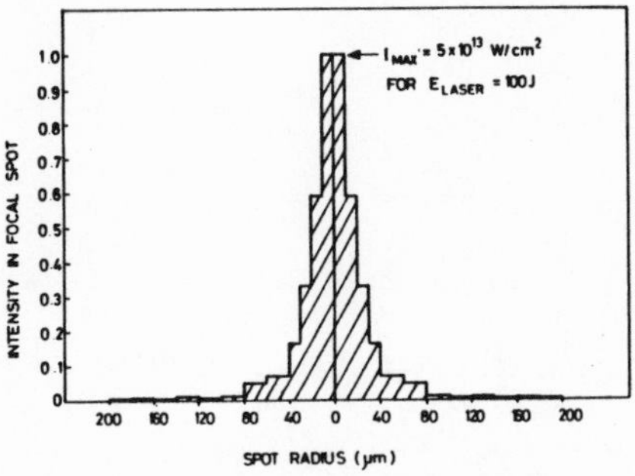

Fig. 19 Intensity versus distance of the Sprite KrF beam in the focal plane of f: 10 focussing optics

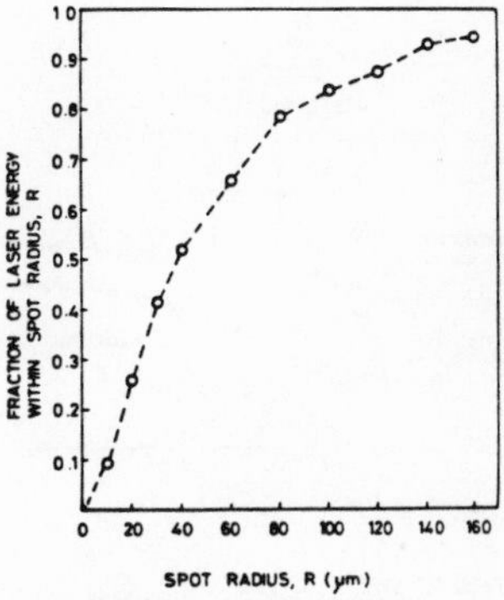

Fig. 20 Energy profile of the focussed Sprite KrF beam

After allowing for various losses as the laser beam propagates from the laser laboratory to the target chamber there is approximately 100J of energy incident on target which corresponds to a peak power of 5 x 10^{13} W/cm^2. From Fig 20 we can see that 50% of the laser beam energy falls within a circle of 80 μm diameter in the target plane for an average power density within this region of 2 x 10^{13} W/cm^2.

The first experiments using the KrF laser facility have been concerned with the generation of X-rays from laser produced plasmas. We have investigated the generation of X-rays of energy >1 keV for X-ray lithography applications by focussing the laser onto Al targets(29). X-ray spectra from the laser-produced Al plasma were recorded using a flat crystal spectrometer and Fig 21 is a single shot spectrum showing strong emission of He-like and H-like lines. The strongest line is emitted at 0.776 nm with a pulse duration of 20 ns and has been estimated to be sufficiently bright to expose X-ray resists on a single shot. The X-ray emission in this line is approximately 70 mJ/steradian.

We have also used the KrF laser to generate soft X-rays by focussing the beam onto plastic targets. Here we are interested in generating

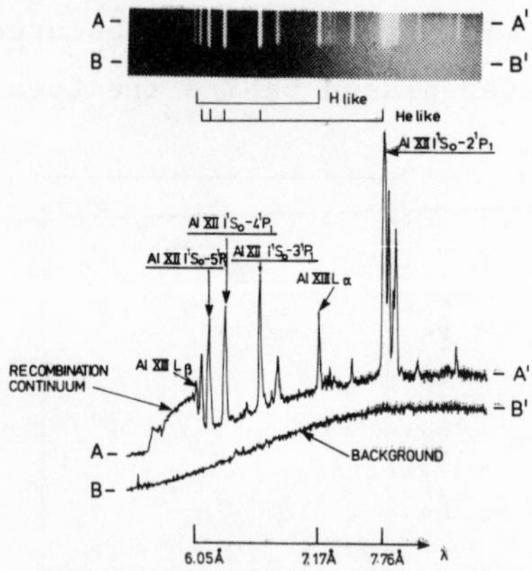

Fig. 21 X-ray spectrum from KrF laser-produced plasma using Al target[29]

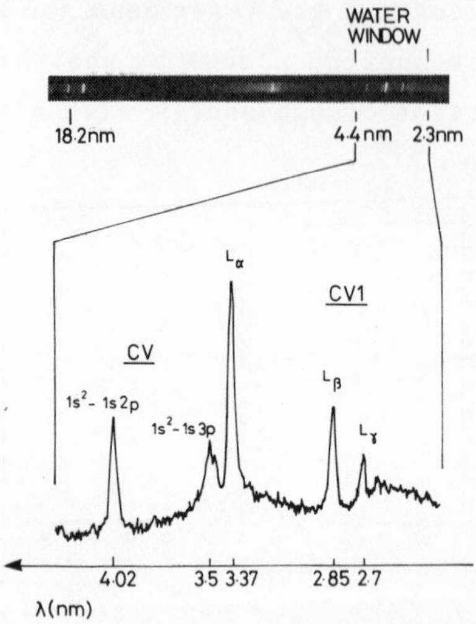

Fig. 22 Soft X-ray spectrum from KrF laser-produced plasma using Mylar target. Mylar = $(C_{10}H_8O_4)_n$

X-rays in the wavelength range 2.3nm - 4.4nm (540eV - 280eV) for application to X-ray microscopy of living biological specimens. The VULCAN glass laser at the CLF has previously been used for this type of research[30] and we are interested in using the KrF laser because of its higher repetition rate and the potentially higher conversion efficiency of 249nm laser light to X-rays. A typical laser plasma X-ray spectrum from a plastic target using Sprite is shown in Fig 22 where we see that line emission from H-like and He-like carbon falls within the spectral range of interest. We have measured an X-ray flux of $10mJ/cm^2$ in the 2.3nm-4.4nm wavelength range at a distance of 5 cm from the plasma which is sufficient to produce X-ray images of biological specimens by contact printing onto an X-ray resist with a single laser shot. Fig 23 shows an image of a fresh foxglove epidermal hair taken by this method. The image shown in Fig 23 was produced by observing the developed X-ray resist using an optical

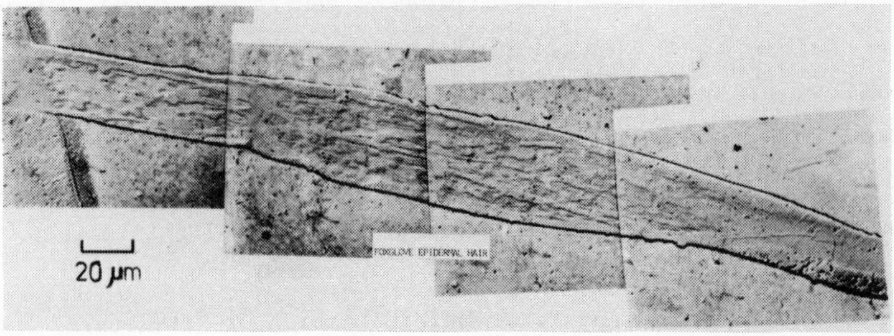

Fig. 23 Image of biological specimen recorded in X-ray resist by contact printing (experimental results by Dr T Stead, Botany Dept, Royal Holloway College, London)

microscope. The potential advantages of using X-ray microscopy are,

(1) Higher resolution than with optical microscopy by observing the resist with an electron microscope

(2) Can observe thicker samples than an electron microscope.

The spatial characteristics of the laser-produced plasmas have been measured using an X-ray pinhole camera to record hard X-ray images of the plasma and two such images taken of a Mylar target plasma are shown in Fig 24. The dimensions of these images correspond to the

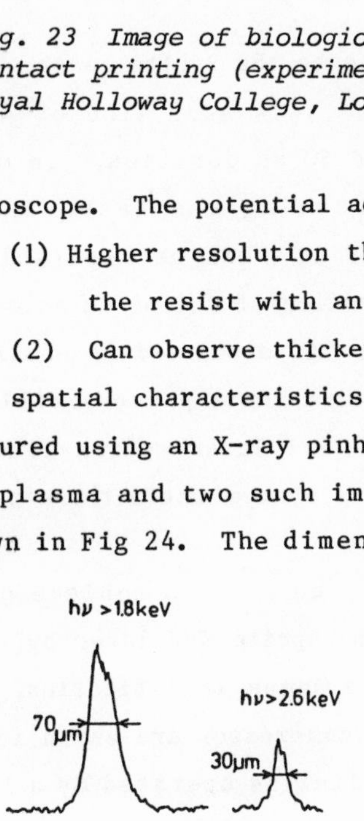

Fig. 24 Microdensitometer traces of X-ray pinhole camera images of KrF laser-produced plasma using a Mylar target

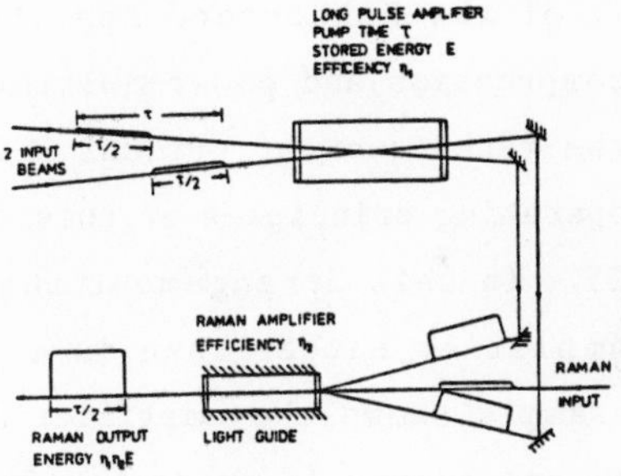

Fig. 25 Schematic diagram of a KrF laser pulse compressor using optical multiplexing and beam combination in a light-guide Raman amplifier

hottest part of the plasma and the image size of 30-70 μm agrees well with the laser spot size on target of 40 μm as shown in Fig 19.

The X-ray measurements presented above represent the first

observations of high temperature plasmas generated by the 249nm KrF laser. We see that even with the pulse duration of 50 ns it is possible to produce high intensities on target of 5×10^{13} w/cm^2 because of the good optical quality of the laser beam. In the near future we plan to apply the laser to the study of other areas of laser plasma interaction physics more closely related to ICF interests.

3.2 Pulse Compression and Power Multiplication of KrF lasers

We have seen above that for target experiments the Sprite laser can at present generate in excess of 10^{13}W/cm^2 in a spot size of 40 μm diameter from a pulse of 100J energy and 50 ns duration. In order to fully utilise this laser for addressing the laser matter interaction physics of relevance to ICF it will be desirable to compress the laser energy into a pulse duration of $<$ 10ns duration and achieve a power of $> 10^{14}$ W/cm^2 on target. As discussed in section 1 it is not possible to achieve this by using a short pulse oscillator followed by an amplifier chain because the KrF laser medium is not a storage medium (like Nd:glass is) and such a system would therefore be very inefficient.

At the Rutherford Appleton Laboratory we plan to achieve pulse compression and power multiplication of the Sprite KrF laser by using the techniques of Optical Multiplexing and Raman Amplification. The operating principles of this type of pulse compressor are shown in Fig 25. In this arrangement the KrF gain medium is operated as a laser amplifier rather than as a laser oscillator with mirrors. In the example shown the amplifier is pumped by an e-beam pulse of duration τ and energy is extracted from the amplifier by two input pulses of duration $\tau/2$. The delay between pulses is $\tau/2$ and each pulse takes a different angled path through the amplifier such that optical delay lines can be used to direct the pulses in synchronism to pump a Raman amplifier. This amplifier, containing a Raman active gas like methane (CH_4) or hydrogen (H_2), behaves like an optically pumped gain medium with a pump time of $\tau/2$ and energy is extracted from the Raman amplifier by a single pulse of duration $\tau/2$ at the 1st Stokes wavelength of 268 nm in CH_4 or 277 nm in H_2.

In the example shown an overall pulse compression of a factor of 2 is achieved with a power gain of $2\eta_1 \eta_2$ where η_1 is the extraction efficiency of the KrF amplifier and η_2 is the efficiency of the Raman system. Remembering that the final output from this laser will be used for target work we can list the advantages of using the Raman amplifier stage as follows.

(1) Many KrF beams are combined into one Raman beam simplifying focussing onto target.

(2) Multiple pump beams of poor optical quality are converted into one diffraction limited Raman beam.

(3) No ASE pre-pulse on target.

(4) Raman amplifier uses a static fill and no cooling.

In order that the full benefit of (1) - (4) can be obtained it is necessary that high power gains can be achieved ie η_1 and η_2 should both be approximately 100%. η_1 is determined mainly by geometric considerations and is expected to be ~ 90%. Recently we have carried out an experiment to measure η_2 under conditions of multiple KrF beam pumping[31]. Multiple overlapping pump beams are obtained as shown in Fig 26 by reflecting the beam from the Sprite laser off a flat mirror which consists of 56 individual segments. This gives 56 KrF beamlets each of dimension 1" x 1" and these are overlapped in a 1.8 m long light guide which is immersed in the Raman gas in a high pressure cell. Use of a light guide provides a long region of uniform gain in the Raman medium.

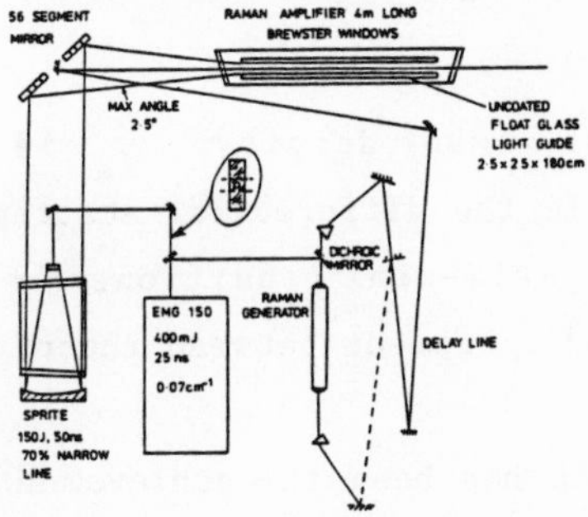

Fig. 26 Experimental arrangement for investigating the light guided forward Raman amplifier[31]

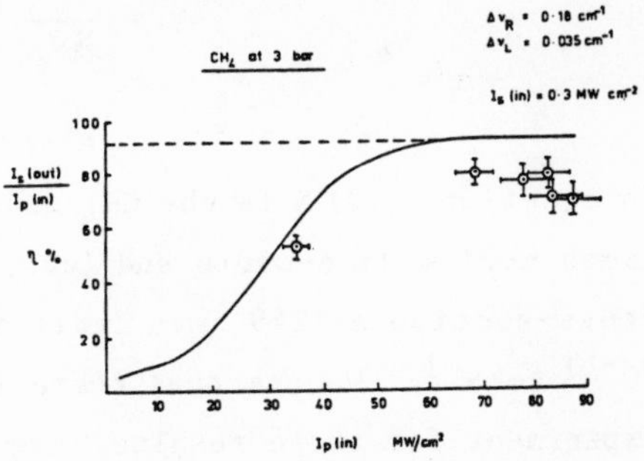

Fig. 27 Power conversion efficiency versus KrF pump power in a Methane Raman amplifier[31]

We have measured the extraction efficiency of the CH_4 Raman amplifier by injecting a 1st Stokes pulse into the input of the waveguide and measuring the power gain down the amplifier. The results are shown in Fig 27 where we have plotted the 1st Stokes output intensity I_s (out) divided by the KrF input pump intensity I_p (in) versus pump input. For these experiments Sprite was operated as an injection locked line-narrowed laser producing pulses of up to 60 J in 50 ns with a line width of $0.035 cm^{-1}$. The results show that an extraction effficiency of 80% is achieved and we have also found that the beam quality of the Stokes wave is maintained after amplification.

The solid curve shown in Fig 27 is calculated from the expression,

$$I_s(\text{out}) = I_s(\text{in}) \frac{\exp\{\gamma I_p(\text{in})L\}}{1 + \omega_p I_s(\text{in})\exp\{\gamma I_p(\text{in})L\}/\omega_s I_p(\text{in})} \tag{3.1}$$

where ω_p and ω_s are the pump and 1st Stokes frequencies and L is the length of the Raman gain medium. γ is the Raman gain coefficient given by,

$$\gamma = \frac{\lambda_s^2}{\hbar\omega_s} \frac{N}{\pi\Delta\nu_R} \left(\frac{d\sigma}{d\Omega}\right)_{249} \tag{3.2}$$

In equation (3.2) N is the CH_4 molecular number density, $\Delta\nu_R$ is the Raman medium line-width and $(d\sigma / d\Omega)$ is the differential scattering cross-section at 249 nm. Under our experimental conditions $\gamma = 6 \times 10^{-10}$ cm W^{-1}. We see that there is good agreement between theory and experiment for these results.

The crucial result of this experiment has been the achievement of high Raman conversion efficiency. Using such amplifiers it should therefore be possible to obtain pulse compression and power multiplication of KrF lasers and we are at present constructing a system for use on the Sprite laser. The Sprite e-beam pump duration is approximately 65 ns and therefore we plan to multiplex the amplifier using 8 beams of 8 ns duration to extract for the full pump time. These 8 pulses will be combined in a Raman amplifier to produce a single beam of 8 ns duration at the 1st Stokes wavelength with an energy > 150J. Such a laser will be a very powerful system for laser plasma interaction studies and will be available in 1986/87.

3.3 Improved efficiency of e-beam-pumped KrF lasers

In this final section of my lecture notes I briefly describe some work that is underway to try and increase the operating efficiency of large e-beam-pumped KrF lasers. We have already seen in section 2.3 that the intrinsic efficiency of e-beam-pumped KrF lasers is typically 10% and that this falls far short of the maximum possible value of 25% given by the quantum efficiency. We have also seen that the lost efficiency is well accounted for as being dissipated in approximately equal amounts in kinetic losses and medium absorption losses. As well as these inefficiencies related to the physics of the laser medium there is also the electrical inefficiencies connected with the conversion of stored electrical energy into deposited e-beam energy. In Sprite for example this reduces the laser intrinsic efficiency from 10% to an overall machine efficiency of approximately 1.5%.

In order that the overall efficiency of KrF lasers can be increased it is necessary therefore to address three problem areas viz

(1) Improve power supply efficiency

(2) Reduce kinetics losses in the laser medium

(3) Reduce internal absorption losses in the laser medium

Improvement of power supply efficiency is perhaps the easiest problem to solve. In the Sprite laser for example the stored energy in the Marx bank (see Fig 12) is converted into deposited e-beam energy with a conversion efficiency of only 15%. From a study of energy losses in the system it seems feasible that design improvements (eg use a low inductance Marx bank to charge the PFLs directly) could lead to a power supply efficiency of >50% which would increase the laser efficiency to >5% overall.

The problem areas listed under (2) and (3) above are more difficult to solve and are being addressed by a number of researchers. In general, attempts are being made to devise operating parameters (eg pump rate, gas mix, etc) which simultaneously improve the kinetic efficiency of the laser medium and reduce internal absorption. In this work extensive use is made of computer kinetics codes to try and predict optimum operating conditions an approach which is justified in view of the fact that experimental results to date have been very accurately modelled by such codes(12,15). The overall conclusion drawn from this theoretical work is that improved laser intrinsic efficiency should be achieved by using lower pressure (1 atm) laser gas mixtures with high krypton concentrations(32).

As an example of the predicted performance Fig 28 shows the calculated variation of the intrinsic efficiency of an e-beam-pumped KrF laser as a function of Kr concentration in the gas mixture. We see that a high intrinsic efficiency of ~ 17% is predicted for a Kr concentration of 50% and a total gas pressure of approximately 1 atmosphere. High e-beam current densities are used in order to keep the specific output energy of the laser (in joules per liter) at a level which can be achieved with argon-rich gas mixtures operating at 2 atm pressure.

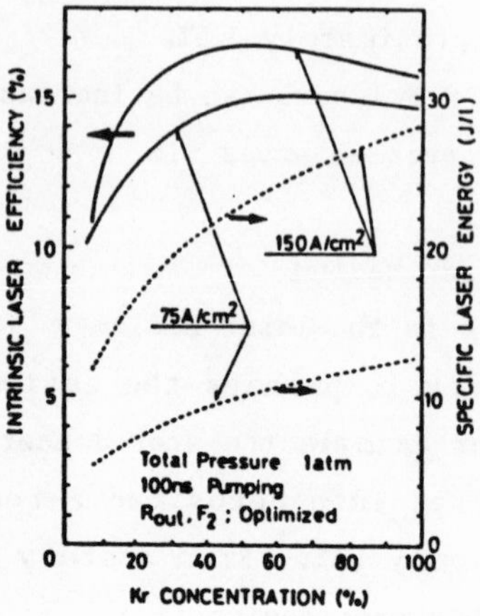

Fig. 28 Theoretical dependence of intrinsic KrF laser efficiency and specific laser energy on the Kr concentration for a 1-atm mixture of $Kr/F_2/Ar$(32)

As we see the low pressure, Kr-rich mixes offer improved laser efficiency. There is also the additional advantage that there is no static pressure load on the windows of the laser cell and thus the thickness of the fused silica windows in such a laser can be kept to a minimum. This will result in a considerable financial saving for very large lasers for ICF where window apertures in the 1-2m range will be necessary.

The results shown in Fig 28 obviously represent a major departure from the operating regime previously used in e-beam-pumped KrF lasers(15,20) where high pressure Ar-rich mixtures were found to be optimum. The underlying physical reasons why low pressure Kr-rich mixtures have the potential for higher efficiency are,

(1) Low Ar concentration gives reduced Ar_2^+ formation. Ar_2^+ is a strong absorber at 249 nm

(2) Low pressure reduces KrF^* losses by quenching processes

(3) Low pressure reduces formation rate of Kr_2F^* which absorbs at 249nm

(4) Use of high e-beam pump rate maximises laser cavity internal flux so stimulated emission competes favourably with kinetics losses

A typical high efficiency KrF laser mixture will consist of 49% argon, 50% Kr and ~1% F_2 and the laser will operate with low medium absorption and high intracavity flux. The high intracavity flux will be achieved by using high e-beam pump rates and low output coupling optics on the laser resonator where optimum performance will be achieved at output coupling > 50% as a direct result of reduced intracavity absorption. In KrF lasers with Ar-rich mixtures the optimum output coupling is ~80% because of high internal losses.

The predicted operating conditions with Kr-rich mixtures are expected to give rise(33) to an additional advantage as a result of the high intracavity fluxes that will be used. At high flux levels it seems feasible that the absorption due to Kr_2F^* can be saturated. The saturation intensity of the 249nm absorbing transition in Kr_2F^* is

$$I_{sat} = \frac{h\nu(249\text{nm})}{\sigma_a(Kr_2F)\ \tau_{coll}} \tag{3.3}$$

where $\sigma(Kr_2F)$ is the absorption cross-section and τ_{coll} is the collision dominated life-time of the absorbing state. Using

$\sigma_a(Kr_2F) = 4 \times 10^{-18} cm^2$ and $\tau_{coll} = 25$ ns[15] we obtain $I_{sat} = 8$ MW/cm^2. Achievement of intracavity fluxes greater than this value will therefore result in a strong reduction of Kr_2F^* absorption and should increase the efficiency over the value of 17% already predicted above. In fact improved laser performance at high intracavity fluxes has already been reported[20] in work with normal Ar-rich gas mixes at RAL where a power efficiency of 15% was observed at an intracavity flux of ~ 9 MW/cm^2.

To date there have been few reports of KrF laser efficiency measurements with Kr-rich mixtures. Some preliminary results have however been reported[34] which do indicate improved laser performance under these conditions and further experiments are underway at RAL to measure laser efficiency under widely varying operating conditions.

4
CONCLUSIONS

These lecture notes are intended to present the reader with a brief overview of high power rare gas halide laser technology. The emphasis has been on e-beam-pumped KrF lasers because to date these are the most powerful systems available. These are also the systems which have been the subject of the most intensive study at RAL and are therefore the systems with which the author has most experience. Many of the topics discussed in these notes are however also applicable to other e-beam-pumped RGH lasers (eg XeF, XeCl) and even to discharge excited systems.

In view of the brevity of these notes there have been a number of important topics left undiscussed eg UV optics fabrication, gas handling facilities for halogens, etc. These omissions have been due to the authors desire to explain fully the underlying physics of the excimer gain medium at the expense of neglecting some areas of laser technology.

It is interesting to look to the future to speculate on

developments that might take place in say the next 5-10 years. The emphasis at RAL will be on the construction of very high power (> 1 TW) short pulse KrF laser systems for application to laser plasma studies in such areas as ICF, X-ray lasers and laser plasma X-ray source applications. The KrF gas laser is well suited to this type of work because of its high efficiency, good beam quality, high repetition rate and short wavelength.

In the USA the emphasis will be on the construction of very large lasers of >100 kJ output and >5% overall efficiency. While this work is at present aimed at fulfilling the requirement of an ICF driver there is a new technological goal on the horizon viz. the so called Strategic Defence Initiative (SDI)[35]. Rare gas halide lasers for SDI uses will operate with long pulses of >500 ns, very high energy and good beam quality. The rare gas halide lasers are well suited to this application where in particular the short operating wavelength means that diffraction limited beams can propagate over very long distances in space with minimum beam expansion. The short wavelength RGH lasers are also absorbed most efficiently on targets which results in the maximum impulse delivery as is required for target destruction. It will be interesting to see how this new technological goal will effect the direction of laser development and laser matter interaction research over the next few years.

ACKNOWLEDGEMENTS

The author has drawn the material for these lecture notes from a number of sources, as evidenced by the list of references, with many contributions coming from colleagues at RAL and UK universities. I wish in particular to acknowledge the efforts of the following individuals:

Laser and Pulsed Power Design M J Shaw, C B Edwards, F Kannari, D Craddock, F S Gilbert, J Boon, E Madrazsek, S Hancock, I N Ross, E M Hodgson, J Partanen, B A Omar, H T Medhurst, D Baker, D Wood

KrF Laser Target Experiments Y Matsumoto, Y Owadano, R Eason, I C E Turcu, O Willi, G Kiehn, A Ridgeley, A Michette, T Stead

Finally I wish to thank the senior management of Laser Division Rutherford Appleton Laboratory who have consistently supported KrF laser research from the initial experiments to demonstrate high KrF laser efficiency in 1979[20] through to the establishment of the first KrF target facility in 1984[29]. During this period the Laser Division was managed initially by A F Gibson and P R Williams and presently by M H Key and P T Rumsby.

REFERENCES

1. M F Golde and B A Thrush, Chem Phys Lett, 29, 486 (1974)
2. J E Velazco and D W Setser, J Chem Phys 62, 1990 (1975)
3. S K Searles and G A Hart, Appl Phys Lett 27, 243 (1975)
4. J J Ewing and C A Brau, Appl Phys Lett, 27, 350 (1975)
5. E R Ault, R S Bradford and M L Bhaumik, Appl Phys Lett 27, 413 (1975)
6. J M Hoffman, A K Hays and G C Tisone, Appl Phys Lett, 28, 538 (1976)
7. H Egger, T S Luk, K Boyer, D F Muller, H Pummer, T Srinivasan and C K Rhodes, Appl Phys Lett 41, 1032 (1982)
8. Topics in Applied Physics Vol 30, Excimer Lasers, ed C K Rhodes (Springer-Verlag)
9. J C Swingle, L G Schlitt, W R Rapoport, J Goldhar and J J Ewing, J Appl Phys 52, 91 (1981)
10. C B Edwards, F O'Neill, M J Shaw, D Baker and D Craddock, Excimer Lasers 1983, AIP Conf Proc No 100 (New York:AIP) pp59-65
11. T H Dunning and P J Hay, J Chem Phys 69, 134 (1978)
12. F Kannari, A Suda, S Yamaguchi, M Obara and T Fujioka, IEEE J Quantum Electron, QE-19, 232 (1983)
13. R G Caro, M C Gower and C E Webb, J Phys D: Appl Phys 15, 767 (1982)
14. D C Lorents, Physica 82C, 19 (1976)

15. C B Edwards and F O'Neill, Laser and Part Beams 1, 81 (1983)
16. A Mandl, Phys Rev A3, 251 (1971)
17. R K Steunenberg and R C Vogel, J Am Chem Soc 78, 901 (1956)
18. C B Edwards, F O'Neill and M J Shaw, Appl Phys Lett 38, 843 (1981)
19. M J Shaw, Appl Phys B30, 5 (1983)
20. C B Edwards, F O'Neill and M J Shaw, Appl Phys Lett 36, 617 (1980)
21. C B Edwards, F O'Neill and M J Shaw, J Phys E: Sci Instrum 18, 136 (1985)
22. R P Little and W T Whitney, J Appl Phys 34, 2430 (1963)
23. R K Parker, R E Anderson and C V Duncan, J Appl Phys 45, 2463 (1974)
24. K H Kingdon and H E Tanis, Phys Rev 53, 128 (1938)
25. C M Slack and L E Ehrke, J Appl Phys 12, 165 (1941)
26. J A Nation, Part Accel 10, 1 (1979)
27. R J Noer, Appl Phys A28, 1 (1982)
28. S C Clark, D C Emmony, B A Omar and M J Shaw, to be published
29. Y Matsumoto, M J Shaw, F O'Neill, J P Partanen, M H Key, R W Eason, I N Ross, E M Hodgson and Y Sakagami, Appl Phys Lett 46, 28 (1985)
30. R J Rosser, K G Baldwin, R Feder, D Bassett, A Cole, and R W Eason, J Microscopy (1985) to be published
31. M J Shaw, J P Partanen, Y Owadano, I N Ross, E M Hodgson, C B Edwards and F O'Neill, to be published
32. F Kannari, A Suda, M Obara and T Fujioka, Appl Phys Lett 45, 305 (1984)
33. F Kannari, private communication
34. E T Salesky and W D Kimura, Paper Th R6-1, IQEC conference (1984)
35. Articles on SDI have recently been published by G Yonas and W K H Panofsky, Physics Today, 38, 24 (1985).

XUV AND X-RAY LASERS

G.J. PERT

Department of Applied Physics, University of Hull, Hull, HU6 7RX

1.
INTRODUCTION

Since the invention and demonstration of lasers in 1959, there has been rapid development from the initial 6000-7000 Å devices to cover a spectral range from microwaves to ultra-violet. Devices operating over this wide range use both atomic and molecular transitions pumped by a variety of sources, chemical, photon, electrical and gas dynamic. In each case for efficient operation the power source must be matched in both the specific energy and the rate to the laser medium. Within this wavelength range photon energies do not exceed 10 eV, and characteristic lifetimes are typically longer than 1 ns. These values closely match those generated by conventional sources such as electrical discharges. Techniques for pumping present-day lasers thus use well established technology. This convenient underlying physics has enabled a very wide variety of lasers to be developed, to the extent that devices are available to provide coherent radiation over the entire spectrum from about 2000 Å to 100 μm.

However the progressive development of laser action to shorter wavelengths is beset by fundamental problems. These arise from two fundamental physics constraints. Firstly the probability of radiative transition scales rapidly with frequency, roughly to the fourth power. Secondly below about 1000 Å the photon energy exceeds valence electron ionisation energies. In consequence the technology of conventional lasers using valence electron transitions in a neutral or weakly ionised medium enclosed in a reflecting cavity is no longer applicable. The pump source must be matched to transitions of 1-10 ps, and energies of 100 eV at 100 Å, using as lasant either a highly ionised plasma, or inner shell transitions without the advantage of an external cavity. This imposes much more severe technological contraints which are only just being successfully overcome.

The value of a coherent source of X-rays in many fields has been realised for many years. Most of the applications proposed extend either present day X-ray technology, or involve optical techniques at ultra-short wavelengths. Typical of these are phase contrast X-ray microscopy and, possibly, crystal holography. In practice, however, the earliest applications are likely to be more mundane as powerful well-resolved X-ray sources for microscopy and radiography. Numerous uses for such sources in medicine, crystallography, non-destructive testing and metallurgy can be readily envisaged. More recently military applications of X-ray lasers have received considerable attention.

In the early 1970s there was considerable speculation on approaches to X-ray laser design[(1)]. By about 1975 the technological constraints referred to earlier were clearly apparent, and much of the activity ceased. To those remaining in the field it was clear that a less ambitious programme with progressive development, as state-of-the-art technology allowed, was likely to prove more rewarding. It was apparent that by careful design systems lasing in the XUV region around 100 Å could be constructed. Furthermore by investigating appropriate schemes scalability into the X-ray region could be developed.

The technological constraints on short wavelength laser operation follow from very fundamental physical principles, as outlined above.

The discussion can be developed in a quantitative manner, which enables the magnitudes of the experimental parameters to be identified. The gain per unit length of a population inverted medium is given by:

$$\alpha = \zeta \frac{\lambda^2}{8\pi} \frac{A}{\Delta\nu} g_3 \left\{ \frac{n_3}{g_3} - \frac{n_2}{g_2} \right\} \tag{1.1}$$

where ζ is the line shape factor, A the spontaneous transition probability, λ the wavelength and $\Delta\nu$ the frequency width of the line, and (n_3, g_3) and (n_2, g_2) are the population density and statistical weight of the upper and lower laser states respectively. In a steady state system power is supplied by the external pump source, converted and dissipated by lasant, and finally absorbed by some internal or external dump. We may easily identify a lower bound for the required input power by evaluating the power radiated by spontaneous emission on the laser transition itself. Thus the power loss per unit area of cross-section of the laser is

$$P = n_3 \ell\, A\, h\nu \geqslant \frac{8\pi\hbar}{\zeta c^2} \nu^4 (\alpha\ell) \frac{\Delta\nu}{\nu} = \frac{4 \times 10^{21}}{\lambda(\text{Å})^4} (\alpha\ell) \left|\frac{\Delta\lambda}{\lambda}\right| \quad \text{W/cm}^2 \tag{1.2}$$

where ℓ is the length of medium. In practical devices additional losses will increase the input power demand by at least an order of magnitude over this bound. Not only must this power be supplied by the external source, but the medium itself must be able to survive the dissipation without its own destruction.

In contrast to conventional lasers only two laboratory sources are at present known which can in principle deliver the pump power densities $\geqslant 10^{15}$ W/cm^2 required by an X-ray laser namely a subsidiary laser or a particle beam: we shall not discuss a third alternative, a nuclear explosion, as its suitability for non-military applications must be extremely limited. In the less restrictive XUV spectral region ($\geqslant 100$ Å) we may add fast high voltage discharge devices such as the superfast pinch.

The problem is further complicated by the absence of simple transmitting or reflecting media at X-ray wavelengths so that the constraints cannot be relaxed by allowing the laser signal to build up in a resonant cavity, or distributed feedback system. In particular it follows from the elementary theory of metals that the minimum wavelength, λ_{min}, which can be reflected from a metal at an angle, θ, is

$$\lambda_{min} = 3.33 \times 10^{14} N^{-\frac{1}{2}} \sin\theta \quad (1.3)$$

where N is density of electrons (/cc). For glass $N \simeq 7.8 \times 10^{23}$/cc, and for gold $N \simeq 46.6 \times 10^{23}$/cc. Thus at XUV wavelengths one must use metals at grazing incidence for mirrors in order to obtain high reflectivity[(2)]. At X-ray wavelengths crystal reflectors may be used but again at non-normal incidences. Under such conditions the construction of a cavity is extremely complex[(1)]. The recent development of multi-layer reflectors in this wavelength range may eventually solve this difficulty but for the present their reflectivity is low[(3,4)]. The absence of a cavity implies that the laser must operate in a weakly coherent travelling wave, or amplified spontaneous emission mode. This form of operation in which the spontaneous fluorescence is amplified by stimulated emission down a rod shaped medium requires a gain-length product $(\alpha\ell) \sim 20$ (i.e. an overall amplification $e^{\alpha\ell}$ of about 100 db). Such a device produces a beam of divergence typically $\sim d/\ell$, where d is the rod diameter and ℓ its length. If this is matched to the diffraction limit, λ/d, the beam will have coherence across its wavefront. The optimum diameter is thus

$$d = \sqrt{\lambda\ell} \quad (1.4)$$

If the diameter is less than this value diffraction losses will reduce the effective gain length to $\ell' \simeq d^2/\lambda$.

Since the total power $P \simeq pd^2$ it is clear that if the total pump power is to be kept within reasonable limits (say 100 TW) then $d \leqslant 10^{-3} \lambda(\text{Å})^2$ cm, and therefore $\ell \leqslant 10^{-6} \lambda(\text{Å})^3$ cm, assuming $\Delta\lambda/\lambda \simeq 10^{-4}$. Hence we may estimate that the minimum inversion density $\Delta n \simeq 6 \times 10^{21}/\lambda(\text{Å})^4$/cc assuming the oscillator strength of the lasing transition at unity.

It is clear from the preceeding discussion that the principal limitation on constructing an X-ray laser is the high pump power required. In principle we may satisfy this condition either directly in which case the external pump must itself satisfy the power condition, or indirectly when the system is pumped to a relatively long lived reservoir state, which may be slowly filled, to be transferred into the upper laser state by some appropriate trigger. The latter approach relaxes the condition on the external pump by using the reservoir energy, but is clearly limited to short pulse action only. Two examples, namely metastable doubly excited state pumping and recombination during expansion, will illustrate this effect.

Any further analysis of the scaling of X-ray lasers is specific to the actual scheme proposed. As an example consider the behaviour of an optical transition scaled similarly along an iso-electronic sequence of ions. The characteristic energies scale as λ^{-1} and the lifetimes as λ^2. The temperature thus scales as λ^{-1} and, if the line is Doppler broadened, the line width $\Delta\lambda \sim \lambda^{\frac{1}{2}} M^{\frac{1}{2}}$, where M is the ion mass. Since the medium is a long thin rod the diameter scaling is determined by the characteristic velocity and time. Hence using (1.4) and the fact that $\alpha\ell \simeq$ const

$$d \sim \lambda^{\frac{3}{2}} M^{-\frac{1}{2}}, \quad \ell \sim \lambda^2 M^{-1} \text{ and } \alpha \sim \lambda^{-2} M \qquad (1.5)$$

Since from (1.1) the gain $\alpha \sim n\lambda^2/\Delta\lambda$, the density, n, and power density, p, scale as:

$$n \sim \lambda^{-\frac{7}{2}} M^{\frac{1}{2}}, \quad p \sim \lambda^{-\frac{9}{2}} M^{-\frac{1}{2}} \qquad (1.6)$$

It is clear that the resultant system is a high gain, high density device of small dimensions. Although the actual scaling laws vary from system to system the general conclusions remain unchanged: indeed the assumptions made in the above analysis are more general than is immediately apparent for dissipation of the waste pump energy will almost always heat the medium to give a Doppler broadened line, and the rod diameter must be limited by the hydrodynamic disassembly time.

The lasing must take place either on inner shell transitions of a neutral or weakly ionised atom, or on an optical transition of a highly stripped ion. We shall call these systems respectively non-plasma and plasma schemes and consider them separately.

An alternative approach for the generation of coherent short wavelength radiation is by frequency multiplication of existing visible or near ultra-violet lasers[5,6]. This method is, at present, the only way in which such radiation can be generated, and demonstrations of coherent beams with wavelengths down to 380 Å have been made. In these experiments ultra-violet radiation is frequency multiplied in a gas medium using a near resonance to obtain a strong coupling coefficient. Although possessing many attractive features such as practicability, coherence etc. the applications to XUV and X-ray wavelengths are extremely limited by its inherent disadvantages. The gaseous non-linear medium must be relatively tenuous to avoid the strong absorption at these wavelengths; as a result efficiencies are extremely low, and energies of the order of nJ only are typical. The application to short wavelengths is restricted by the need, at present, for a neutral gas as medium to about 300 Å. In the present context it is not considered that frequency multiplication can provide an alternative source to a true X-ray laser: rather it is expected that this approach will fill a complementary role as a source of coherent VUV and XUV radiation.

2.
NON-PLASMA SYSTEMS

These schemes may conveniently be divided into two groups, inner shell excitation and electron channelling.

2.1 Inner Shell Excitation

Inner shell transitions involve a vacancy within a closed shell of a neutral or weakly ionised atom. The vacancy can be created by particle collision or by photo-excitation. The subsequent decay of the hole can involve a variety of processes including radiative, Auger, radiative-Auger and more complicated shake-off and shake-up processes. Inner-shell transitions are thus inherently complex, and in quasi-static mode consume considerable power to maintain the lasing against these losses, and which must be dissipated from the medium if it is to maintain its un-ionised form. Furthermore the pump source energy must be restricted to a narrow band to avoid outer shell removal: in this context it is important to note the favourable wavelength scaling of photo-ionisation cross-sections which makes this process selective to shells whose ionisation energy matches that of the photons.

The basic laser process involves the removal of an inner shell electron to form an inversion with an outer shell vacancy. Superficially it would appear that such transitions would have narrow line widths conferring a marked advantage to this approach. In practice the strong Auger decay leads to widths which are comparable with Doppler widths in typical plasmas giving strong emission at similar wavelengths

The principal interest in these schemes has been directed towards generating K shell holes by photo-ionisation using filtered X-rays produced by a laser produced plasma as a pump. For small atomic numbers, such schemes involving a transition between K and L shells are self-terminating and limited to inversion lifetimes of the order of the K-L shell transition time. For medium atomic numbers Auger decay dominates to form a LL shell vacancy, which is not the lower laser state, and a sustained inversion is possible[(7,8)]. However, maintaining the medium against continuing ionisation and removing the waste energy present design problems of extreme difficulty. Indeed one can readily appreciate that as a result of the photo-ionisation source alone, without Auger losses, one will rapidly deplete the limited number of bound electrons, and strip the ion[(9)]. In this case the shells must be refilled by recombination, and the system becomes closely allied to the recombination scheme discussed later.

A novel proposal[(10)] for generating inversion in the XUV region uses excitation of a level meta-stable against both auto-ionisation and radiation, which is photo-excited to the upper laser level, itself stable against auto-ionisation, in the doubly excited series of neutral lithium. The lower laser state is above ground and emptied both by radiation and photo-ionisation. The relevant term scheme is shown in Fig. 1.

The metastable vacancy level $1s2s2p\,{}^4P^0$ is pumped by X-rays from a laser plasma generated at a high Z target within the heat pipe oven containing the lithium. A tuned dye laser transfers this population to the level $1s2s^2\,{}^2P$ which is the upper state of the lasing transition to $1s^2 2p\,{}^2P^0$ at 207 Å. A gain coefficient of about 0.15/cm is predicted for these experiments[(10)].

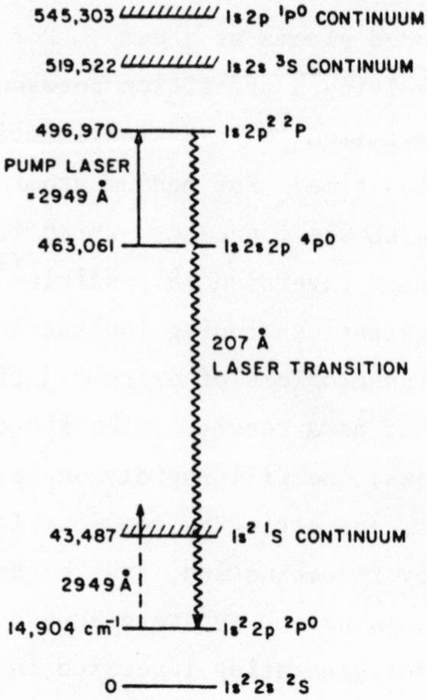

Fig. 1 Energy level diagram for the proposed 207 Å laser in neutral Li, from ref. (10).

2.2 Electron Channelling

A parallel beam of electrons incident on a crystal consisting of parallel lattice planes will undergo Bragg reflection. If the beam is aligned along the crystal planes at an exact Bragg position the beam will be contained within a set of periodic "channels" formed by the lattice planes which act rather like waveguides, Fig. 2.

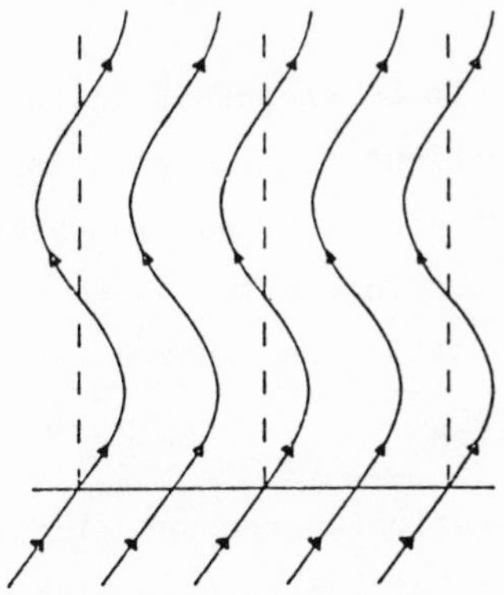

Fig. 2 Flux line paths of electrons incident on a crystal at the Bragg reflecting position from ref. (11).

Clearly the electrons have an oscillatory motion due to the Bragg reflection, which should be accompanied by the emission of radiation, called Bloch wave channelling radiation. The electrons oscillate with a depth periodicity ξ_g whose value varies with electron energy, so that the wavelength is continuously variable (e.g. for gold 100 keV electrons give 114 Å and 10 MeV 0.5 Å). A more detailed study[(12)] shows that the emitted radiation is due to transitions between discrete Bloch wave states, and that since the excitation amplitude of each Bloch wave depends on crystal orientation, population inversions amongst these states may be readily achieved.

Observations of Bloch wave channelling radiation by Swent et al[(13)] are in good agreement with theory[(12)]. However, detailed analysis of gain and input electron parameters are not yet available. Clearly good monochromaticity of the input electron beam is essential. Other practical limitations must involve the damage to the crystal resultant from the fundamental dissipation limit (1.2). Although, since the device is essentially c.w., a crystal resonant cavity may be used to relax this constraint.

3.
PLASMA SYSTEMS

The remaining approaches to be discussed in this survey involve the pumping of optical transitions in highly ionised species. These may be characterised into two groups, those in which the pumping transition is upwards, collisional excitation and photo-excitation; and those for which it is down, recombination and charge exchange.

3.1 Collisional Excitation

In most collisional excitation approaches electron collisional pumping of a forbidden optical transition, with a high rate, is used to generate a relatively large population in the upper laser state. The lower laser state is selected with a large optical decay rate (to prevent a build-up of population). The system is clearly quasi-steady state, the inversion being present in a state of coronal equilibrium.

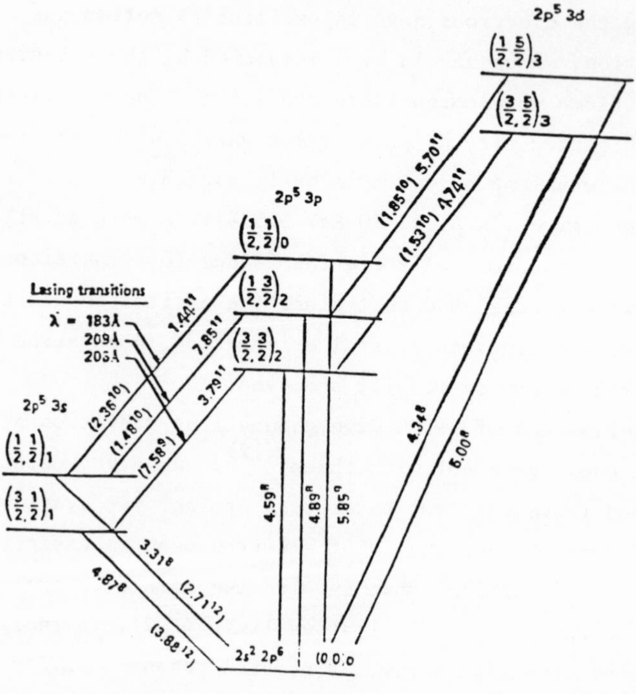

Fig. 3 Calculated term scheme for the levels n = 3 and n = 2 of Se XXV (from ref. 20).

Of particular interest in this regard are ions with ground state configuration $1s^2 2s^2 2p^n$, of which the carbon-like (n = 4)[14] and neon-like (n = 6)[15-17] have received study. The relevant term scheme for the latter is shown in Fig. 3. Pumping occurs as a result of the large monopole electron collisional excitation rate of the transition 2p - 3p. Since the lower laser state 3s has an allowed transition to the ground state the system has a simple three-level form. Detailed calculations are hampered by a lack of accurate atomic data, but estimates show that the gain is strongly temperature and density dependent. One dimension of the plasma must be restricted to allow the line 3s - 2p to be optically thin. Gains as high as 100 cm^{-1} calculated for Ca XI by Vinogradov et al[15] are probably overestimates but values of about 10 cm^{-1} in Se XXV by Hagelstein[17] are more realistic. These gains are obtained in a plasma at about 10^{20}/cc electron density and 150 eV electron temperature, conditions typical of those in a plume produced by a nsec Nd.glass laser pulse incident on a solid target. At longer wavelength transitions mirrors can be constructed and operation in a CW cavity mode may be possible.

Early experiments to check these predictions of gain in Ca XI were made by Ilyukhin et al[18] using a line focussed laser plasma from a sandwich target. Since the predicted wavelengths lay in the range 600-660 Å cavity operation using gold and ruthenium mirrors was possible. Some evidence of collimated emission was obtained, but the experiments were irreproducible and the results inconclusive.

This approach has been successfully applied by workers at Lawrence Livermore Laboratory[19] who in a careful series of experiments have unequivocally demonstrated spontaneous amplified emission using neon-like selenium ions. Initial experiments using laser irradiated selenium coated foils were disappointing, no gain being observed. This was caused, it was believed by severe refraction of the XUV radiation in the large density gradient of the blow-off plasma[17]. The use of thin films (1500 Å) coated with selenium provided a more uniform plasma and eliminated this effect[20].

In these experiments a gain-length product of 6.5 was measured at wavelengths of 207 Å and 209 Å from 3s - 3p transitions in neon-like selenium Se XXV. The plasma was generated by the irradiation of a

1500 Å formavar film coated with a 750 Å layer of selenium using a 0.53 μm laser beam focussed to a line 1 - 2 cm long by 0.2 mm wide at an irradiance of 10^{14} W/cm^2 in a pulse of 450 ps. Amplified emissio was identified by four tests: angular distribution of emitted radiation temporal behaviour, line width considerations and, principally, the variation of intensity with plasma length (Fig. 4). These experiments were all consistent with the generation of gain.

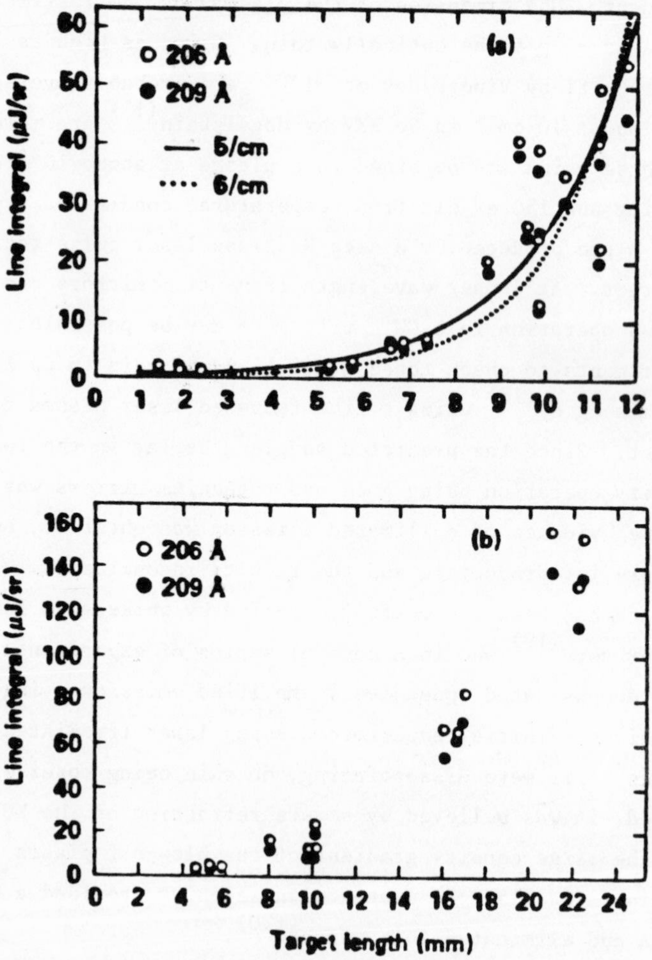

Fig. 4 Variation of total intensity with target lengths for the lines 206 and 209 Å of Se XXV, showing the exponential signal amplification (from ref. 19).

Similar experiments with the selenium replaced by yttrium yielded gain at shorter wavelengths 155 Å and 157 Å in Y XXX.

The experimental results are in conflict with theoretical prediction[20]. The gain is observed on the (J = 2) → (J = 1) transitions at 207 Å and 209 Å instead of the predicted (J = 0) → (J = 1) line at 183 Å. This is in conflict with the expected dominant monopole collision process from the (J = 0) ground state to the (J = 0) excited state. This discrepancy is not yet understood.

Extension of this approach to shorter wavelengths using the 2s - 2p transitions of helium-like ions is disappointing[14]. Thus any progress towards shorter wavelengths may occur along the iso-electronic sequence, but is likely to involve higher densities and temperatures, not conveniently matched to an easily generated plasma.

3.2 Photo-excitation

The concept of using a strongly emitting plasma of one species to selectively pump a second colder one of different composition has been considered for many years. The pumping radiation must be strongly emitted by the source, and is usually the resonance line of an appropriate ion. The difficulties of this method are to find a sufficiently well-matched transiton between pump and lasant, and to obtain good coupling between the plasmas.

The use of near resonant transitions in hydrogen-like and helium-like species was first investigated by Vinogradov et al[21] who identified several possible candidate pairs. Amongst these the sodium-neon pair (Na X $2^1P - 1^1S$ and Ne IX $4^1P - 1^1S$) at 11.00 Å and silicon-aluminium (Si XIII $2^1P - 1^1S$ and Al XII $3^1P - 1^1S$) at 6.650 Å have been further examined[17][21-24]. The wavelength mismatches are respectively 0.001 Å and 0.025 Å. Although the mismatch can to some extent be tolerated if the pump line is broad, the pumping efficiency is then reduced. As a result the silicon-aluminium scheme is unlikely to be practicable unless motional Doppler shifts can be introduced to compensate for the mismatch. Further surveys at Livermore and KMS have identified a number of other well-matched pairs and have accurately measured their separations[25].

A number of experimental arrangements have been proposed. Experiments at Livermore have used gas cell targets containing the lasant enclosed in a foil of the pump material. Irradiation by the Novette laser was used to generate a line pump plasma, whose X-ray emission both photo-ionised and pumped the lasant. This scheme has the merit that the laser ions are relatively cold, and therefore have a narrow line width. A number of candidate transitions have been investigated without success, but the best matched is believed to be fluorine-manganese (F IX 1s - 3p and Mn XXII $2p^2\ {}^1D_2 - 2p3d\ {}^1F_3$)(25).

The neon-sodium scheme is being investigated by workers at NRL(26) (frozen neon layers) and Rutherford Laboratory(27) (ion implanted neon foils) using a laser produced plasma as both pump and lasant. To date only preliminary results have been obtained.

3.3 Charge Exchange Pumping

These schemes propose to mix an ionised gas with a second neutral or weakly ionised background. The gases are chosen so that a near resonant charge exchange from the neutral atom populates an upper level of the ion. The scheme has extremely favourable scaling by virtue of the selective, and strongly coupled nature of the resonant charge exchange process. Two distinct variations are considered. In the first charge exchange directly populates the upper laser level, for example:

$$C^{6+} + Ar^{2+} \rightarrow C^{5+}\ (n = 3) + Ar^{3+}$$

leading to a CVI Balmer α transition(25). In the second approach the mechanism is less specific, charge exchange populating highly excited levels of the ion, which then radiatively cascade to the ground state(29 In many cases the distribution of transition rates within the term scheme is such that population inversions form during the cascade. This approach is very similar to the recombination scheme to be discussed later, but in view of the much larger (by a factor of ~ 10^4) charge exchange cross-section compared to that of recombination is extremely attractive. However, the onset of hydrodynamic (shock) and radiative heating of the background gas, before the arrival of the ionised medium greatly limit the density at which the interaction can take place, and therefore the overall gain. Population inversions observed in CVI and CV have been ascribed to such charge exchange cascades by workers at NRL(32,31) but unfortunately at too low a density (10^{15}/cc) for significant gain to occur.

3.4 Recombination Lasers

The most promising approach to a laboratory XUV laser at present seems to be the recombination laser. Population inversion generated in this way is a familiar feature in many laser produced plasmas, and significant gain has been measured. The inversion is achieved during the cascade through the excited states of highly charged ions following recombination into high lying states. Two basic approaches can be envisaged, requiring different forms of the ion term scheme. If the principal de-excitation process in the cascade is via electron collisions, the inversion forms between a pair of well separated levels within a group of closely spaced ones. In contrast with radiative decay the inversion generally forms between a pair of levels, the lower of which is well separated from its next lower state, for example, the 2p state of hydrogen. The first approach has been successfully applied by workers at Bell Labs in the visible region of the spectrum[32], whereas the latter has been used in the XUV with emphasis on hydrogen-like ions[33], it being relatively easy to produce a high fractional population of the parent ion by stripping the medium. In order to achieve significantly large gain within the recombination cascade, it is necessary to first ionise the medium, and then rapidly cool it so that the dominant recombination is into high lying states rather than directly into the ground state. This, the basic problem, may be accomplished in a number of ways of which adiabatic cooling by expansion[34,35] and radiative heat loss[36] are the most promising.

Radiative cooling has been principally studied by workers at Princeton University and predicted to give significant gain in large low density systems. In this proposal a long lived plasma containing ions of the lasant (e.g. carbon) in a mixture with ions of heavier atoms is initially pulse heated to a sufficiently high temperature to fully strip the lasant ions. The complex atomic structure of the ionised heavy atoms gives rise to rapid emission of radiation, thereby cooling the plasma. Experiments involve a magnetically confined plasma which is initially heated by a CO_2 laser pulse, to give a measurable demonstration of gain.

Initial results using gas fill gave only small gain on CVI, H_α 182 Å. Subsequent experiments have been performed with solid targets, either carbon discs or thick fibres. In particular using carbon fibres 100 μm diameter, 2 cm long, irradiated by 10.6 μm laser pulses of about 300 J lasting 20 ns, gain-length products of up to 6.5 have been measured on the CVI H_α transition at 182 Å. Population inversions have also been measured on lithium-like ions. In future experiments it is planned to use aluminium overcoats to enhance the radiation loss[37].

Radiative cooling has been attributed as the principal heat loss in experiments at the University of Rochester[38]. In this attempt to repeat the Livermore experiment discussed earlier, gain-length products of up to 4 were observed on the CVI H_α line at 182 Å, but no lasing was seen on the selenium lines at 207 Å or 209 Å. The carbon originates from the 1000 Å formavar substrate on which a 750 Å selenium layer was deposited. The foil was irradiated by 0.351 μm laser pulse of irradiance 7×10^{13} W/cm^2 and duration 600 ps along a line 0.1 mm wide and up to 13.6 mm long. The gain was measured from the variation of signal amplitude with plasma length.

There have been several observations of gain either measured directly or indirectly from expansion cooled recombination systems. Using slab targets irradiated with Nd.glass laser pulses of 10 ns duration Elton and co-workers[39] at NRL have indirectly estimated a gain of 2%/cm at 520 Å on the hydrogen-like carbon lines CVI, P_α.

Jaeglé and associates[40] at Orsay have in a series of experiments measured gain-length products of up to about 2 on the lithium-like Al XI line 3d - 5f at 105.7 Å. In this work a 1.06 μm laser pulse of 40 J duration from 5 - 25 ns is used to form a line plasma 50 μm wide and up to 2 cm long. The gain is measured from the variation of signal amplitude with length.

Irradiation of a simple formvar foil target containing 15% oxygen with two beams of the Novette laser at relatively low power yielded population inversion and gain on the H_α transition of O VIII at 102.4 Å[41]. The gain with $\alpha\ell$ of about 0.5 over 1 cm was observed by varying the length of line irradiated from 0.27 to 1.27 cm on a foil 2300 Å thick in a focus 2 mm by 1.4 cm with a pulse of 500 J in 100 ps. This gain was ascribed to recombination following adiabatic cooling, and is probably associated with the enhancement of gain in low concentration dopants[42] in such systems.

Gain-length products of up to 5 on CVI H_α at 182 Å have been measured by the Hull group[43] during the expansion of thin carbon fibres of 4 μm diameter following heating by a 1.06 μm laser pulse focussed to a line 40 μm wide by 2 mm long. Since the underlying principles of all these approaches is similar we examine the principles and details this latter experiment more extensively.

The basic conditions necessary for generating inversion during recombination can be simply identified. The plasma must be first rapidly heated, and then cooled over timescales of the order of the characteristic ionisation time: for CVI ions at density of $\sim 10^{21}$ ions/cc this implies times of order 100 ps. These times are characteristic of the duration of mode-locked pulses from a Nd.glass laser, and of the expansion of plasma of appropriate temperature $\sim 10^{-3}$ cm in size. In practical terms these parameters closely match those obtained by heating carbon fibres of a few microns diameter with an appropriate laser pulse, a conclusion confirmed by detailed computer modelling[34,35]. In this system the inversion is achieved at sufficiently high density ($\sim 10^{19}$ ions/cc) to give an inversion density $\sim 10^{15}$/cc with gain coefficients ~ 10/cm, sufficient to measure in a laboratory experiment. A direct observation of gain at 182 Å in such a plasma formed by an expanding laser heated carbon fibre has been made[43,44].

In this experiment two spectrographs (Fig. 5) were used to simultaneously measure the spontaneous and amplified spontaneous emission from a 2mm length of heated carbon fibre. Signal amplifications of up to 30 times of the CVI H_α line along the fibre were measured and the inferred gain/length parameters are shown in Fig. 6. The strong variation with laser energy shown is in good agreement with computer modelling of the interaction. The need for extreme care in cross calibrating the spectrographs in any experiment of this type should be noted.

It was found that the measured plasma parameters were only weakly dependent on the fibre diameter, provided it was not too large ($\lesssim$ 6 μm). This was inferred to be due to an incomplete burn of the fibre, the actual mass participating in the gain generation being strongly dependent on the absorbed laser energy. This result has been confirmed in subsequent experiments, and the mass/energy relationship deduced for

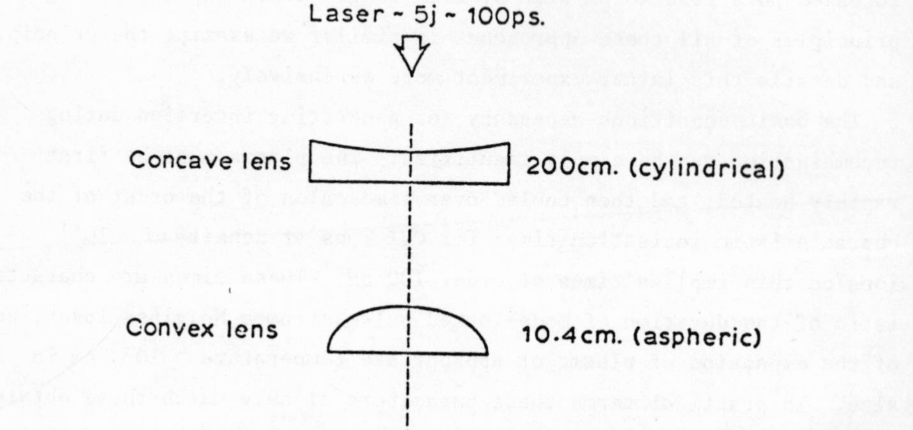

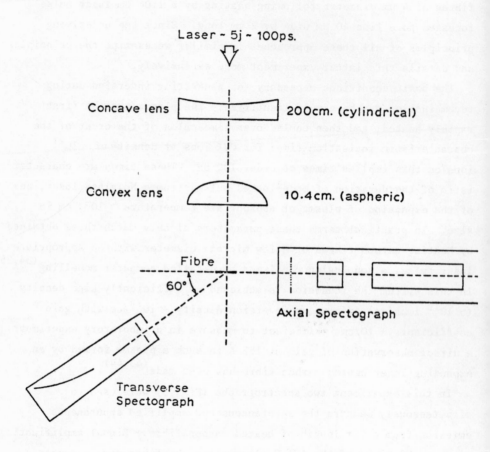

Fig. 5 Diagram of the apparatus used in experiments to measure gains from laser heated carbon fibres (from ref. 44).

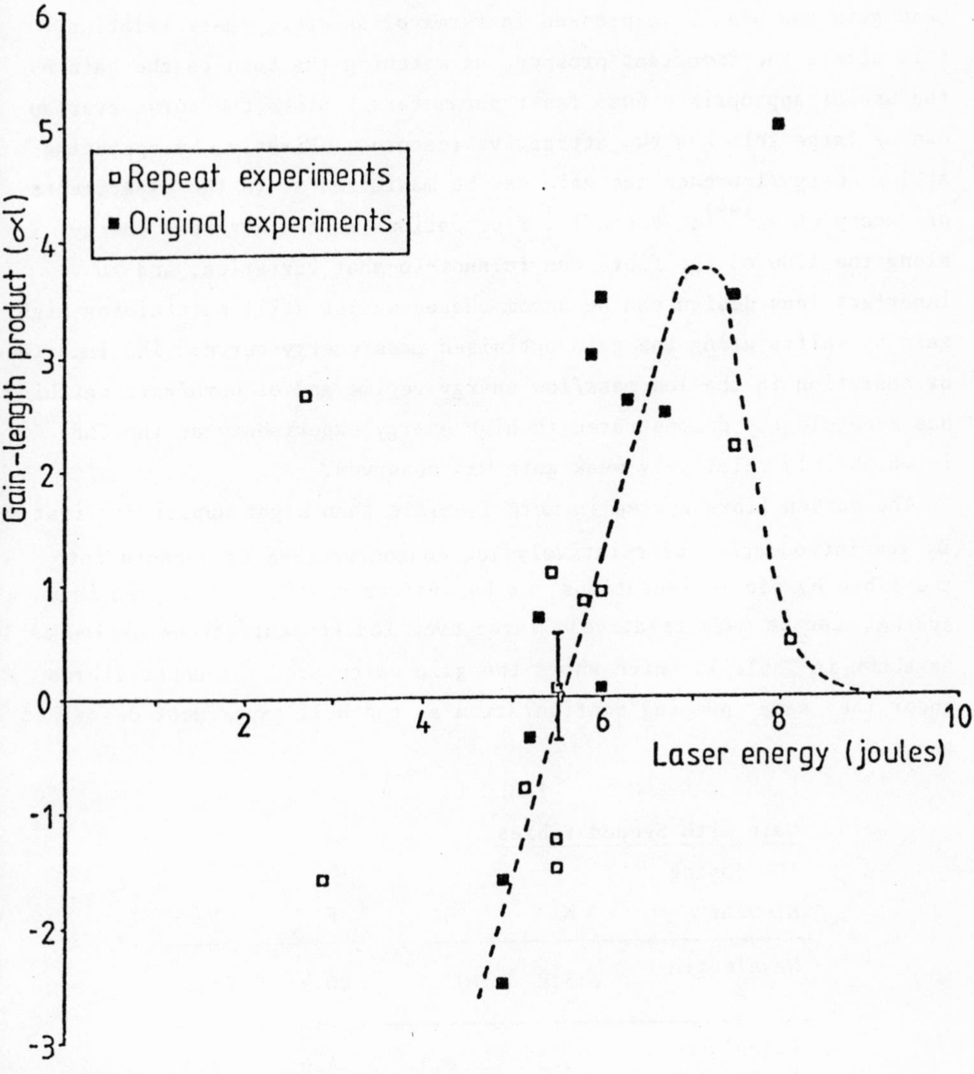

Fig. 6 Plot of the measured gain-length product (αl) at 182 Å from laser heated fibres as a function of the input laser energy[12].

a number of pump laser configurations[45]. Since the condition for peak gain can also be expressed in terms of an energy/mass relation, this offers the important prospect of matching the burn to the gain by the use of appropriate pump laser parameters. Since the curve overlap can be large this has two attractive features. Firstly, by operating at low energy/low mass the gain can be maximised as in the experiments of Jacoby et al[44]. Secondly, fluctuations in the absorbed energy along the line of the fibre due to shot-to-shot variation, and to imperfect lens design, can be accommodated whilst still maintaining high gain by shifts along the gain-optimised mass/energy curve. The importance of operation in the low mass/low energy regime and of burn/gain matching has been clearly demonstrated in high energy experiments at the CLF[45] in which only relatively weak gain was observed.

The carbon fibre system is more flexible than might appear at first sig By the introduction of relatively low concentrations of dopants into the fibre by ion implantations, or by surface coating, the gain in such systems can be made relatively large even for concentrations as low as 10 as shown in Table I, which shows the gain calculated for doped fibres under the same pumping configuration as the Hull experiment described above.

TABLE I

Gain with Seeded Fibres

10% doping

Element	N	O	F	Ne
Wavelength Å	133.8	102.4	80.9	65.5
Calculated Gain /cm	8.3	8.5	4.6	1.4

This approach has been confirmed by an experimental observation of gain on N VII, H_α at 133.8[42] in these same experiments with gain-length product of about 3. The large gain, comparable with that in the carbon base, arises despite the low concentration of the impurity as a result of the corresponding weakening of L_α opacity trapping whilst the electron density is maintained by the background carbon ions. As a result the gain is initiated earlier at higher density, and therefore correspondingly higher gain.

It is likely that significantly enhanced cooling can be provided by radiation to improve the fibre gain, and therefore the overall efficiency of the device. Low concentrations of elements of mass number about 30 (calcium, selenium) may be added without severe mass penalty yet substantially increase the radiation heat loss rate. Care must, however, be taken to ensure that the radiation does not pump the strong lower state decay in the lasant as this would destroy the inversion.

3.5 Scaling to Shorter Wavelengths

One of the attractive features of devices which use a lasing transition on which the principal quantum number changes is that wavelength decreases rapidly with the ionic charge Z allowing scaling along the iso-electronic sequence to generate shorter wavelengths. These scalings are easily developed for expansion dominated devices, whose behaviour is similar along the sequence.

It is readily shown that as a function of the laser frequency ν, the photon energy $E \sim \nu$, and the spontaneous emission probability $A \sim \nu^2$. Hence for similar expansion histories all individual energies must scale as ν and all times as ν^{-2}. If the ion mass is M, it follows that the expansion velocity $v \sim Z^{\frac{1}{2}}\nu^{\frac{1}{2}}M^{-\frac{1}{2}}$ and the width of the plasma $a \sim Z^{\frac{1}{2}}\nu^{-\frac{3}{2}}M^{-\frac{1}{2}}$. In addition the Doppler line width $\Delta\nu \sim \nu^{\frac{3}{2}}M^{-\frac{1}{2}}$. To proceed further it is necessary to know the scaling of density. This arises from a detailed study(34) of the atomic physics within the plasma from which it follows that for hydrogenic systems, the atomic rate coefficients scale in conformity with the above forms if the population density $\sim Z^{10}$ and electron density scales as Z^7. The initial ion density $n \sim Z^6$. Hence the gain coefficient $\alpha \sim Z^{10}\nu^{-\frac{3}{2}}M^{\frac{1}{2}}$.

Applying these results to hydrogenic systems where $\nu \sim Z^2$ and $M \sim Z$ we obtain

Gain scales	as	$Z^{7.5}$
Density	as	Z^6
Time	as	Z^{-4}
Duration	as	Z^{-3}
Energy	as	Z^2
Wavelength	as	Z^{-2}

The rapid scaling of density with wavelength introduces problems in proceeding to shorter wavelengths. The present design of carbon fibre systems contains a degree of pre-expansion whilst the laser heating and ionisation take place, the matching to the similar adiabat occuring at density of approximately 1/100 solid density. Using the scaling laws it follows that the expansion must start at solid density for aluminium systems, and must be pre-compressed for longer atomic number elements.

The requirements to operate at so near solid density without pre-expansion for $Z > 12$ introduces further constraints on the diameter of the target, in order that ionisation be completed before the expansion is initiated[46]. As a result it is necessary to operate on lower gain adiabats than for carbon, and to sacrifice some of the strong gain scaling for practical feasibility. Detailed numerical calculations[47] have shown that viable devices to generate gain at 38.7 Å can be designed using variants of aluminium foils as targets.

4. CONCLUSIONS

Several experiments observations of gain in the wavelength range 100-200 Å have now been made. It is generally accepted now that stimulated emission and population inversion can be generated in two ways, electron collisional excitation, and recombination. Several other approaches have been proposed, but none have yet shown experimenta realisation.

Although gain has now been convincingly demonstrated, practicable laser action has not yet been achieved. To do so gain length products of the order 20 or more are required, necessitating the present systems to be substantially increased in length. In principle no new technological constraints are involved, although improved optics and allignment techniques may have to be developed.

Present devices are extremely inefficient. As a result their application is likely to be very limited, and confined to a few large facilities. There is therefore need to discover more efficient schemes and to optimise existing ones. The development of scaling laws is therefore a matter of some urgency. Such relations have been identifie for the carbon fibre device: a reasonably optimised system compatible with the present pump configuration can be designed. The extension of the scalings to other recombination schemes will no doubt lead to further improvements in efficiency.

The wavelengths generated in present devices are too long for widespread application. The longest wavelengths likely to be widely used fall below the carbon edge at about 40 Å. Fortunately such wavelengths appear achievable with aluminium recombination devices. Such lasers must represent the goal of present research. Present-day devices demonstrate proof of principle, and establish that XUV and soft X-ray lasers can be designed, developed and constructed. In view of their initial history of insubstantial claims, unrealistic proposals and loss of confidence, the present position represents a notable technological step forward.

REFERENCES

1. R.W. Waynant and R.C. Elton, Proc. I.E.E.E., 64, 1059 (1976).
2. J.A.R. Samson, "Techniques of Ultra-violet Spectroscopy", Wiley, New York (1967).
3. T.W. Barbee jr., "Low Energy X-ray Diagnostics", Ed. D.T. Attwoo and B.L. Henke, Amer. Inst. Phys. Conf. Proc. No. 75, 131 (1981).
4. E. Spiller, "Laser Techniques in the Extreme Ultraviolet", Ed. S.E. Harris and T.B. Lucartorto, Amer. Inst. Phys. Conf. Proc. No. 119, 312 (1984).
5. J. Reintjes, R.C. Eckardt, C.Y. She, N.E. Karangelen, R.C. Elton and R.A. Andrews, Phys. Rev. Lett. 37, 1540 (1976), Appl. Phys. Lett., 30, 480 (1977).
6. M.H.R. Hutchinson, C.C. Ling and D.J. Bradley, Opt. Comm., 18, 203 (1976).
7. Y.L. Stankewich, Sov. Phys. Dok., 15, 356 (1970).
8. R.C. Elton, Appl. Opt. 14, 2243 (1975).
9. P.S. Axelrod, Phys. Rev. A, 13, 376 (1976), Phys. Rev. A, 15, 1132 (1977).
10. S.E. Harris, Opt. Lett. 5, 1 (1980).
11. C.J. Humphreys, Elec. Micros. 4, 68 (1980).
12. R.G. Caro, J.C. Wang, J.F. Young and S.E. Harris, "Laser Techniques in the Extreme Ultraviolet", Ed. S.E. Harris (1984) an T.B. Lucatorto, Amer. Inst. Phys. Conf. Proc., No.119, 417.
13. R.L. Swent, R.H. Pantell, M.J. Alguart, B.L. Bennan, S.D. Bloom and S. Datz, Phys. Rev. Lett., 43, 1723 (1980).
14. L.J. Palumbo and R.C. Elton, J. Opt. Soc. Am. 7, 32 (1977).
15. Quant. Elec. 5, 59 (1977), Sov. Phys. J. Quant. Elec. 7, 32.
16. A.V. Vinogradov and A.N. Shlyptsev, Sov. Phys. J. Quant. Elect. 10, 754 (1980).
17. P.L. Hagelstein, Plasma Physics, 25, 1345 (1983).
18. A.A. Ilyukhin, G.V. Peregudov, E.N. Ragozin, I.I. Sobelman and V.A. Chirkov, Sov. Phys. J.E.T.P., Lett., 12, 535 (1978).
19. D.L. Matthews, P.L. Hagelstein, M.D. Rosen, M.J. Eckart, N.M. Ceglio, A.W. Hazi, H. Medecki, B.J. MacGowan, J.E. Trebes, B.L. Whitten, E.M. Campbell, C.W. Hatcher, A.M. Hawryluk, R.L. Kauffmar

L.D. Pleasance, G. Raunbach, J.H. Scofield, G. Stone and T.A. Weaver, Phys. Rev. Lett., 54, 110 (1985).

20. M.D. Rosen, P.L. Hagelstein, D.L. Matthews, E.M. Campbell, A.W. Hazi, B.L. Whitten, B. MacGowan, R.E. Turner, R.W. Lee, G. Charatis, G.E. Busch, C.L. Shepard, P.D. Rockett and R.R. Johnson, Phys. Rev. Lett., 54, 106 (1985).
21. A.V. Vinogradov, I.I. Sobelman and E.A. Yukov, Sov. Phys. J. (1975).
22. B.A. Norton and N.J. Peacock, J. Phys. B, 8, 989 (1977).
23. W.E. Alley and G. Chapline, J.Q.R.S.T., 27, 257 (1982).
24. J.P. Apruzese, J. Davis and K.G. Whitney, J. Appl. Phys., 53, 4020 (1982).
25. D. Matthews, P. Hagelstein, M. Rosen, R.L. Kauffman, R. Lee, C. Wang, H. Medecki, M. Campbell, N. Ceglio, G. Leipelt, P. Lee, P. Drake, L. Pleasance, L. Seppla, W. Hatcher, G. Rambach, A. Hawryluk, G. Heaton, R. Price, R. Ozarski, R. Speck, K. Manes, J. Underwood, A. Toor, T. Weaver, L. Coleman, B. DeMartini, J. Auerbach, R. Turner, W. Zagotta, C. Hailey, M. Eckart, A. Hazi, B. Whitten, G. Charatis, P. Rockett, G. Busch, D. Sullivan, C. Shepard, R. Johnson, B. MacGowan and P. Burkhalter, "Laser Techniques in the Extreme Ultraviolet", Ed. S.E. Harris and T.B. Lucartorto, Amer. Inst. Phys. Conf. Proc., No.119, 25 (1984).
26. R.C. Elton, R.H. Dixon and T.N. Lee, "Laser Techniques in the Extreme Ultraviolet", Ed. S.E. Harris and T.B. Lucartorto, Amer. Inst. Phys. Conf. Proc., No.119, 489 (1984).
27. J.G. Lunney, R.E. Corbett, M.J. Lamb, C.L.S. Lewis, P. McCavana, L.D. Shorrock, S.J. Rose and F. Pinzong, Opt. Comm., 50, 367 (1984).
28. J.F. Seely and W.B. McKnight, J. Appl. Phys., 48, 3691 (1977).
29. A.V. Vinogradov and I.I. Sovelman, Sov. Phys. J.E.T.P., 36, 1115 (1973).
30. R.J. Dixon and R.C. Elton, Phys. Rev. Lett., 38, 1072 (1977).
31. R.J. Dixon, J.F. Seely and R.C. Elton, Phys. Rev. Lett., 40, 122 (1978).
32. W.T. Silfvast and O.R. Wood, III, J. de Physique, 41, C 9, 439 (1980).
33. R.J. Dewhurst, D. Jacoby, G.J. Pert and S.A. Ramsden, Phys. Rev. Lett., 37, 1265 (1976).
34. G.J. Pert, J. Phys. B, 9, 3301 (1976).

35. G.J. Pert, J. Phys. B., 12, 2067 (1979).
36. S. Suckewer and H. Fishman, J. Appl. Phys., 51, 1922 (1980).
37. S. Suckewer, C.H. Skinner, H. Milchberg, C. Kune and D. Voorhus, Phys. Rev. Lett., 55, 1753 (1985).
38. J.F. Seely, C.M. Brown, W. Feldman, M. Richardson, B. Yaakobi and W.E. Behring, Opt. Comm., 54, 289 (1985).
39. R.C. Elton, J.F. Seely and R.H. Dixon, "Laser Techniques for Extreme Ultraviolet Spectroscopy", Ed. T.J. McIlrath and R.R. Freeman, Amer. Inst. Phys. Conf. Proc., No.90, 277 (1982).
40. P. Jaeglé, G. Jamelot, A. Carillon, A. Khisnick, A. Sureau, H. Guennou, "Laser Techniques in the Extreme Ultraviolet", Ed. S.E. Harris and T.B. Lucartorto, Amer. Inst. Phys. Conf. Proc., No.119, 468 (1984).
41. D.L. Matthews, E.M. Campbell, K. Estabrook, W. Hatcher, R.L. Kauffman, R.W. Lee and C.I. Wang, Appl. Phys. Lett., 45, 226 (1984)
42. G.J. Pert and L.D. Shorrock, "Quantum Electronics and Electro-optics", Ed. P.L. Knight, J. Wiley and Sons (Chichester), 269 (198
43. D. Jacoby, G.J. Pert, S.A. Ramsden, L.D. Shorrock and G.J. Tallents Opt. Comm., 37, 193 (1981).
44. D. Jacoby, G.J. Pert, L.D. Shorrock and G.J. Tallents, J. Phys. B, 15, 3557 (1982).
45. G.J. Pert, L.D. Shorrock, G.J. Tallents, R. Corbett, M.J. Lamb, C.L.S. Lewis, E. Mahoney, R.B. Eason, C. Hooker and M.H. Key, "Laser Techniques in the Extreme Ultraviolet", Ed. S.E. Harris and T.B. Lucartorto, Amer. Inst. Phys. Conf. Proc., No.119, 480 (1984).
46. G.J. Pert, Plasma Physics: in press.

DIAGNOSTICS OF LASER PRODUCED PLASMAS

T.A. Hall
Department of Physics
University of Essex, Colchester CO4 3SQ, U.K.

1.
INTRODUCTION

1.1 Types of diagnostic and parameter range

Diagnostics are concerned with finding out what is happening inside a plasma. To do this we can use either electromagnetic waves or particles. These waves or particles can originate inside the plasma itself, or from a subsidiary source and the absorption, reflection, scattering or energy loss, is measured as they pass through the plasma.

In a simple laser produced plasma experiment - take for example a spherical microballoon implosion - the range of parameters is very large indeed and no single diagnostic is going to be useful in all regions. Densities and temperatures range from 10^{18} cm^{-3} and a few eV up to 10^{26} cm^{-3} and many keV.

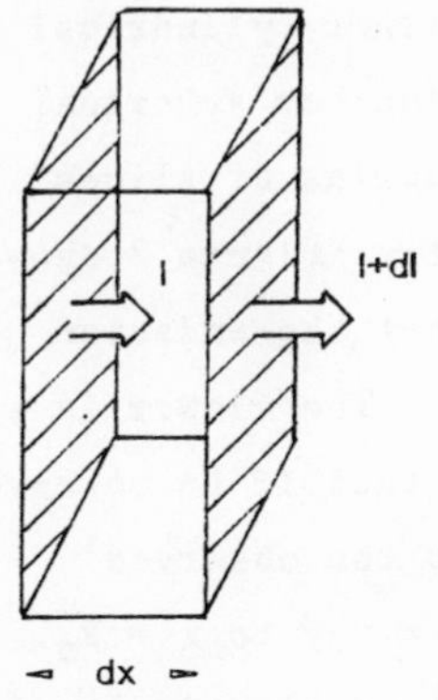

Fig. 1.

In the coronal plasma wave diagnostics can use visible or near visible EM waves where their frequency is greater than the plasma frequency, whereas only much shorter wavelengths, in the X-ray region, can escape from the dense plasma core. Radiation emitted from the dense core, even if it is at a higher frequency than the plasma frequency ($\lambda < \lambda_{crit}$), may still be strongly absorbed before it can leave the plasma.

1.2 Emission and absorption of radiation

In general, in any given region of plasma, a wave of frequency, ω, and irradiance, I_ν, will be attenuated by absorption and enhanced by emission. Figure 1 represents a slab of plasma with radiation incident from the left.

In passing through this <u>thin</u> slab of thickness, dx, the irradiance will be enhanced by $j_\nu(x)dx$ (emission) and reduced by $k_\nu(x)I_\nu(x)dx$ (absorption). Thus

$$\frac{dI_\nu(x)}{dx} = j_\nu(x) - k_\nu(x)I_\nu(x) \qquad (1.1)$$

where j_ν and k_ν are called emission and absorption coefficients, respectively. Equation 1.1 can readily be solved in three special cases: these are

(a) when there is negligible absorption ($k_\nu = 0$)

or (b) when there is negligible emission ($j_\nu = 0$)

or (c) when the plasma is uniform, ie. when k_ν and j_ν do not depend upon x.

This latter case will be dealt with further in section 4.1.3c. The analysis in either of the first two cases is very similar and, by way of an example, we will consider case (a) further.

1.3 <u>Abel inversion</u>

Most laser produced plasma experiments have either cylindrical symmetry (as in the case of single beam illumination) or spherical symmetry. In this latter case we can consider a series of slices through the sphere, each having cylindrical symmetry. Figure 2 shows such a slice of outer radius R. Let us suppose that the emission coefficient j_ν is only a function of the radius r. The viewer is some distance away in the x direction and he will note that if he observes the plasma section at a height y, contributions to the observed intensity will come from parts of the plasma from r = y to $r = R_0$. Thus when y = 0 one observes the sum of contributions of emission from all points across the diameter and when $y \simeq R_0$ only outer sections of the slab will contribute.

The irradiance at a height y can thus be written

$$I_\nu(y) = 2\int_{x=0}^{x=x_0} j_\nu(r)dx = 2\int_y^R \frac{rj_\nu(r)dr}{(r^2-y^2)^{\frac{1}{2}}} \qquad (1.2)$$

This is one form of Abel's integral equation. If $I_\nu(y)$ is known as a continuous function, then this equation can be inverted

$$j_\nu(r) = -\frac{1}{\pi}\int_r^R \frac{I'_\nu(y)dy}{(y^2-r^2)^{\frac{1}{2}}} \qquad (1.3)$$

where

$$I'_\nu(y) = \frac{dI_\nu(y)}{dy}$$

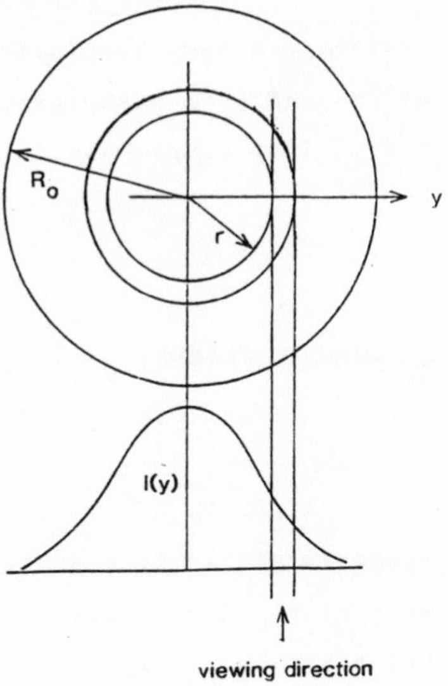

Fig. 2. Schematic diagram to illustrate Abel inversion.

More generally in laser produced plasmas, the value of I_ν is known at various values of y, y_n, and we wish to find j_ν at various values of r, r_m. Equation 1.2 can now be expressed as a sum

$$I_\nu(y_n) = \sum \alpha_{mn} j_\nu(r_m)$$

this can be inverted to give

$$j_\nu(r_m) = \sum \beta_{mn} I_\nu(y_n) \qquad (1.4)$$

A straightforward evaluation of the coefficients β_{mn} results in an inversion which is unduly dependent on the accurate measurement of the outer values of $I_\nu(y_n)$. These are the values that are often least well known. By weighting the values of $I_\nu(y_n)$ various other values of β_{mn} can be found. One of the earliest methods was due to Nestor and Olsen(1), but a preferred method is due to Bockasten(2), which is more accurate and direct. If the data is noisy, then a method of Deutsch and Beniaminy(3) is reported to give big improvements over Bockasten's method.

1.4 Time resolution

In most laser produced plasma experiments carried out recently, the laser pulse length is typically a few nanoseconds, down to a few tens of picoseconds, and many of the plasma processes change on a time scale faster than this. Quite often a particular effect can be effectively 'frozen' by making use of the behaviour of the plasma, but perhaps more often an ultra-fast camera is needed. The image converter streak camera has found wide use in laser produced plasma experiments and different versions of it are sensitive to a wide spectral range, from the near infrared to the X-ray region.

Inverse sweep speeds of down to ~ 30 psec/mm and time resolutions of ~ 5 psec are possible with these cameras(4). Various examples of streak records will be given during the rest of this course and in the other courses.

2.
RADIATIVE DIAGNOSTICS OF THE CORONAL PLASMA

2.1 Range of diagnostics

In this section we will consider measurements of electron density based on plasma refractivity and measurements of electron density, temperature and various other parameters which can be inferred by studying the light reflected or scattered back from the plasma at, or around, the fundamental or harmonics of the incident laser light.

2.2 Refractive index of a plasma

The dispersion relation for an infinite EM wave passing through a uniform plasma is can be written in the form

$$\mu = \frac{c}{v_p} = \frac{ck}{\omega} = \left(1 - \frac{\omega_p^2}{\omega^2}\right)^{\frac{1}{2}} = \left(1 - \frac{n_e}{n_c}\right)^{\frac{1}{2}} \tag{2.1}$$

where ω and k are the angular frequency and wavevectors of the wave respectively. The quantity ω/k is the phase velocity of the wave, v_p, and μ is the refractive index of the plasma and n_e and n_c are respectively the electron density and critical electron density for a wave of frequency ω. If we have a plasma of length ℓ, the change in optical path of any probe beam due to the plasma will be $\mu\ell$ and the corresponding change in phase will be $k\mu\ell$.

In practice the plasma will not be uniform and μ will be a function of position, ℓ, through the plasma. The phase change, δ, will now be given by

$$\delta = k \int \mu(\ell) d\ell \tag{2.2}$$

This phase change can be measured using an <u>interferometer</u> but two other techniques are available to observe changes in μ. The <u>Schlieren</u> effect results from a deflection of light and depends upon $d\mu/dy$ rather than μ. <u>shadowgraphy</u> relies upon $d^2\mu/dy^2$. Of these three techniques, interferometry will more easily yield quantitative results but if $d\mu/dy$ and $d^2\mu/dy^2$ become large then interferograms can be difficult to interpret.

In all of these techniques the plasma motion is 'frozen' by using a very short laser pulse for the interferometer. This pulse is often only a few tens of picoseconds long and may be a part of the main laser pulse. In order to penetrate to dense regions of the plasma, the probe beam is often frequency doubled or quadrupled.

2.3 <u>Interferometry</u>

The most commonly used interferometer in laser produced plasmas is due to Nomarski[(5)] which was modified by Benattar et al[(6)]. It has a compact design, (Figure 3) is insensitive to small movements and

Fig. 3. Schematic diagram of Normarski interferometer after Benattar[(6)].

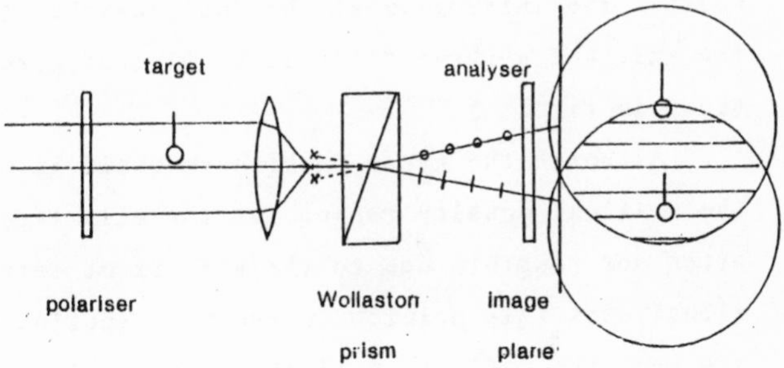

suitable to the geometrical constraints of laser produced plasmas. The incident probe laser light is polarised at 45° to the vertical and is collimated onto the position of the target ensuring that a much larger region is illuminated than is occupied by the plasma. The target is imaged onto the camera by the lens. Two overlapping images are produced on the camera by the Wollaston prism. These do not normally interfere with each other because they are of orthogonal polarisations. However, the second polariser set at 45° to the vertical recombines the two beams and allows interference between them. Figure 4 shows an interferogram(7) taken using this type of interferometer. This interferogram was obtained using frequency doubled probe of a neodymium laser. The plasma resulted from the

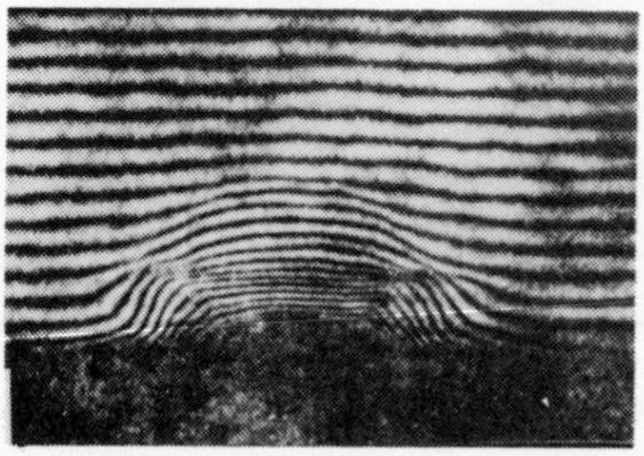

Fig. 4. Interferogram taken with .53 μ light of plasma produced by double laser pulse(7). The laser light is incident from above.

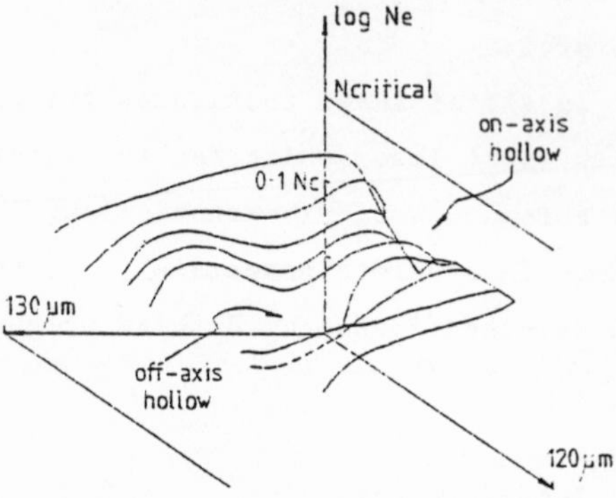

Fig. 5. Three dimensional plot of electron density showing dip in density on axis.

illumination of a plane solid target by two pulses a few nanoseconds apart. The inferogram can be Abel inverted for various positions along the axis and a three dimensional plot of electron density obtained, as shown in figure 5.

Although the green probe beam could in principle propagate beyond the critical density region for the main laser pulse (1.06μ) this is often not possible due to the effects of refraction. Figure 8 illustrates this point. In the next section on the Schlieren effect we see that the deflection of the beam, θ, is given approximately by

$$\theta \sim \frac{\ell}{2n_c} \frac{dn_e}{dz}$$

If θ is large then the light will not be collected by the optics of the system and no fringe would be observable. Thus the highest density that can be measured depends on the aperture of the collecting optics, the probe wavelength (n_c is the probe critical density), the path length through the plasma and density gradient.

2.4 The Schlieren effect

In this technique we use the fact that in a density gradient the probe beam will be deflected. Figure 6 represents a Schlieren system.

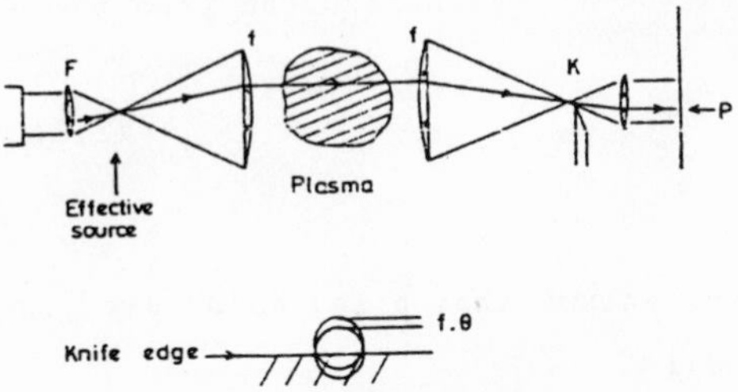

Fig. 6. Schematic of Schieren system[8].

Fig. 7. Schlieren image of gold coated microballoon

Light (usually from a laser) is focussed to a point and subsequently collimated by a second lens. A third lens, similar to the second brings the light to a second focus. In the absence of the plasma, a knife edge is brought up to the second focal point so that it just obscures the beam. A fourth lens is arranged to produce an image of the plasma on the screen but, with the knife edge in position, a completely dark field is observed. If an object (in our case a plasma) deflects the light upwards it will pass by the knife edge and will produce a bright area on the screen which will correspond to the image of the region where the ray was deflected from. Figure 7 shows a Schleiren[9] **photograph of a microballoon at an irradiance of 2 x 10^{13} Wcm^{-2}, 0.5 nsec after the 1.5 nsec laser pulse has finished. The bright regions occur where the rays have been deflected past the knife edge. This photograph shows very clearly the striations in the coronal plasma called jets. It is not difficult to show that**

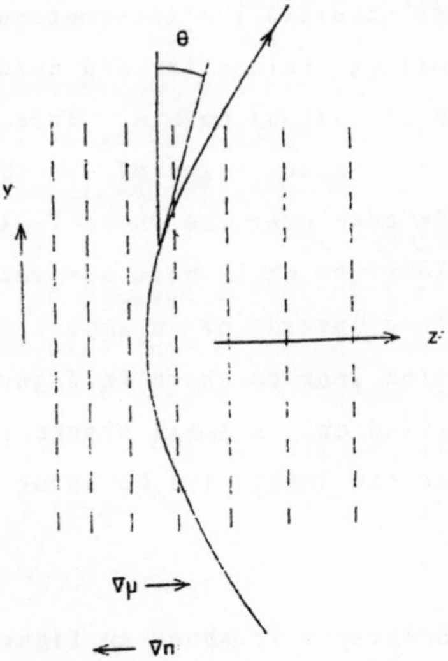

Fig.8. Deviation of a ray in a density gradient.

for a ray travelling at an angle θ to a refractive index gradient that

$$\frac{d\theta}{dy} = \frac{1}{\mu}\frac{d\mu}{dz}$$

where the angle and directions refer to figure 8. Thus after passing through a length, ℓ, of plasma

$$\theta = \int_0^{\ell} \frac{1}{\mu}\frac{d\mu}{dz} \cdot dy \qquad (2.3)$$

In order to obtain estimates we may assume that μ and dμ/dz are constant with y and hence, using 2.1

$$\theta = \frac{\ell}{\mu}\frac{d\mu}{dz} = -\frac{\ell}{2n_c}\frac{dn}{dz} \qquad (2.4)$$

To be observable, θ (or a component of θ) must be in a direction normal to the knife edge. Thus, the images observed represent density gradients in this direction.

It is quite difficult to get accurate quantitative information from Schlieren images but we can get approximate values for ℓ/μ dμ/dz by considering the brightness of the image at various points. This can be estimated by considering a medium with a constant value of ℓ/μ dμ/dz in the direction perpendicular to the knife edge over the whole field. This medium will act like a prism and deflect the whole beam upwards. Figure 6b shows how the beam focus is shifted upwards by an amount fθ where f is the focal length of the collecting lens as shown in figure 6a. For optimum sensitivity $f\theta \simeq d$. If $f\theta \ll d$ only a small change in brightness will occur whereas if $f\theta \gg d$ then the image will be saturated.

2.5 Shadowgraphy

The experimental arrangement for shadowgraphy is shown in figure 9 and is extremely simple[8]. The collecting optics and knife edge of the Schlieren system are omitted. Changes in the brightness of the shadow image depend now on variations of dμ/dz i.e. upon $d^2\mu/dz^2$ (in laser produced plasmas there will almost always be a region where the

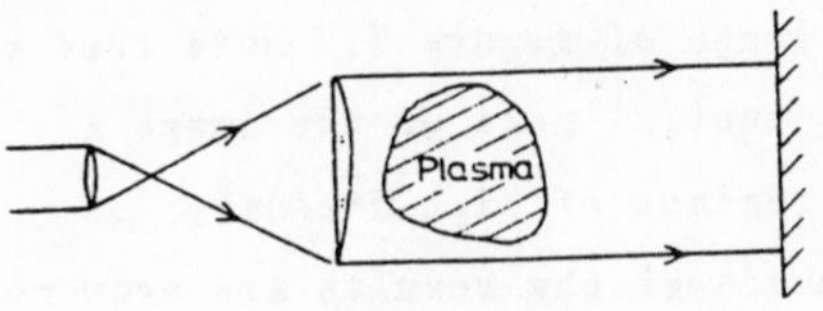

Fig. 9. The shadowgraph technique[8].

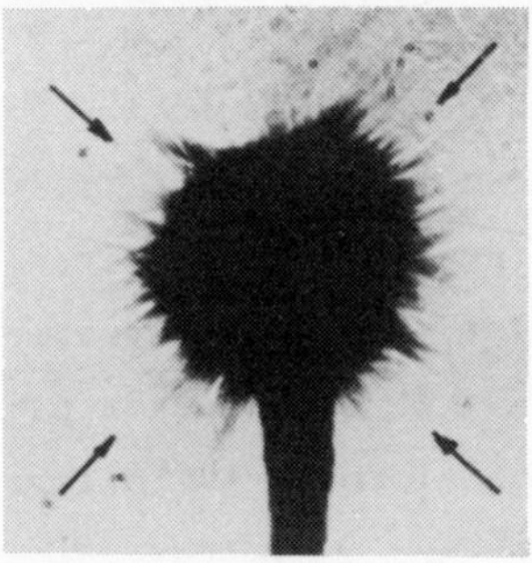

Fig. 11. Shadowgraph of a gold coated microballoon similar to Fig. 7[10].

light is not transmited – for $\omega_p > \omega$ – these will appear dark on shadowgraph). Figure 10 illustrates how a non zero $d^2\mu/dz^2$ can lead to a focussing of the rays. From figure 10b) and (2.4) we see that

$$\theta = \frac{\ell}{\mu_1}\frac{d\mu_1}{dz} \quad \text{and}$$

$$\theta + \Delta\theta = \frac{\ell}{\mu_2}\left(\frac{d\mu_1}{dz} + \frac{d^2\mu}{dz^2} . \Delta z\right)$$

If we assume $\mu_1 \sim \mu_2 \sim \mu$, then

$$\Delta\theta \simeq \frac{\ell}{\mu} \quad \frac{d^2\mu}{dz^2} . \Delta z$$

This leads to a focussing of the rays with a focal length, f, where

$$f = \frac{\Delta z}{\Delta\theta} = \frac{\mu}{\ell . \frac{d^2\mu}{dz^2}}$$

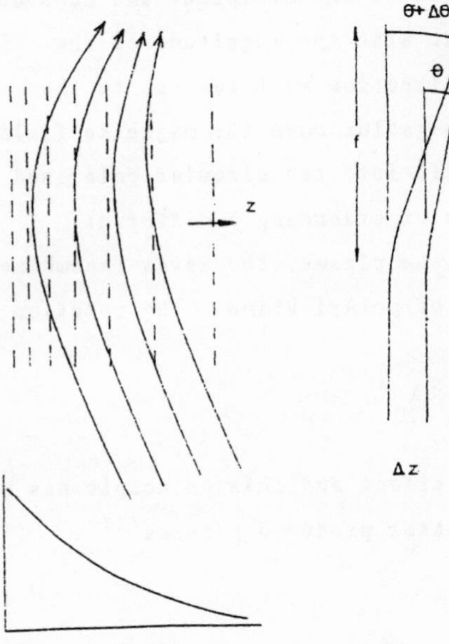

Fig. 10. Focussing of rays in a plasma with finite $d^2\mu/dz^2$.

Thus the focussing power of plasma depends upon the second derivative

of the refractive index. Figure 11 shows a shadowgram[10] under similar conditions to the Schleiren image of figure 7. Note that now the image has a bright field and the central part of the image is opaque and hard to distinguish from regions of high $d^2\mu/dz^2$.

Quantitative interpretation of shadowgraphy results are even more difficult than Schlieren pictures and are only used to indicate the presence of small scale irregularities.

2.6 Magnetic field measurements

In the presence of a magnetic field the simple formula for the refractive index given in equation 2.1a no longer holds and a wave can propagate through the plasma with two orthogonal polarisations. The refractive index for each polarisation is now different and dependent, not only on the electron density, but also the magnitude of the magnetic field and the propagation direction with respect to the magnetic field. In the case of propagation down the magnetic field, an incident plane polarised wave is split into two circular polarised waves with opposite senses, each one experiencing a different refractive index. On emerging from the plasma, the waves recombine to produce a net rotation of the plane of polarisation. The rotation θ is given by

$$\theta = 2.6 \times 10^{-17} \int_o^L \lambda_p^2 n_e B dx.$$

This effect is known as the Faraday effect and this principle has been used to measure magnetic fields in laser produced plasmas[11].

2.7 Harmonic emissions

Intense light emissions from the coronal plasma have been observed in spectral regions at and around the incident laser frequency and its harmonics and sub-harmonics. In certain circumstances these emissions can be used to gain information on the state of the plasma. A description of these emissions is given elsewhere[12] and is beyond the scope of this course. We will, however, look at some aspects of these emissions which will yield diagnostic information.

2.7.1 Emission around the incident laser frequency

If we observe the light reflected or scattered back through the focussing lens of a laser produced plasma and spectrally disperse it, we find that the spectrum is shifted, broadened and often split into several components. Motion of the critical density surface and motion of plasma through the surface can both result in a Doppler shift of the reflected light. Motion of the critical surface towards the observer results in a blue shift and vice versa. A second process is due to Brillouin scattering. In this process an incident photon is converted into a scattered photon and a phonon. The matching conditions for frequency (energy) and wave vector (momentum) require that the scattered photon is directed in the opposite direction to the incident photon (i.e. backscattered) and down shifted by the frequency of the ion wave. These coupling conditions allow this interaction to occur over a wide range of densities and phonon frequencies. The backscattered spectrum from Brillouin scattering is thus generally red shifted and quite broad.

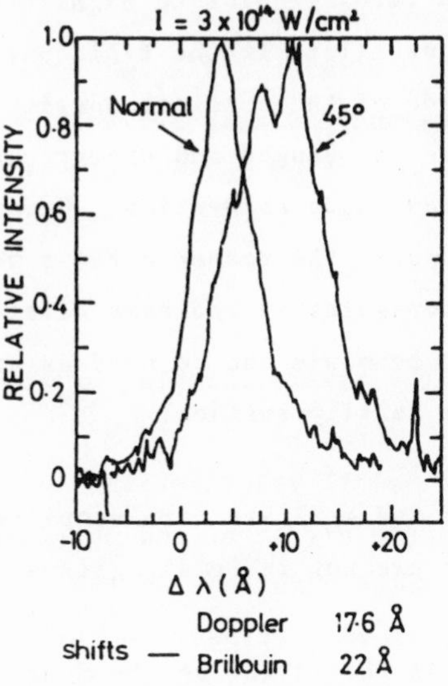

Fig. 12. Brillouin backscatter spectrum at normal and 45° degree incidence.

In general both of these processes occur simultaneously but it is possible to separate the effects if the targets are tilted to the incident laser beam. Figure 12 shows some results(13) which show how the spectrum varies with angle and from this the shifts implied from Doppler and Brillouin are shown. The interpretation of Brillouin spectra is difficult and in general not very rewarding from a diagnostic point of view but the Doppler shift gives an indication of the motion of the critical density surface which can be helpful in determining hydrodynamic flow.

2.7.2 Emission at the second harmonic

Second harmonic emission is usually associated with the process of resonance absorption(14). Resonance absorption occurs when p-polarised light is incident obliquely on a plasma. The light is reflected by the refractive index gradient at densities lower than the critical density and the wave field is evanescent from the turning point to higher densities. In this region the electrons oscillate in the field but are unable to gain energy from it. Either side of the critical density surface the phase of the electron response is changed and energy can be extracted from the wave field. This is resonance absorption. Because the phase response on each half cycle changes, the response has a very strong second harmonic component which propagates in the same direction as the refracted fundamental. The second harmonic can be used as an indicator of the position of the critical density surface.

2.7.3 Emission at $3\omega_L/2$ and $\omega_L/2$

The mechanics for the generation of $3\omega_L/2$ and $\omega_L/2$ are more complicated and there are aspects of the theory which are not in total agreement with the experimental results.

The incident laser photon moves towards the target up the density gradient. At the density at which the local plasma frequency is half

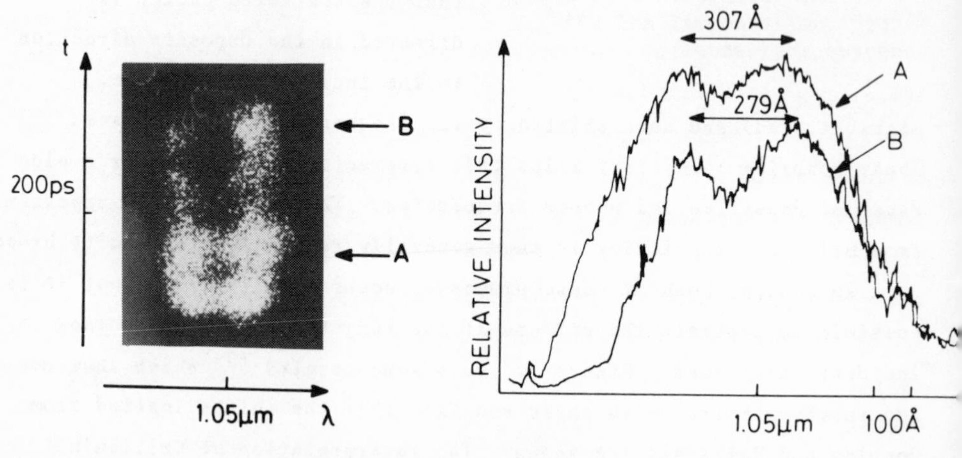

Fig. 13. Spectrum of $\omega_L/2$ emission from a planar target(15).

the wave frequency (at $n = n_c/4$) the incident photon can decay by one of two processes: Raman scattering or the two plasmon decay. In Raman scattering the incident photon of frequency, ω_L, produces a photon and plasmon each of frequency near $\omega_L/2$. The photon will be backscattered and the plasmon forward scattered. In the two plasmon decay instability the incident photon decays to two plasmons one of which propagates in the forwards direction and one in the backwards. The plasmons produced from either process may be reflected from the $n_c/4$ density (if they are travelling inwards), and they may combine with an incident or backscattered fundamental frequency photon to produce a $3\omega_L/2$ photon which, if it is travelling inwards can be reflected from the $9n_c/4$ layer: the number of possibilities is very large.

Experimentally it is often observed that the backscattered $3\omega_L/2$ (and the $\omega/2$) consists of two components, one shifted each side of the exact value of $3\omega_L/2$. Figure 13 shows the spectrum of $\omega_L/2$ emission(15). The emission is time resolved using a streak camera with time moving vertically and wavelength horizontally. In these experiments the laser wavelength was 0.53 μm and the incident energy was 16J in a 800 psec pulse.

As we have seen above, the number of possible wave interactions is large and the interpretation of the data is open to some doubt. Theories of Avrov(16), Barr(17) and Karttunen(18) predict a double peaked structure for the $3\omega_L/2$ emission. In all three cases the separation of the peaks is proportional to the local electron temperature although the constant of proportionality is different. Thus, a measurement of the separation of the peaks of the $3\omega_L/2$ emission could serve as a temperature diagnostic. Unfortunately the temperatures predicted by the theories are much higher (or very much higher - depending on which theory!) than has been measured by other techniques. The reason for this has not been fully explained as yet.

Since the $3\omega_L/2$ emission originates from the $n_c/4$ layer the spatial emission of the radiation can be used to plot the position of this region(19).

3.

PARTICLE DIAGNOSTICS FROM CORONAL PLASMAS

3.1 Plasma expansion

The plasma created at the ablation front moves outwards towards the laser, being heated as it moves. At, or just below the critical density layer the flow velocity is around Mach 1, i.e. just supersonic. Further heating of the plasma will occur around the critical region due to absorption of the laser light but this becomes less effective as the density falls. As the plasma expands further, thermal energy is converted into flow energy and the temperature falls and the Mach number increases, i.e. the plasma becomes strongly supersonic. At some point in the expansion, collisions cease to become effective and the expansion is controlled by internally generated electric fields. These electric fields arise because the electrons initially have a much higher velocity and attempt to expand ahead of the ions. This sets up an electric field in the plasma which holds back the electrons and accelerates the ions. The situation is complicated by the energy distribution of the electrons. In addition to a group of thermal electrons there are a number of very high energy electrons which are created in the laser light absorption region by processes such as resonance absorption and Raman scattering [9].

A considerable amount of theoretical work has been done on this expansion with only limited success (the reader is referred to the review by Sigel[20] for a fuller account).

3.2 Time of flight measurements

The simplest form of ion collector is Faraday cup (usually, this has a grid across the entrance, as shown in Figure 14). Also shown in Figure 14 is the biassing arrangement. The negative voltage applied to the cup ensures that only ions are collected. In order to ensure that any secondary electrons liberated by ion impact are recollected, the Faraday cup is often constructed with longitudinal honeycomb interior. The type of result obtained with such a probe is shown in Figure 15. The form of the results can change dramatically with only marginal changes in the

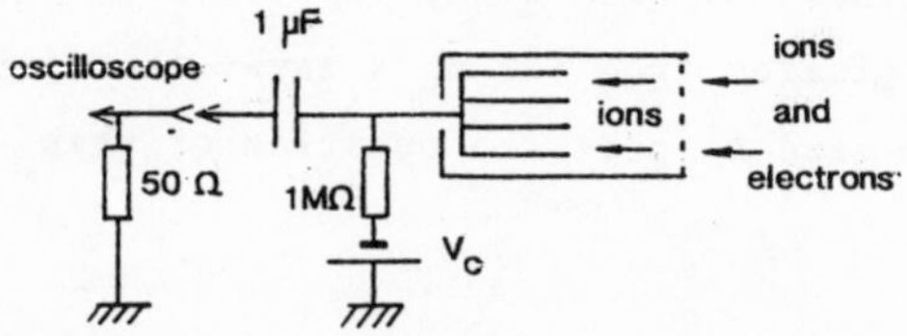

Fig. 14. Faraday cup charge collector and biassing arrangement.

plasma conditions.

3.2.1 Plasmas produced by short laser pulses

For plasmas produced by short laser pulses (< 200 ps) the analytic model of Wickens[22] has had some success. In this model, a one dimensional plasma expansion is driven by electrons with a two temperature distribution for the duration of the laser pulse and it is then assumed that the ions drift to the walls with a constant velocity. The velocity spectrum is found to be of the form

$$\frac{dN}{dv} = K.\ \frac{(n_h + n_c)}{Z}\left[1 - \tfrac{1}{2}\ .\ \frac{n_h n_c}{(n_h/T_h + n_c/T_c)^2}\ .\ \left(\frac{1}{T_c} - \frac{1}{T_h} \right)^2 \right] \qquad (3.1)$$

where n_h n_c T_h and T_c are density and temperature of the hot and cold components of the plasma. n_h and n_c vary through the plasma, but are described by the appropriate Boltzmann density distribution written in terms of the plasma potential .

$$n_h = n_{ho} \exp\ (e\phi/kT_h)$$

and

$$n_c = n_{co} \exp\ (e\phi/kT_c)$$

where n_{ho} and n_{co} are the densities of the hot and cold components of the source plasma. Figure 16 shows an experimental result similar to that shown in figure 15, but plotted semi-logarithmicaly in the form of dN/dv. Also plotted on figure 16 are the model results using 3.1 above for various values of n_{co}/n_{ho} and T_h/T_c.

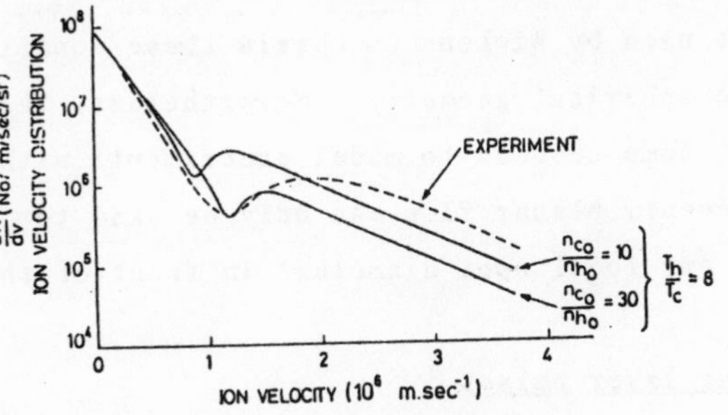

Fig. 16. Experimental and theoretical charge collector velocity spectra.

The dip in the experimental curve which is often observed is qualitatively described by Wickens' model. However, detailed agreement is rarely obtained. Nevertheless, an important observation can be made from the model: for low velocity ions

$$\frac{dN}{dv} \propto \exp - (v/c_c)$$

and for high velocity ions

$$\frac{dN}{dv} \propto \exp - (v/c_h)$$

where

$$c_c = \left(\frac{ZkT_c}{m}\right)^{\frac{1}{2}} \quad \text{and} \quad c_h = \left(\frac{ZkT_h}{m}\right)^{\frac{1}{2}}$$

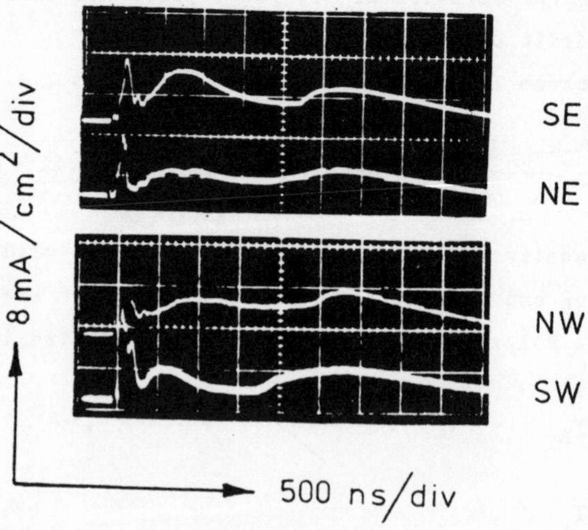

Fig. 15. Four ion collector signals from glass microsphere implosion(21)

Thus the slopes of ln(dN/dv) at low and high velocities can yield values of T_c and T_h.

It is interesting to note that as T_h/T_c increases towards a value of $5 + \sqrt{24}$ (~ 9.9) the velocity in the above equations becomes multiple valued. This is not physically acceptable and is usually attributed to the formation of a rarefaction shock.

It may be that the dramatic changes in probe signals that are observed for very small changes in the laser condition arise because of the occasional formation of such a shock. This has not been verified.

The similarity techniques used by Wickens to obtain these equations cannot strictly be applied to spherical geometry. Nevertheless, the equations have been used with some success to model experiments with spherical symmetry. In any event, planar flow can only be said to exist for short distances (~ one focal spot diameter) in front of the target.

3.2.2 Plasmas produced by long laser pulses

As the laser pulse length is increased the ion spectrum observed by a Faraday cup can change markedly, as shown in figure 17. This change is believed to arise because of a transition to a quasi-steady state flow condition, as opposed to the impulse assumed by Wickens.

For a stationary flow situation the equations of continuity, momentum and energy can be solved in simplified form[24] to yield solutions similar to those shown schematically in figure 18. In this model it is assumed that energy is deposited at the critical radius r_c and is conducted inwards to the target surface at a radius r_p. At some intermediate radius the plasma expansion becomes supersonic. Beyond r_c the plasma velocity increases and the temperature decreases with radius. As $r \to \infty$ then $T \to 0$ and $v \to v_\infty$, where v_∞ is dependent on the laser irradiance and is typically $\sim 4\ c_s$.

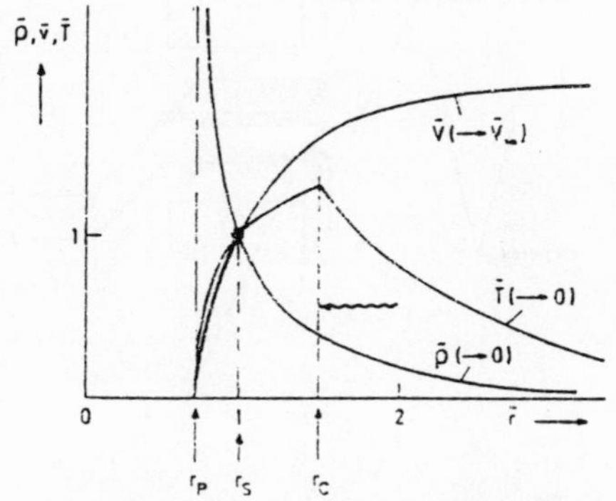

Fig. 18. Density, velocity and temperature against radius in stationary spherical expansion[24].

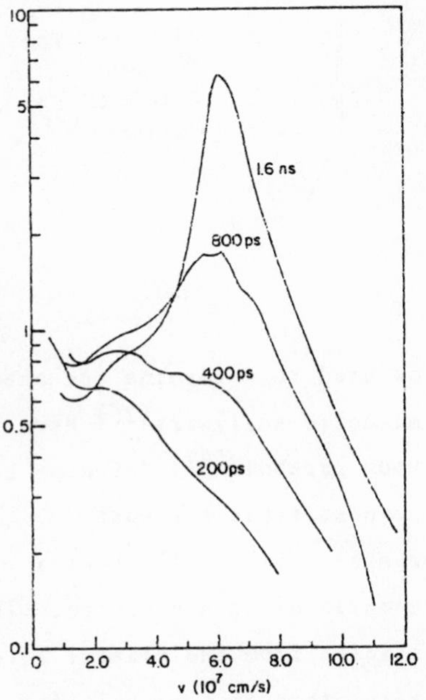

Fig. 17. Ion velocity distributions for various pulse lengths at an irradiance of 10^{14} *W cm*$^{-2}$ [23].

Clearly, in any experiment, truly stationary flow is impossible and for a given length of laser pulse the stationary flow approximation will break down at a particular radius which will depend upon the laser pulse length. The measured ion spectrum will consist of a single peaked spectrum in v with a width determined by the local temperature. This is illustrated in figure 17.

The introduction of a two temperature electron energy distribution can lead to the production of a rarefaction shock, as in the similarity solutions of Wickens, when $T_h/T_c \sim 5 + \sqrt{24}$ and the distribution tends to split into two parts.

3.3 Mass spectrometer analysers

Despite their simplicity, time of flight Faraday cup collectors suffer from a serious drawback: the current that is collected depends not only on the number of ions collected per second, but also their charge state. Recombination and charge exchange processes will often leave a distribution of charge states reaching the detector.

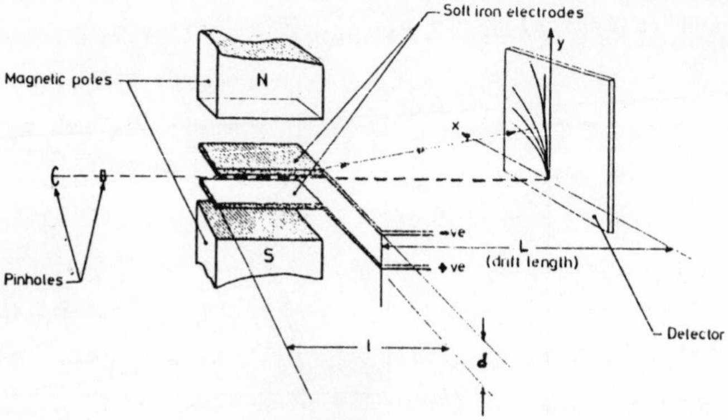

Fig. 19. The Thomson parabola ion analyser[28].

The mass spectrometer detector can be used to determine the charge state. Although electrostatic[25] and magnetic analysers[26] have been used in laser produced plasmas, the Thomson parabola[27] detector has proved suitable for most purposes. Figure 19 shows a schematic of such a device. The particles from the plasma, after passing through a pinhole (to restrict the range of angles), pass through a region of parallel electric and magnetic fields. The particles with a particular ionisation state Z are then deflected to fall along a parabola in the detecting plane. The detector can either be plastic, such as cellulose nitrate or CR39, which is damaged by ion impact, photographic film or a

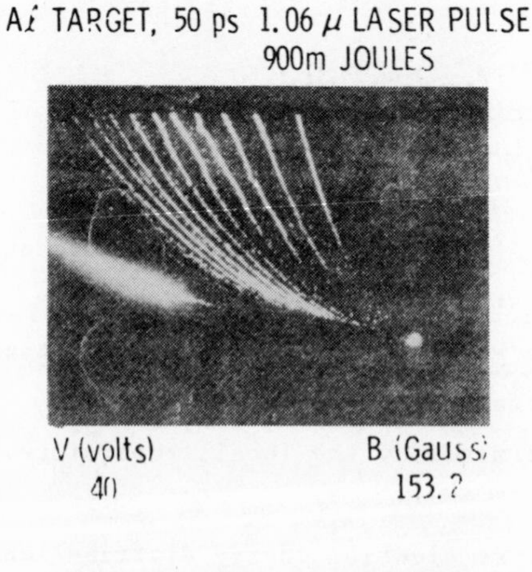

Fig. 20. Thomson parabola output: lower left trace are protons and others are various stages of ionisation of Aluminium[27].

microchannel plate. Fast particles strike the detector plane near the origin and slow particles are deflected and lie further along the parabola. Different parabolae are traced for particles of different Z/m which can usually be identified unambiguously with a particular ionisation state of an ion.

Figure 20 shows some results obtained with a Thomson parabola for an aluminium target[27]. Analysis of the Thomson parabola results shows that the first ions collected are nearly always of hydrogen. Even with targets which ostensibly contain no hydrogen, hydrogen ions are almost always observed unless very special precautions are taken to clean the target surface just before laser impact.

Results for the Faraday cup collector shown in figure 15 show several peaks and it is tempting to assign the peaks to different species of ions. Gurevich[29], however, suggests that the dips are more consistent with a self similar expansion within each ion species similar to that shown for the two component plasma.

3.4 Plasma Calorimeters

In experiments involving fusion, it is often important to be able to measure the absorbed laser energy and the energy that is converted into particles and non-laser radiation. The differential plasma calorimeter has been devised for this purpose[30]. Two identical thin aluminium discs are arranged close together: one of the discs will collect all particles, X-rays and laser light, and the other disc has a glass or optical filter window in front of it so that it only collects the scattered laser light. On firing the laser, the collected radiation and particles causes the temperature of the discs to rise. This is detected by an identical thermopile attached to each one. The difference signal provides the collected plasma energy. This technique has proved very successful: it is easily calibrated with a heating coil, its slow response time (~ 1 sec) means that measurements are taken well after the electrical noise from the laser has disappeared and, since it is relatively inexpensive, large numbers can be sited around the target chamber.

4.

RADIATIVE DIAGNOSTICS OF DENSE PLASMAS

This section will attempt to cover diagnostics using electromagnetic waves which are useful for probing densities beyond critical for the laser frequency. The opacity of these dense regions almost always requires that the wavelengths used are in the XUV (λ~100-10nm) or X-ray (λ<10nm) part of the spectrum. It is fortuitous that in many experiments high temperatures are encountered and most of the resulting emission occurs in this spectral region.

4.1 Spectral emission lines

Figure 21 shows a typical spectrum obtained from a laser produced plasma in the 9 - 14 Å spectral region[31]. In general, in order to be able to use spectra such as this, we must make some very simplifying

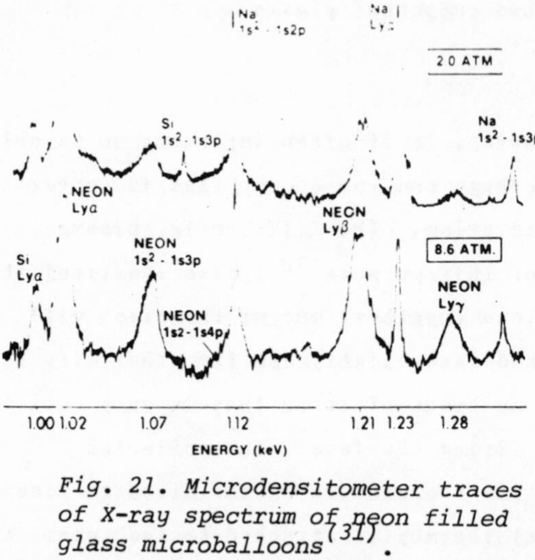

Fig. 21. Microdensitometer traces of X-ray spectrum of neon filled glass microballoons[31].

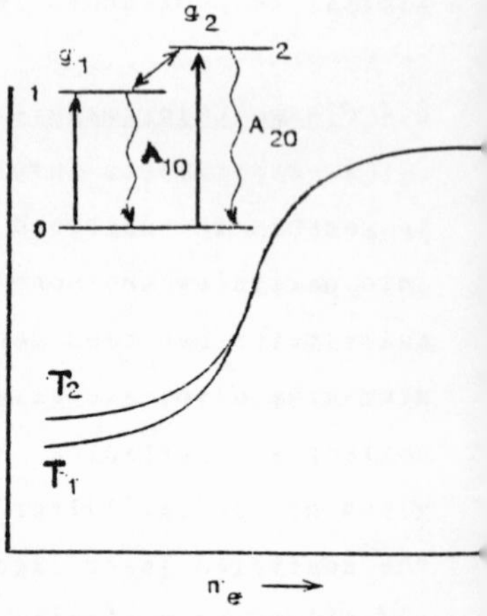

Fig. 22. Three level model ion.

assumptions concerning thermal equilibrium and the uniformity of the plasma conditions in space and time. We will usually assume that the emission region has uniform, constant density and temperature for the whole of the time of emission and that there is no emission from outside this region. Clearly, this situation can never be achieved in practice but it does allow approximate values of the plasma parameters to be found. A more accurate approach is to use the density and temperature profiles from fluid computer codes such as MEDUSA to predict a particular spectral feature and then compare this with the experimental results. If there is good agreement between the two then this is taken as support for the computer predictions. However, many very different density and temperature distributions in space and time can produce the same spectral feature, so one should always be aware of the limitations. In principle, for spherically or cylindrically symmetric distributions with no reabsorption the emissions could be Abel inverted at each wavelength to obtain the local emission. However, in laser produced plasmas the techniques usually used make this type of interpretation impossible.

4.1.1 Line intensities

One of the more obvious features of the spectrum in figure 21 is that different emission lines have different intensities. These intensities depend on the plasma conditions, in particular the density and temperature. In principle measurement of absolute line intensities is possible but this is rarely practicable in laser produced plasmas and so we will restrict ourselves to relative line intensities. How these line intensities vary can best be illustrated by reference to figure 22 which represents a model 3-level atom or ion. We will assume that the separation of levels 1 and 2 is small compared to their separation from the ground state (level 0) and that the spontaneous transition rates A_{10} and A_{20} are very different from each other. Thus at low densities the relative intensities are just determined by the relative collisional excitation rates from the ground state

$$\frac{I_{10}}{I_{20}} = \frac{\langle\sigma_{01}\rangle}{\langle\sigma_{02}\rangle}$$

At high densities, however, there is a complete coupling between the two upper levels as a result of collisions and the relative intensities of the two lines are determined by spontaneous transition rates and the level degeneracies.

$$\frac{I_{10}}{I_{20}} = \frac{g_1 A_{10}}{g_2 A_{20}}$$

In the intermediate region the intensity ratio is dependent on the electron density.

a) Intercombination/resonance line intensities

The ratio of resonance line to intercombination line intensity in the

He-like ions ($1s2p\,^1P_0 - 1s^2\,^1S_0/1s2p\,^3P_1 - 1s^2\,^1S_0$) is a commonly used diagnostic. The lines are usually strong emitters and the ratio is only weakly dependent on temperature. Figure 23 shows the variation

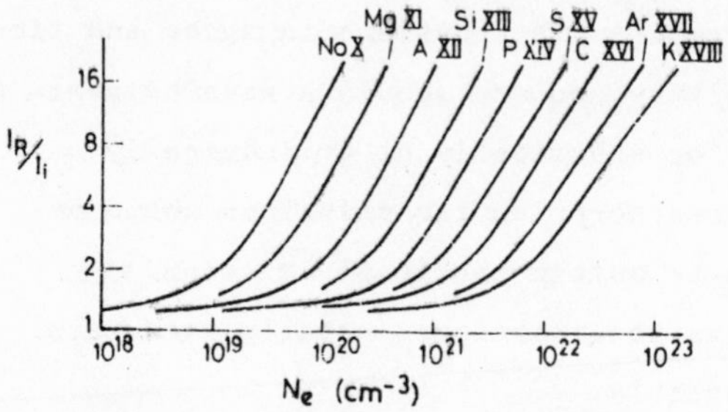

Fig. 23. Intensity ratio of singlet to triplet decays to ground in He-like ions[32].

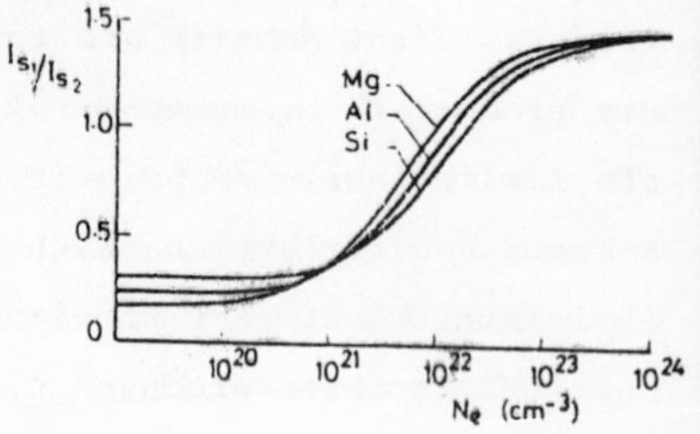

Fig. 24. Line satellite intensity ratio (see text).

of the ratio over a wide range of densities for the He-like ions of various elements[32].

A common difficulty with this technique is that the resonance line is very often optically thick in laser produced plasmas. This can sometimes be overcome by 'seeding' a small fraction of a monitoring element into the plasma. Although figure 22 suggests that this technique will work over a wide range of densities, the dynamics of plasma expansion in the coronal region often result in super cooled ions and recombination becomes the major source of population of the n=2 levels. In such cases the ratio may well drop below unity (this super-cooling is used as a way of producing a population inversion in recombination laser schemes).

b) <u>Line satellites</u>

Direct inner shell excitation or dielectronic recombination can often leave an ion in a doubly excited state. Let us imagine we have a Li-like ion which has one electron excited to n=2 levels and the second excited electron excited to a higher level. The n=2 electron will effectively see a He-like core which is only partially screened by the outer electron. This partial screening results in a reduction of the effective core charge and hence reduces the n=2 state energy. Thus, several satellite lines (depending on the second excited electron's energy level) can exist on the long wavelength side of the He-like resonance lines. A similar situation arises for H-like ions.

The ratio of He-like ions to Li-like ions is strongly temperature dependent and consequently the ratio of the resonance line to satellite line intensity is also strongly temperature dependent. However, this temperature dependence is similar for all satellites of one ion species. Furthermore, the excitation of the second excited electron can be density dependent as shown in figure 22. In these circumstances the relative intensities of some satellites are density dependent and relatively temperature independent. Figure 24 shows such a density dependence for satellites of H-like ions of magnesium[33]. In this figure I_{s1} refers to the transition

$$2p^2\ ^3P_{1,2} \quad - \quad 1s2p\ ^3P_{0,1,2}$$

and I_{s2} refers to the transition

$$1s2p\ ^3P_{0,1,2} \quad - \quad 1s2s\ ^3S,$$

The satellite intensities are generally very much weaker than the resonance line intensities and consequently do not suffer from the problem of opacity. However, they are usually only excited in the region of critical density and tend to be unobservable far away from the region.

c) Other line ratios

The use of line ratio measurements already described to measure electron density can be subject to large errors under certain conditions[34]. However, the choice ratio of lines in different ionisation species can provide a reasonably accurate measurement of the electron temperature. Figure 25 gives the ratio of some H-like to He-lines of an Argon plasma[35]. These intensity ratios are very sensitive to temperature in this range, changing by two orders of magnitude for a factor of two increase in temperature but insensitive to density, changing by only a few percent for a factor of three rise in density.

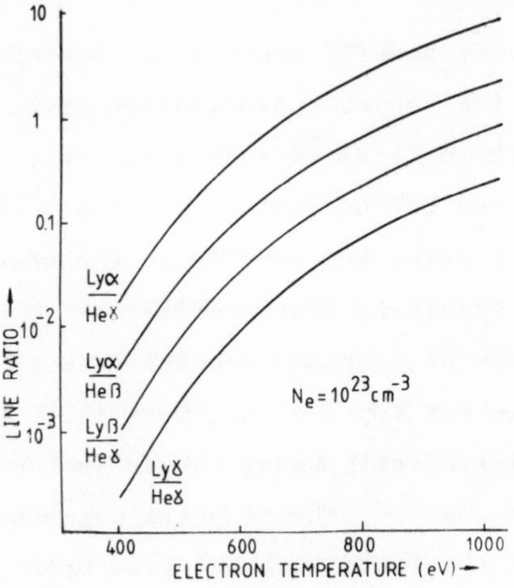

Fig. 25. Intensity ratio of H-like to He-like ions of Argon.

One minor disadvantage with

this technique is that, in practice, the range of temperatures that can be measured with one element (such as Argon in this case) is limited. This is usually overcome by the use of other tracer elements.

4.1.2 Line Profiles

There are three important mechanisms of line broadening in laser produced plasmas, these are:- Stark broadening, Doppler broadening and opacity broadening. Other mechanisms such as power broadening, pressure narrowing and magnetic field effects are usually insignificant. Furthermore, the broadening is usually sufficiently great to completely neglect any natural line width.

a) Doppler broadening

Doppler effects occur because of thermal or mass motions of the plasma. A Doppler shift ($\Delta\lambda/\lambda = u/c$, where u is streaming velocity) is readily observed in the coronal region of the plasma where lines are relatively narrow and directed mass motion is large. Doppler broadening occurs due to thermal motion of the emitters, which for a Maxwellian distribution results in a Gaussian line profile with a width

$$\frac{\delta\lambda}{\lambda} = 4.5 \times 10^{-5}\sqrt{\frac{T}{A}}$$

where T is in eV and A is the atomic mass of the ion.

b) Stark broadening

Theories of stark broadening in a plasma usually make use of one of two approximations. The first is called the impact approximation where it is assumed that the perturbation to the emitter is sudden and only lasts for a time which is short compared to the time of emission. This assumption is similar to pressure broadening and results in a change of phase of the emitting waveform. The resulting line profile is Lorentzian. However, the simple theory of pressure broadening does not work well for electron impacts. Different types of emitters (i.e. H-like or He-like ions) behave very differently under the influence of the electric fields introduced by the impact. The response depends not only on the energy levels involved in the transition but also upon neighbouring levels. For H-like ions the effect is linear in electric field but for all other ions the effect is quadratic. Statistical averaging over many impacts results in a broadened, unshifted line for

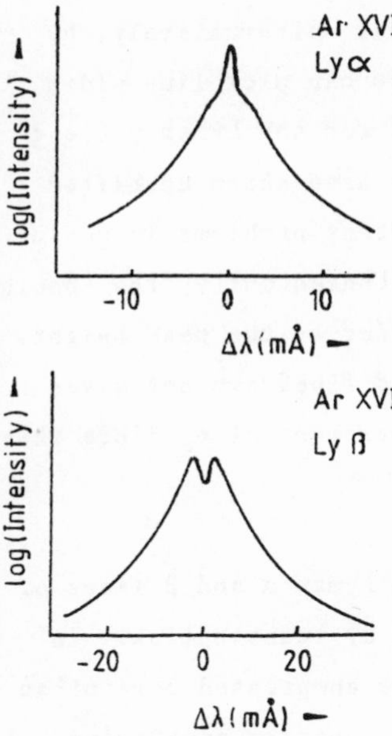

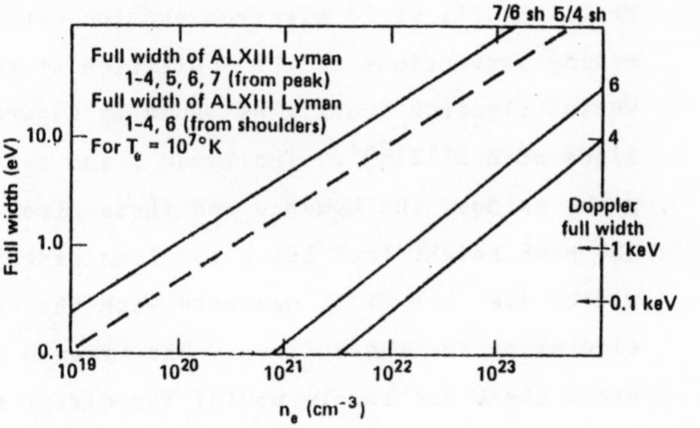

Fig. 27. Stark broadened full width at half maximum of Al XIII Lyman series lines.

Fig. 26. Line shapes of Stark broadened Lyα and Lyβ lines of Ar XVIII.

H-like ions but in general for all other ions the line is broadened and shifted (usually to longer wavelengths). Because of the much higher velocities of electrons than ions in a plasma, the impact approximation applies to broadening by electron impact rather than to ion impact. A detailed description of the impact approximation is given by Griem(36).

The second approximation that is often used is called the quasi-static approximation. This is quite the opposite of the impact approximation insofar as the perturber is considered stationary with respect to the emitter during the time of emission. The line profile is obtained by a statistical average over all perturbers. As one may imagine this approximation works best for relatively slow moving ions. It was first tackled by Holtzmark [see reference 36] and is often called Holtzmark broadening. The line profiles predicted by Holtzmark's theory depend very much on the line in question. For example the Lyman-α profile has a central spike whereas Lyman-β profile has a central dip as shown in figure 26. In these curves electron impact broadening is also included. Since impact broadening is dependent on the electron density and quasi-static broadening depends on the ion density of the perturbing ion species, careful fitting of

profiles will yield electron and ion densities. Alternatively, by making assumptions of the ionisation state one can plot line width versus electron density as shown in figure 27 for the Lyman γ δ ε ξ lines of A XIII[38]. The Lyman γ and ε lines have sharp unshifted peaks as does the Lyman α and these often present problems in measuring the peak height (and hence the line width). Consequently, the shoulder width, i.e. the width measured with the shoulder as the peak height, is also given for those lines. The Lyman α and β lines are not given since these are rarely useful for direct measurement of n_e since they are very often opacity broadened.

c) Opacity broadening

In many laser produced plasma experiments the Lyman α and β lines of any modest atomic number target are broadened by reabsorption. In figure 21 the profiles the Ne Lyman-α from the compressed core of an imploded microballoon illustrate the effect of opacity broadening. At 2.0 atmospheres filling pressure only the sharp central peak has been 'lost' due to reabsorption but at a higher fill pressure of 8.6 atmospheres the line has the characteristic 'flat top' shape which arises from reabsorption over a wider spectral range.

It is possible to make use of this opacity broadening to make an estimate of the average ρr of the emitting ion. This type of measurement is of considerable significance to thermonuclear fusion, since for the fuel,

$$\rho r > 1 \text{ gm cm}^{-2}$$

is the Lawson criterion.

The equation of radiative transfer (equation 1.1) in one dimension can be transformed to optical depth, τ, rather than physical depth. We define

$$\tau = \int_0^x k_\nu dx \quad \text{and} \quad d\tau = k_\nu dx$$

In LTE we can write $B = j_\nu / k_\nu$ where B_ν is the Planck function.

Thus

$$\frac{dI_\nu}{d\tau} = B_\nu - I_\nu$$

For a uniform slab this can be integrated to give

$$I_\nu = B_\nu(1 - e^{-\tau})$$

The absorption coefficient k_ν, integrated over the whole spectral line is related to the number of absorbers in the lower energy state N_1 by:-

$$\int k_\nu d\nu = \frac{e^2}{4\varepsilon_o mc} \cdot N_1 f$$

where f is the line oscillator strength.

Again assuming k_ν constant in space.

$$\int \tau d\nu = x \int k_\nu d\nu = \frac{e^2}{4\varepsilon_o mc} \alpha N f x$$

where α now is the proportion of total number of absorbers, N, in the ground state.

However

$$MNx = \langle\rho R\rangle$$

and $\tau = \tau_0 P_\nu$ where P_ν is the original line profile normalised to unit height and τ_0 the optical depth at line centre.

Since $\int P_\nu d\nu \simeq \Delta\nu_{1/2}$ the half width of the line, then

$$\langle\rho R\rangle \simeq \frac{4\varepsilon_o mMc}{e^2 \alpha f} \cdot \Delta\nu_{\frac{1}{2}} \tau_o$$

This technique has been used to find the $\langle\rho R\rangle$ of a compressed neon gas filled microballoon(39). (On the same spectra the line profiles of the optically thin lines were used to obtain ρ).

4.1.3 <u>Crystal and grating spectrographs</u>

Grating spectrographs used near to grazing incidence are useful in the range $20Å < \lambda < 500Å$. Because of the small solid angle presented by the grating at the entrance slit, the early spectrometers were difficult to align. A more recent instrument, the GML-5, (40) has overcome many of these problems and is capable of resolving powers $\lambda/_{\delta\lambda} > 4 \times 10^3$ at $\lambda = 100Å$. This versatile instrument has a precision ground 5m (Rowland) circle on which the entrance slit, grating and detector can be moved to vary the angle of incidence. Detection can be either photographic or photoelectric. This instrument has been used

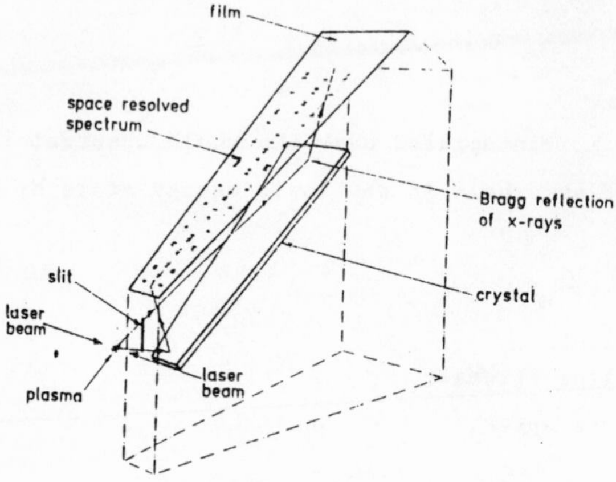

Fig. 28. Miniature space resolving X-ray spectrograph.

extensively on the CVI X-ray laser experiments at the CLF[41].

In most compression experiments wavelengths <20Å must be used to reduce opacity effects. In this spectral region crystal spectrometers are almost universally used. A simple crystal spectrograph suitable for use with laser produced plasmas is shown in figure 28. The plasma source is located ~20mm in front of the entrance slit of the spectrometer. X-rays from the source pass through the slit and strike the crystal at an angle of incidence dependent on the position along it. X-ray wavelengths which satisfy the Bragg condition are strongly reflected whereas other wavelengths are not. Thus the spectrum is spread out along the film plane. If the entrance slit width is reduced to be much smaller than the source size, then some spatial resolution is obtained perpendicular to the dispersion direction. A thin foil is usually placed over the entrance slit to reduce visible and U.V. light entering and fogging the film.

The variation of angle of incidence that is possible with such an instrument depends on the closeness of the plasma and crystal length. With 100J laser pulses the blast damage to the entrance foil usually becomes severe if the distance from the plasma is much closer than ~20mm. For higher energy laser pulses the spectrograph must be moved further away. Crystal lengths are typically 20-50mm and the range of angles of incidence are ~30°-60°. Thus from the Bragg condition

$$n\lambda = 2d\sin\theta$$

the wavelength range is ~±35% about the central wavelength in the first order. Different spectral regions may be covered by using a crystal with a different 2d value. Table 4.1 lists some commonly used crystals

Table 4.1

Crystal	orientation	2d(A)	Resolving Power	Integrated Reflectivity
LiF	(422)	1.652	–	–
Ge	(220)	4.00	4500	$2x10^{-4}$
Ge	(111)	6.532	3000	10^{-4}
PET	(002)	8.74	3000	10^{-4}
ADP	(101)	10.64	–	$5x10^{-5}$
Beryl	(100)	15.954	–	–
Mica	(002)	19.84	1000	$5x10^{-5}$
TlAP	(100)	25.9	–	–
lead myristate		79.9	200	$8x10^{-4}$
carbon/Tungsten – sputtered layers		40-200	20	$2x10^{-3}$

and pseudo-crystals and some of their properties. The spectrum shown in figure 21 was obtained using a TlAP crystal.

One disadvantage of the flat crystal spectrometer is that its light gathering power (or X-ray gathering power) is limited. Considerable improvement can be obtained using a curved crystal. A format suitable for laser produced plasmas is the Von-Hamos type(42) shown schematically in figure 29. The crystal is bent to form part of a cylinder and the source and image points (for various wavelengths) all lie on the axis of the cylinder. Light gathering powers > 100 times that of a flat crystal are possible

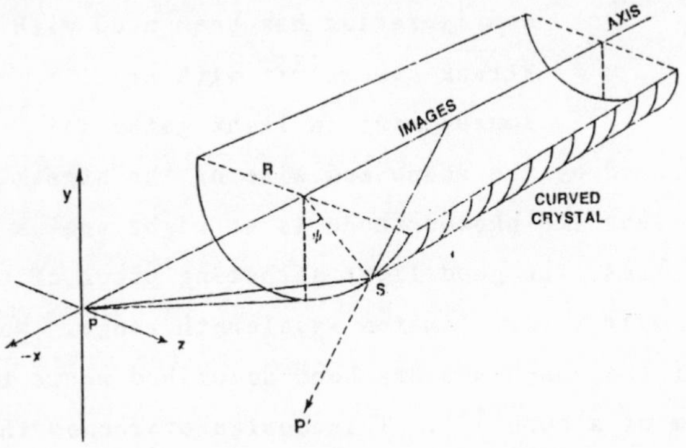

Fig. 29. Von Hamos curved crystal spectrograph geometry.

with this instrument but off axis aberrations tend to restrict the spatial resolving power.

Two dimensional spatial resolution at a single wavelength can be achieved by bending a thin crystal onto a spherical mirror substrate[43] and using this crystal mirror near to normal incidence in same way that an optical mirror would be used except that only wavelengths satisfying the Bragg condition are reflected.

4.1.4 Time resolved spectroscopy

Crystal spectrometers can be readily coupled to a streak camera to yield time resolved spectra, an example of which is shown in figure 30. This streaked spectrogram shows the emission from an alumninium target overlaid with a layer of SiO_2. From the delay in the appearance of the Al lines at 7.17 Å and 7.80 Å the time taken for the laser to burn through the SiO_2 can be found[44].

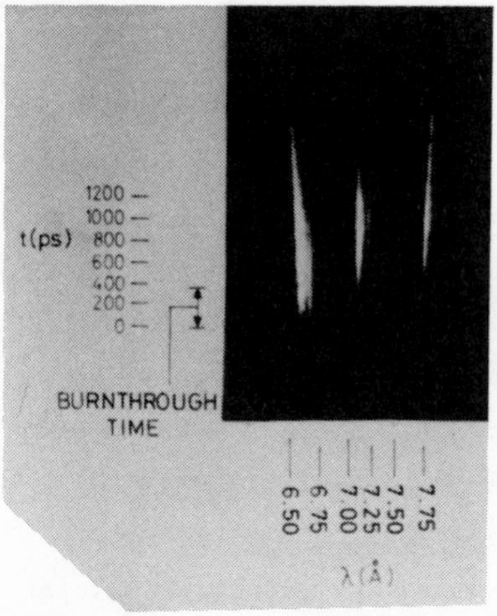

Fig. 30. Streaked spectrograph showing burnthrough of SiO film on Al.

Although a streak camera is generally more sensitive than film, the dispersion of the X-rays in wavelength and time requires a bright source when using a flat crystal disperser. The Von Hamos configuration has been used with a streak camera[35] with an improvement in light gathering power. The constraints imposed by the shape and size of the streak camera, however, determine that the photocathode is at right angles to the Von Hamos image line. Thus, the good light gathering power of the Von Hamos can only be used over a very limited wavelength range. More recently a modified form of the Von Hamos has been described where the crystal is bent in the form of a cone[45]. This design overcomes the problem of limited spectral range whilst maintaining a high light gathering power.

4.2 X-ray continuum measurements

In addition to emission line spectra already described, crystal spectrometers often show a low level continuum. This continuum is either bremsstrahlung or recombination and can provide a very useful diagnostic of temperature. For targets with moderate to high atomic number recombination continuum is dominant, whereas at low atomic number and high temperatures, bremsstrahlung dominates. The attraction of using continua as a diagnostic is because its power spectrum can be expressed in a relatively simple analytic form. In the spectral region $h\nu > hT$ the logarithmic slope of the spectrum is just proportional to $- h\nu/kT$. Figure 31 shows a plot of this slope taken from the spectrum obtained with a crystal spectrograph(46). However, it is not necessary to use a high resolution dispersive element to obtain the temperature from continuum emission: very much cruder techniques are possible which lend themselves very readily to photoelectric recording.

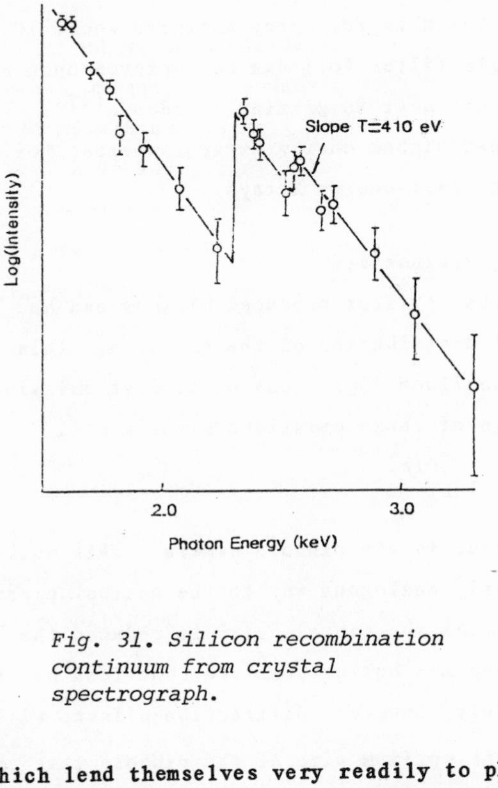

Fig. 31. Silicon recombination continuum from crystal spectrograph.

Thin metal or plastic foils will absorb different regions of the X-ray spectrum and by using an array of such foils it is possible to divide up the spectral region of interest. The X-rays that are transmitted through the foils can be detected using either vacuum photodiodes, PIN semiconductor diodes or plastic scintillator-photomultiplier combinations. The PIN diodes are quite sensitive to X-rays with energies <5keV but since the sensitive intrinsic layer is ~100μm thick much harder X-rays tend to pass straight through. At higher X-ray energies, plastic scintillators with photomultipliers are more effective.

For very bright X-ray sources a more sophisticated technique is possible: X-rays from the plasma first pass through a prefilter foil and strike a second fluorescer foil. The fluorescence produced in this foil is further filtered by a second absorbing filter and the relatively narrow band of X-ray energies that are transmitted are detected by a scintillator and photomultiplier[47]. The spectral discrimination obtained in this way is much superior to the single foil technique and is particularly suitable for X-ray energies above 10 keV.

At lower energies the single filter foil can be improved upon by reflections from polished mirrors near to grazing incidence[47]. These reflections discriminate against higher energy X-rays, whereas foils generally discriminate against lower energy X-rays.

4.3 Spatially resolving X-ray diagnostics

Useful data on many aspects of laser produced plasmas can be obtained just from the spatial distribution of the emission. This is particularly the case in microballoon implosions where most emission occurs at stagnation. An image of these emissions gives a good representation of the compressed core.

4.3.1 Pinhole cameras

The simplest X-ray imaging device is the pinhole camera. This works in an exactly analogous way to the optical pinhole camera. As the pinhole size is reduced the image becomes better resolved, but less bright. Ultimately, however, diffraction effects will limit the optimum size of the pinhole that can be used. Mack[48] has shown the optimum diameter to be

$$d = \left(\frac{3.6\lambda v}{1 + m} \right)^{\frac{1}{2}}$$

where v is the image distance and m the magnification. A typical image produced by a pinhole camera is shown in figure 32. This image resulted from a six beam implosion of a

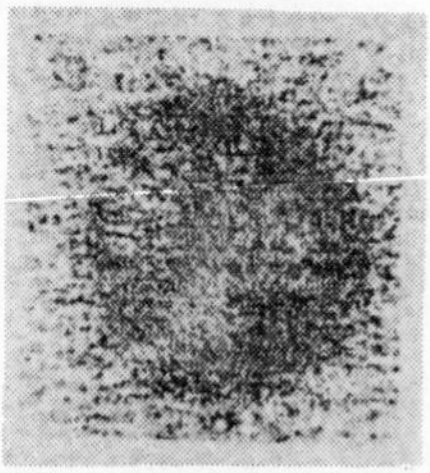

→ 102μm ←

Fig. 32. X-ray pinhole camera photograph of glass microballoon filled with 2 Bar of Argon.

glass microballoon filled with 2 Bar of Argon(49). The imprint of the six beams is seen around the original diameter of the microballoon, together with a very bright central core. Resolutions of ~ 5 μm are possible with short working distances.

The use of several pinholes each with different filter foil can give an indication of the temperature of the emitting region. An accurate determination is not possible, however, since the filter foils still let through a wide wavelength range and the sensitivity of the recording film varies with wavelength. Under normal circumstances it is not easy to unscramble the two effects.

4.3.2 X-ray mirror cameras

Apart from concave crystal mirrors already described, polished non-crystalline materials only reflect X-rays near to grazing incidence. A spherical mirror used near to grazing incidence produces a very astigmatic image. This astigmatism can be mostly corrected for by reflecting the X-rays off a second similar grazing incidence mirror set at perpendicular to the first. This type of microscope is known as a Kirkpatrick-Baez(50) and can achieve 3.5 μm resolution at much greater working distances than a pinhole camera (important with very high power laser systems) and with greater light gathering power.

An axisymmetric mirror pair consisting of coaxial hyperboloid and ellipsoid surfaces has a much higher light gathering power still. This Wolter type microscope(51) has been constructed with a resolution of ~4μm and very high light gathering power. The tolerance and surface finish on these mirrors is very exacting and is usually the factor determining the ultimate resolution obtained.

4.3.3 Coded aperture imaging

A different approach to X-ray imaging uses coded apertures, the simplest of which is a simple Fresnel zone plate(52). The production of zone plates for the visible region is trivial, but this is not the case for the X-ray region. Because of the much shorter wavelength the X-ray zone plates are very small and must be fabricated from some X-ray absorbing material, such as gold. Free standing gold zone plates have been made (see section 5.3.4), but for X-ray work they are usually made on thin plastic substrates. Unlike its optical counterpart, the X-ray zone plate does not image the source in the conventional way, but rather produces a magnified shadow of the zone plate which is modified by the source size and shape. From this image the original source can be reconstructed, either optically using a laser, or numerically by computer. Resolutions of 1-2 μm have been obtained by this technique and the light gathering power is quite high, although the working distance is rather short.

The zone plate coded imaging system can introduce artifacts into the reconstructed image. Artifact-free images can be produced using uniformly redundant arrays (URA's)[53], which have similar light gathering powers, but can only be reconstructed by computer.

4.3.4 Space and time resolved imaging

The pinhole camera, K-B or Wolter microscopes can readily be coupled to an X-ray streak camera. With a slit in front of the photocathode one can obtain a streaked image of a section through the two dimensional image produced by the microscope. This technique has found particular favour in radiography and will be described more fully in section 4.4.

Two dimensional time resolved images have been obtained using a high speed X-ray framing camera[54]. This camera produces several sequential frames and has opening time ~ 200 psec.

4.4 Radiography

This technique uses a subsidiary X-ray source produced by an extra laser beam. This source is arranged to be of much higher brightness in the spectral region of interest than the target under observation.

4.4.1 Pulsed radiography

The radiographic technique was pioneered by Key and coworkers[55] at the Rutherford Appleton Laboratory and has proved extremely successful in diagnosing relatively cool dense implosions and the acceleration of thin foils. The technique relies upon the fact that ablative type implosions of low Z microballoons emit X-rays very much less copiously than medium to high Z materials under similar or higher irradiance. Figure 33 shows schematically the arrangement used by Key: a brass backlighting target is illuminated by a short pulse (~ 100 psec) and produces copious X-rays of energy ~ 1.5 keV. The shadow of the

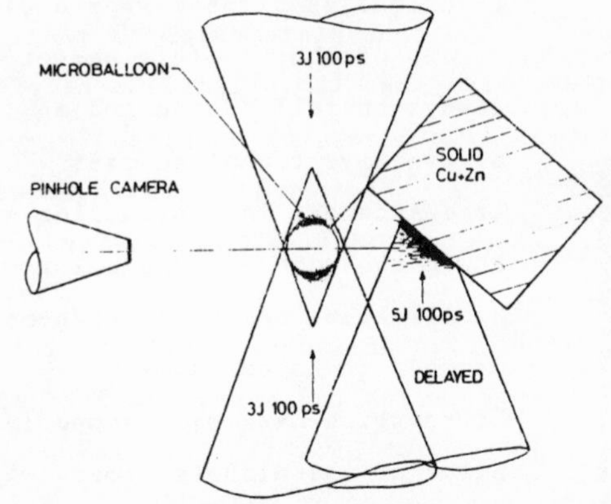

Fig. 33. Experimental configuration for pulsed X-ray radiography.

microballoon is viewed against this background by a pinhole camera. The backlighting source is typically a few mm's, and the pinhole camera a few cm's, away from the microballoon). In these experiments the high pressure gas fill produced a relatively long implosion time and the short backlighting pulse was successively delayed on separate laser shots to produce a sequence of radiographs, as shown in figure 34. The densitometer traces shown in figure 34 were changed to opacity profiles and these opacity profiles were then compared with

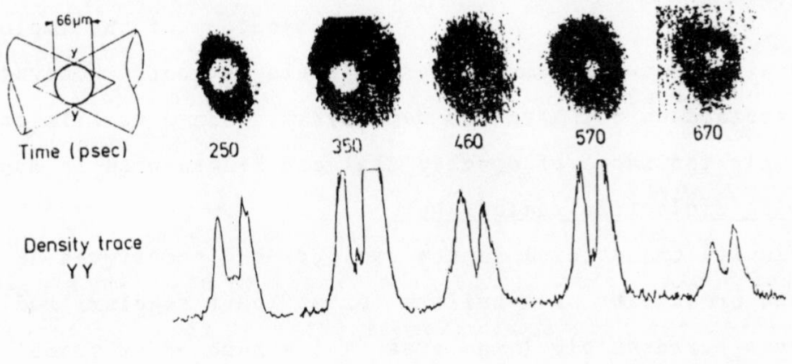

Fig. 34. A series of pulsed radiographs with delays shown from peak of implosion pulse.

those generated from the computer code MEDUSA. In the computer code, it was assumed that the absorption was predominantly due to K shell absorption of He-like oxygen in the balloon wall, and neon in the gas fill. These results corresponded to a compressed density of 1 gm cm^{-3} in the compressed core.

4.4.2 <u>Streaked radiography</u>

To obtain a sequence of radiographs such as shown in figure 34 the

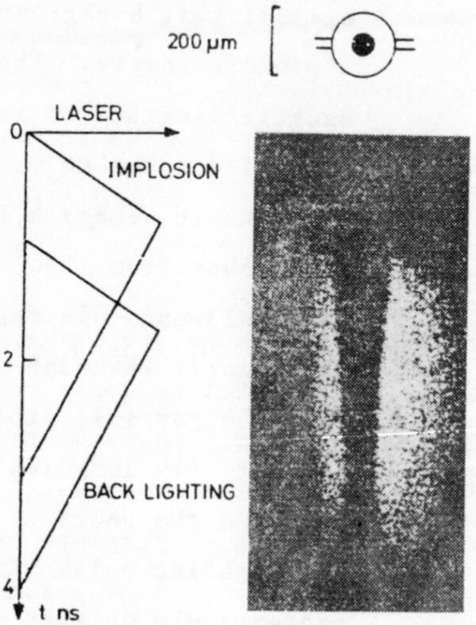

Fig. 35. Streaked radiograph of a plastic coated glass microballoon.

technique requires a reproducible series of shots. This cannot always be relied upon and an alternative technique uses a streak camera in conjunction with the pinhole camera and a long backlighting pulse. A refinement on this technique uses Kirkpatrick-Baez microscope in place of the pinhole camera which gives higher light gathering power and a better working distance(56). Figure 35 shows a streak record of a microballoon implosion. This streak record clearly shows the collapse of the microballoon and enables the trajectory of the implosion to be compared with computer simulations on a single shot. However, the streak camera does not have the same dynamic range as film and consequently the range of opacity that can be measured is much less.

4.4.3 Point projection radiography

One difficulty encountered in the radiographic techniques described so far is the production of a uniform (or at least regular) radiographic source over a reasonably large area (a few hundred μm diameter). Defocussing the laser beam onto the backlighting target quite often produces a number of bright spots of X-ray emission rather than a large uniform source. An alternative technique uses a very bright neo-point source of X-rays and uses this to cast a shadow of the microballoon onto the film or streak camera <u>without</u> a pinhole(57). This point projection radiography requires that the X-ray emission from target is very much weaker than the backlighting source, otherwise the detector will be uniformly fogged from the target emission. Also, at high magnification, refraction of the X-rays in the target plasma begins to become important.

This technique has also been used with a dispersing crystal(57) to produce monochromatic X-ray radiography. The use of monochromatic backlighting X-rays greatly simplifies the analysis of radiographic results and, provided a measurable transmission is observed at every point across the target, allows an Abel inversion of the transmission profile to be made and hence obtain the radial distribution in density directly without recourse to computer modelling.

4.5 Super dense plasma diagnostics

The interest in stellar interiors has continued now for many years, but it is only fairly recently that these degenerate, strongly coupled, plasmas have been created in the laboratory. The diagnosis of these plasmas is still very much in its infancy. One recent experiment[58] used colliding shocks produced in 40 μm thick slabs of parylene by irradiating both sides of the slab simultaneously with defocussed laser beams. These experiments produced a plasma with coupling parameter $\Gamma \sim 6$ and observations were made on the shift of the chlorine K-edge using a Bismuth backlighting source and a crystal spectrometer coupled to a streak camera. (The chlorine was introduced as a dopant to the central region of the parylene.) The experiments showed a red shift of the K edge of ~ 10 eV which was compared with various theoretical models.

5.
DIAGNOSTICS WITH FUSION PRODUCTS

5.1 Fusion Reactions

The nuclear reaction of particular interest to I.C.F. are :-

$$D + T \rightarrow \alpha\ (3.5\ \text{MeV}) + n\ (14.1\ \text{MeV})$$

$$D + D \rightarrow {}^3He\ (0.82\ \text{MeV}) + n\ (2.45\ \text{MeV}) \quad \text{) approx}$$

$$\searrow\ T\ (1.01\ \text{MeV}) + p\ (3.02\ \text{MeV}) \quad \text{) 50\% each}$$

$$D + {}^3He \rightarrow \alpha\ (3.6\ \text{MeV}) + p\ (14.7\ \text{MeV})$$

The reaction rates N_R for these processes varies strongly with temperature[59] $\quad N_R = n_1 n_2 \langle\sigma v\rangle$

where n_1 and n_2 are the densities of the two fusing components, σ is the cross section and v the relative velocity. The quantity $\langle\sigma v\rangle$ is usually represented as an average over a Maxwellian distribution for an ion temperature θ_i.

Since reactions will occur for a range of impact velocities, the reaction products will also have a distribution in energies [see ref. 20]. For the D-T reaction the energy spread is :-

$$\Delta E = 177\ \theta_i^{1/2}\ (\text{keV})$$

and is the same for both reaction products. In any future power station the number of D-T neutrons produced per implosion will be ~ 10^{20} but, with current devices, the numbers that have been recorded are in the range 10^5 - 10^{13} and most diagnostics so far have been developed to observe in this range.

Most of the techniques that have been developed are described in the review by Sigel(20) and will be only briefly mentioned here. An exception to this is some recently developed diagnostics using an improved track detector plastic CR 39.

5.2 Neutron measurements

In any laser produced implosion, in addition to any possible thermonuclear particles, an intense electromagnetic pulse (EMP) is produced. This can cause severe difficulties in measuring the effects of single particles and several techniques have been devised to ensure that the actual measurement is delayed until after the EMP has died away.

5.2.1 Activation techniques

This technique relies on the fact that neutrons are absorbed by certain elements, which will then decay. It is the decay of the activated nucleus that is measured.

Two materials have been used with success: the first is the silver activation

$$Ag^{109} + n \rightarrow Ag^{110} \rightarrow Cd^{110} + \beta^-$$
$$Ag^{107} + n \rightarrow Ag^{108} \rightarrow Cd^{108} + \beta^-$$

The silver decay has a half life of ~ 24 sec and the β decay is usually measured in situ in the target chamber.

The second material is copper

$$^{63}Cu + n \rightarrow \ ^{63}Cu + 2n$$
$$\searrow \ ^{62}Ni + \beta^+$$
$$\beta^+ + e \rightarrow 2\gamma$$

The half life ^{62}Cu is ~ 10 min: enough time for the copper sample to be removed from the chamber and placed in a measuring system where the γ rays are detected by coincidence techniques.

5.2.2 Time of flight measurements

14 MeV neutrons have a velocity of ~ 5.1 x 10^7 $msec^{-1}$ and the velocity spread for a 10 keV plasma ~ 10^6 $msec^{-1}$ (see section 5.1). Thus a detector 50 m from the source will detect the neutrons ~ 100 nsec after the implosion and arrival times will be spread over ~ 20 nsec. A plastic scintillator and photomultiplier has a time resolution of ~ 3 nsec and is thus readily able to measure this spread. Measurement of the arrival time spread thus will yield the ion temperature. Such a system has been used at LLNL on the SHIVA system for neutron yields in excess of 10^8 [see ref. 20].

5.2.3 Target activation

In most experiments that have been conducted to date, the neutrons escape from the target without any significant interaction. However, as the ρR of the fuel and tamper are increased the neutrons can activate the target material. If this material is collected then the proportion of activated nuclei will yield the ρR of the pellet (usually the tamper). ^{28}Si is present in glass microspheres and is activated to ^{28}Al by neutron capture. This decays by β decay back to ^{28}Si with a half life of 2-24 min. The number of activated nuclei in the tamper is given by

$$B^* = 2.1 \times 10^{-3} \langle \rho R \rangle_T N_n$$

where N_n is the neutron yield and $\langle \rho R \rangle_T$ is the mean ρR of the tamper (in gm cm^{-2}). Experiments have been carried out using this technique [see ref. 20] where the target debris was collected inside a hollow plastic sphere.

5.2.4 Time of emission measurements

As more powerful lasers become operational the production of neutrons from ablative implosions becomes more feasible and the detailed time history of the neutron emission becomes important. Recently, a neutron streak camera has been described which should allow measurements of the time of emission of the neutrons[59]. This measurement is important because the highest neutron yields appear to occur under conditions where a precursor shock would occur. It is possible that the neutrons are emitted at the time of arrival of the shock rather than at the main compression. This situation is ultimately undesirable since shock preheating will reduce the compressions that can be attained.

5.3 Charged particle measurements

5.3.1 Time of flight measurements

The spread of velocities due to the fusion ion temperatures is also reflected in the charged particle emission from D-T and D-D and D-^{3}He reactions and a time of flight measurement is also possible on α particles and protons. Their slower speed means that α particle TOF systems are more compact but, unlike neutrons, the flight path must be evacuated. It is possible to focus the α particles to increase the collection angle and to deflect them in order to discriminate against interference. Unlike neutrons, however, even at modest values of ρR the α's can interact with fuel and tamper producing a slowing down and a further spread in energies. Irregularities in the pellet ρR will result in a further spread in detected energies. For high irradiance shots the sheath electric field generated around the pellet corona can actually accelerate the α's[60].

5.3.2. Relative yield measurements

The temperature dependence of σv for the various reactions is markedly different. By varying the fuel composition so that two reactions take place simultaneously and measuring the relative numbers of the reaction particles, it is possible to calculate the fuel temperature. The pairs of reactions most commonly used are D-T/D-D where the α to proton ratio is measured and the D-D/D-^{3}He where the ratio of the 3MeV protons to the 14.7MeV protons is measured.

5.3.3 Range measurements

Charged reaction products are slowed down on their passage through the core as mentioned in 5.3.1. In the absence of other effects, this slowing down can be used as a measure of the ρR of the material through which it passes. These types of measurements are greatly facilitated by the use of plastic track detectors. In this type of detector the charged particle is stopped within the plastic leaving a trail of damage behind it. Subsequent etching of the plastic will reveal the damaged tracks since etching occurs faster along the tracks than for the bulk material. Careful control over the plastic quality and etching conditions enables not only the number of charged particles to be measured but also their energy.

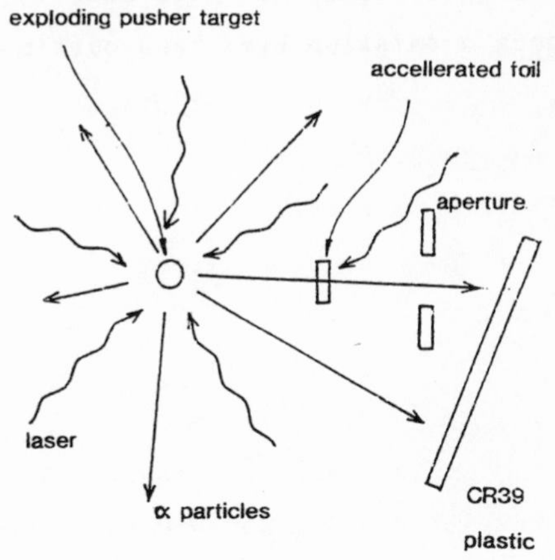

Fig. 36. Schematic of α-particle backlighting experiment.

Early experiments were conducted with cellulose nitrate film but this has been superceded by CR39 plastic. One difficulty in measuring ρR by this technique is that the range of α's in hot matter is different from that in cold material and without a knowledge of the temperature, the results are ambiguous

An elegant use of CR39 track detector has been recently carried out at the CLF by Evans et al.(61) in which very short scale length Rayleigh-Taylor breakup of a thin foil has been monitored by using CR39. Figure 36 shows a set up of the experiment. α particles are emitted from an exploding pusher target with minimal slowing down. Some will pass through a plastic lollipop foil which is accelerated by a subsidiary laser beam. The breakup of the foil involves parts of it becoming very thin and some parts even thicker(62). This can be detected by measuring the range of residual energies of the α-particles on exiting from the foil. This technique works for wavelengths smaller than the resolution limits of any imaging system.

5.3.4 Imaging systems

Implosions with low tamper/fuel ρR not only allow α-particles and protons to escape without slowing down but also without significant deflection. They can thus in principle be imaged using a pinhole camera. However, the number of α's from this type of implosion is

usually much too small for a pinhole camera. However, the larger collecting angle of the zone plate coded aperture has made such measurements possible. Images of core α-emission have been obtained with ~3μm resolution [see ref. 20].

REFERENCES

1. O.H. Nestor and H.N. Olsen, SIAM Review 2, 200 (1960)
2. K. Bockasten, J. Opt. Soc. Am. 51, 943 (1961)
3. M. Deutsh and I. Beniaminy, Appl. Phys. Lett. 41, 27 (1982)
4. Imacon 675, Hadland Photonics
5. M.G. Nomarski, J. de la Phys. et la Radium 16, 95 (1955)
6. R. Benattar et al., Rev. Sci. Instrum. 50, 1583 (1979)
7. L.M. Newell, Ph.D. Thesis, University of Essex (1984)
8. D.E. Evans, Plasma Physics Summer School, Culham (1973)
9. Z.Q. Lin, P.T. Rumsby, O. Willi, S. Sartang, section 3.2, Rutherford Appleton Laboratory, Laser Division (RAL) Ann. Rep.(1981)
10. O. Willi, P.T. Rumsby, W.D. Luckin, C. Hooker, A. Raven and Z.Q. Lin, section 3.7, RAL Ann. Rep. (1982)
11. O. Willi, P.T. Rumsby and C. Duncan, Optics Comm. 37, 401 (1981)
12. T.P. Hughes, Laser Plasma Interactions, SUSSP, pp 1-90 (1979)
13. M.D. Rosen et al. LLNL Report UCRL-82146 (1978)
14. N.S. Erokin, S.S. Moiseev, V.V. Mukhin, Nuclear Fusion 14,333 (1974)
15. E. McGoldrick, S.L.M. Sim. R. Turner, section 4.2 RAL Ann Rep.(1983)
16. A.I. Avrov et al. Zh. Eksp. Teor. Fiz. 72, 970 (1977)
17. H.C. Barr, section 8.3.3. RAL Ann. Rep. (1979)
18. S.J. Karttunen, Laser and Particle Beams, 3,2,157 (1985)
19. V. Aboites, S.L.M. Sim, D. Barrett,I. Ross, section A2.2.3 RAL(1984)
20. R. Sigel, Laser Plasma Interactions, SUSSP, pp. 661-709 (1979)
21. H. Miles et al. section 3.3. RAL Ann. Rep. (1978)
22. L.M. Wickens, J.E. Allen and P.T. Rumsby, Phys. Rev. Lett. 41, 243 (1978)
23. M.K. Matzen and R.L. Morse, Phys. Fluids 22 (4), 654 (1979)
24. S.J. Gitomer, R.L. Morse and B.S. Newberger, Phys. Fluids 20, 234-238 (1977) and B. Besserides, D.,W. Forslund, E.L. Lindeman, Phys. Fluid 21, 2197 (1978)
25. R. Decoste and B.H. Ripin, Rev. Sci. Instrum. 48, 232-236 (1977)
26. H.G. Ahlstrom 'Diagnostics of inertial confinement fusion experiments', LLNL Rep. UCRL 79894 (1977)
27. M.A. Gusinow, M.M. Dillon, G.J. Lockwood and L.E. Ruggles 'The Thomson parabola ion analyser' Sandia Labs. Rep. SAND78-0336 (1978)

28. T.J. Goldsack, J.D. Kilkenny, S. Sartang, W.T. Toner and S. Veats section 4.4. RAL Ann. Rep. (1980)
29. A. Gurevich, D. Anderson and H. Wilhelmsson, Phys. Rev. Lett. 42, 769–772 (1979)
30. LLNL Ann. Rep. section 3.5.2. UCRL-50021-76 (1976)
31. B. Yaakobi, D. Steel, F. Thorsos, A. Hauer and B. Perry, Phys. Rev. Lett. 39, 24, 1526 (1977)
32. A.V. Vinogradov, I. Yu Skobelev and E.A. Yukov, Kvant Elektron (USSR), 2, 1165 (1975)
33. V.A. Boiko, S.A. Pikuz and A. Ya. Faenov, J. Phys. B, 12, 1889 (1979)
34. N.J. Peacock 'Laser Plasma Interactions' SUSSP, pp.711–806 (1979)
35. B.J. McGowan, Ph.D. Thesis, University of London (1982)
36. H.R. Griem 'Plasma Spectroscopy' McGraw-Hill (1964)
37. J.T. O'Brien and C.F. Hooper, Phys. Rev. A 5, 867 (1972)
38. R.W. Lee, J.D. Kilkenny, R.L. Kauffman and D.L. Matthews, section 7.6, RAL Ann. Rep. (1983)
39. B. Yaakobi et al. Phys. Rev. A 19, 3, 1247 (1979)
40. R.J. Speer, Space Science Instrum. 2, 463 (1978)
41. G.J. Pert, L.D. Shorrock, G.J. Talents, M.H. Key, G.L.S. Lewis, E. Mahoney and J.M. Ward, section 6.2, RAL Ann. Rep. (1981)
42. L. Von Hamos, Z. Kristallogr, 101, 17 (1939)
43. N.G. Basov, Yu. Mikhailov, G. Sklizkoz and S. Fedotov, Laser Thermonuclear Installations. Radiotechniques series, Volume 25, Viniti, Moscow (1984)
44. P. Cunningham et al. section 4.4. RAL Ann. Rep. (1981)
45. T.A. Hall, J. Phys. E. 17, 110 (1984)
46. M.H. Key et al. Paper IV.9, 11th ECLIM Oxford (1977)
47. H.G. Ahlstrom, L.W. Coleman, F. Rienecker Jr., V.W. Slivinsky, J. Opt. Soc. Am. 68, 1731 (1978)
48. J.E. Mack and M.J. Martin 'The photographic process', McGraw-Hill, New York (1939)
49. J.D. Kilkenny, B.J. McGowan and P.T. Rumsby, section 5.5 RAL Ann. Rep. (1983)
50. F. Seward et al. Rev. Sci. Instrum. 47, 464 (1976)
51. LLNL Ann. Rep. UCRL-50021-76 (1976)

52. N.M. Ceglio, D.T. Attwood and E.V. George, J. Appl. Phys. 48, 1566 (1977)

53. E.E. Fenimore, T.M. Cannon, D.B. Van Hulsteyn and P. Lee, Appl. Opt. 18, 945 (1979)

54. N. Finn, T.A. Hall and E. McGoldrick, Appl. Phys. Lett. 46, 731 (1985)

55. M.H. Key, C.L.S. Lewis, J.G. Lunney, A. Moore, T.A. Hall and R.G. Evans, Phys. Rev. Lett. 41, 1467 (1978)

56. M.H. Key et al. Paper WA7 Topical meeting on Inertial Confinement Fusion, San Diego (1980)

57. C.L.S. Lewis, J. McGlinchey, S. Saadat, R.E. Corbett, R.W. Eason, C. Hooker, D. Bassett, section A2.3 RAL Ann. Rep. (1984)

58. D.K. Bradley, J. Hares, A. Rankin and S.J. Rose RAL Rep. No. RAL-85-020 (1985)

59. H. Niki 'Advances in Inertial Confinement' p.478 Osaka (1984)

60. Y. Gazit, J. Delettrez, T.C. Bristow, A. Entenberg and J. Soures, A.P.S. Meeting, Plasma Phys. Div., Nov. 12-16 (1979)

61. P.M. Evans, A.P. Fews, A. Cole, C. Edwards, C.J. Hooker, D. Pepplar, W.T. Toner, J. Wark, section A2.4, RAL Ann. Rep. (1985)

62. R.L. McCrory, L. Montierth, R.L. Morse and C. Verdun, Phys. Rev. Lett. 46, 336 (1981)